AF831010

Polysaccharides in Medicinal and Pharmaceutical Applications

Edited by

Valentin Popa

SMITHERS
*i*Smithers

*i*Smithers – A Smithers Group Company

Shawbury, Shrewsbury, Shropshire, SY4 4NR, United Kingdom
Telephone: +44 (0)1939 250383 Fax: +44 (0)1939 251118
http://www.ismithers.net

First Published in 2011 by

***i*Smithers**

Shawbury, Shrewsbury, Shropshire, SY4 4NR, UK

©2011, Smithers Rapra

All rights reserved. Except as permitted under current legislation no part of this publication may be photocopied, reproduced or distributed in any form or by any means or stored in a database or retrieval system, without the prior permission from the copyright holder.

A catalogue record for this book is available from the British Library.

Every effort has been made to contact copyright holders of any material reproduced within the text and the authors and publishers apologise if any have been overlooked.

ISBN: 978-1-84735-436-5 (hardback)
978-1-84735-437-2 (paperback)
978-1-84735-438-9 (ebook)

Typeset by Integra Software Services Pvt. Ltd.
Printed and bound by Lightning Source

Dedicated to the memory of Severian Dumitriu

The idea of bringing out this book was that of one of my best, Severian Dumitriu, of the University of Sherbrooke, Canada, who unfortunately passed away 1 year ago. Keeping in mind his previous contributions to a number of books in the field of polysaccharides in the capacity of scientist as well as editor, we decided together with Daniela, his beloved daughter, and Helene Chavaroche, project coordinator, to continue with and finalise this book, and dedicate it to his memory.

Severian Dumitriu was a research professor in the Department of Chemical Engineering, University of Sherbrooke, Quebec, Canada. He was editor of several books including *Polymeric Biomaterials*, Marcel Dekker, New York, USA, 1994; *Polysaccharides in Medicinal Applications*, Marcel Dekker, New York, USA, 1996; *Polysaccharides: Structural Diversity and Functional Versatility*, Marcel Dekker, USA, 1998; *Polymeric Biomaterials*, second edition, Marcel Dekker, New York, 2002; *Polysaccharides: Structural Diversity and Functional Versatility*, second edition, Marcel Dekker, New York, USA, 2005. He was the author or coauthor of 170 scientific papers and book chapters in the field of polymer and cellulose chemistry, polyfunctional initiators, and bioactive polymers held 15 international patents. Dr. Dumitriu received his BS (1959) and MS (1961) degrees in chemical engineering and a PhD (1971) in macromolecular chemistry from the Polytechnic Institute of Iasi, Romania. He was awarded the Romanian Academy Prize in 1998 for his pioneering research in bioactive polymers. He was a member of the American Chemical Society. Upon completing his doctorate, he worked with Professor G. Smets at the Catholic University of Louvain, Belgium; later, he was research associate at the University of Pisa, Italy; the Hebrew University Medical School, Jerusalem, Israel; and the University of Paris-South, France.

Contents

Contributors

Isabelle Bataille

Inserm, U698, Bio-ingénierie de Polymères Cardiovasculaires; and Institut Galilée, Université Paris 13, 99 Av. J.B. Clément, 93430 Villetaneuse, France

Amélie Bochot

UMR CNRS 8612, Physicochemical, Pharmaceutical Technology and Biopharmacy, School of Pharmacy, University Paris-Sud 11, France

Bradley S. Campbell

Biotechnology Research Centre, La Trobe University, Bendigo, Victoria 3552, Australia

Frédéric Chaubet

Inserm, U698, Bio-ingénierie de Polymères Cardiovasculaires; and Institut Galilée, Université Paris 13, 99 Av. J.B. Clément, 93430 Villetaneuse, France

Dominique Duchêne

UMR CNRS 8612, Physicochemical, Pharmaceutical Technology and Biopharmacy, School of Pharmacy, University Paris-Sud 11, France

Jennifer H. Elisseeff

Department of Biomedical Engineering, Whiting School of Engineering, Johns Hopkins University, Baltimore, MD 21218, USA

Thomas Heinze

Centre of Excellence for Polysaccharide Research, Institute of Organic Chemistry and Macromolecular Chemistry, Friedrich Schiller University of Jena, Humboldtstrasse 10, D-7743 Jena, Germany

Nicholas Andrew Hoenich

Institute of Cellular Medicine, Newcastle University, Newcastle on Tyne NE2 4HH, UK

Jan Jezek

Institute of Organic Chemistry and Biochemistry, Flemingovo nam. 2, 166 10 Prague, Czech Republic

Didier Letourneur

Inserm, U698, Bio-ingénierie de Polymères Cardiovasculaires; and Institut Galilée, Université Paris 13, 99 Av. J.B. Clément, 93430 Villetaneuse, France

Anne Meddahi-Pellé

Inserm, U698, Bio-ingénierie de Polymères Cardiovasculaires; and Institut Galilée, Université Paris 13, 99 Av. J.B. Clément, 93430 Villetaneuse, France

Cătălina Anişoara Peptu

Technical University Gheorghe Asachi, Faculty of Chemical Engineering and Protection of the Environment, D. Mangeron Street, 71A, 700050, Iasi, Romania

Marcel Popa

Technical University Gheorghe Asachi, Faculty of Chemical Engineering and Protection of the Environment, D. Mangeron Street, 71A, 700050, Iasi, Romania

Valentin I. Popa

Gheorghe Asachi Technical University of Iasi, Faculty of Chemical Engineering and Environmental Protection, Blvd. Mangeron No.71, Iasi, 700050, Romania

Mark T. Quinn

Immunology and Infectious Diseases, Montana State University, Bozeman, MT 59717, USA

Milan Reinis

Institute of Molecular Genetics, Videnska 1083, 142 20 Prague 4, Czech Republic

Igor A. Schepetkin

Department of Veterinary Molecular Biology, Montana State University, Bozeman, MT 59717, USA

Kerstin Schlufter

fzmb GmbH, Research Centre for Medical Technology and Biotechnology, Geranienweg 7, D-99947 Bad Langensalza, Germany

Frank Schmid

School of Agriculture, Food and Winem University of Adelaide, Glen Osmond, South Australia, 5065, Australia

Stephanie Schubert

Centre of Excellence for Polysaccharide Research, Institute of Organic Chemistry and Macromolecular Chemistry, Friedrich Schiller University of Jena, Humboldtstrasse 10, D-7743 Jena, Germany

Jaroslav Sebestik

Institute of Organic Chemistry and Biochemistry, Flemingovo nam. 2, 166 10 Prague, Czech Republic

Robert J. Seviour

Biotechnology Research Centre, La Trobe University, Bendigo, Victoria 3552, Australia

Iossif A. Strehin

Department of Biomedical Engineering, Whiting School of Engineering, Johns Hopkins University, Baltimore, MD 21218, USA

Catharine Le Visage

Inserm, U698, Bio-ingénierie de Polymères Cardiovasculaires; and Institut Galilée, Université Paris 13, 99 Av. J.B. Clément, 93430 Villetaneuse, France

Preface

In line with current focus on a sustainable economy, polysaccharides have received tremendous attention, making many headlines. The sources of polysaccharides are very diverse. The polysaccharides can originate from bacteria, fungi, algae and higher photosynthetic plants. At cellular level, polysaccharides represent either the reserve compounds in cytoplasm, or structural components of the membrane or cell wall of organisms. The chemistry and assembly of polysaccharides have direct influence on their methods of extraction and purification. Polysaccharides with varying physicochemical properties can be extracted at relatively low cost and can be chemically modified to suit specific needs for pharmaceutical and medicinal applications. Native and modified plant-, algae- and mushroom-derived polysaccharides have found to exhibit various beneficial pharmacological properties including immunomodulatory anticoagulant, anticancer, wound-healing, antihyperglycaemic and lipid-reducing properties. They also find use in drug delivery.

The aim of this book is to provide a comprehensive overview of research that has been carried out in the field of applications of polysaccharides in medicine and pharmacy covering the achievements, challenges and future needs.

The following applications of cellulose, which is the world's most abundant natural polymer, are proposed. Oxidised regenerated cellulose originating from plants is used in the form of woven cotton gauze dressing for wounds. Its haemostatic effect and action as an adhesion barrier has been exploited for many years and this application continues to be widespread. Cellulose is recognised as an excellent wound-dressing material because of the high water-holding capacity, elasticity, conformability, mechanical strength and porosity. In addition, it provides a barrier against external bacteria, and it is easy to sterilise and mould *in situ*. Furthermore, cellulose has been applied for over decades as a material for haemodialysis. Haemodialysis is a rather complex process since it involves more mechanisms than diffusion, for example, the adsorption of biomolecules on the surface. In order to minimise responses to other biological materials existing in blood, the active surfaces were coated with poly(ethylene glycol). Another approach to minimise the activation of polymorphonuclear leucocytes is functionalisation of the cellulose membrane with vitamin E, which reduces oxidative stress. Cellulose has also found application as culture material and postoperative adhesion barrier and scaffolds in tissue engineering and skin substitution. The structure of the cellulose matrix used depends strongly on its medical application. For example, wound dressings are nanoporous materials that retain moisture during the healing process, whereas cellulose as artificial skin substitute should possess high porosity in order to facilitate cell growth within the cellulose scaffold. Depending on its nano- and microstructured architectures, versatile characteristics of the biopolymer cellulose can be achieved and addressed to perform specific functions. Consequently, cellulose is a promising and broadly applicable material not only (as commonly known) in the paper and textile industries but also in the areas of medical and pharmaceutical devices, among many others. Membranes based on cellulose had historically been used for blood purification, and although they do continue to be thus used, in recent years, their use has declined in favour of membranes manufactured from synthetic material blends. The chapter on the applications of cellulose for blood purification reviews mechanisms for solute and water transport across membranes and discusses the biocompatibility of the material as well as the reasons for the shift away from its use in blood purification. As the clinical application of cellulose for blood purification processes declines, new applications for its use are emerging. Of these, the most promising appears to be the use of cellulose synthesised by *Acetobacter xylinum* as a wound-healing system.

Hemicelluloses are the second most abundant polysaccharides in nature after cellulose. The most important properties of the hemicelluloses include controllable biological activity, biodegradability and their ability to form hydrogels. These properties make these polysaccharides exceptionally suitable for improving lipid metabolism and mineral balance, protection against colon cancer, reducing risk of heart disease, improving body health and immunity, and formulation of drug delivery systems. Thus hemicelluloses can be used in nonmodified or modified form as emulsifiers, suspending agents and disintegrants, sustaining agents in tablets and as gelling agents. The most interesting utilisations of hemicelluloses are as controlled drug release, vehicles for incorporating antioxidants, antifungal and antimicrobial agents, colours and different nutrients.

Fungal exopolysaccharides are synthesised under the appropriate culture conditions. They include heteropolysaccharides, which are often complex and composed of different sugar monomers. Some of them have strong antioxidant character and are bioactive, and many have unique physical, chemical and biological properties of interest, for example, as immunomodulators, antitumoral agents, as effective agents in treating a range of infections microbially mediated diseases and in lowering blood pressure and cholesterol in humans. Their properties are correlated with their structure, which depends on the microorganism and the conditions of production. The elucidation of biosynthetic pathways and their regulation, which is possible by genetic manipulation, will make feasible the production of such polysaccharides on large scale.

Pullulan is an example of a fungal polysaccaharide that is commercially available. Due to the close resemblance of its features with those of dextran, some studies have been carried out with pullulan to develop blood-plasma substitutes. Pullulan can be easily derivatised in order to extend its applications by introducing different chemical groups (e.g., carboxymethyl, sulfate) on the backbone. Thus it has been possible to develop different systems based on hydrogel, films and multilayer formation for drug delivery. Pullulan was sulfated with the aim to obtain new alternatives to heparin. Pullulan hydrogels were proposed for use in creating drug delivery systems in the form of nano- and macrogels.

Botanical polysaccharides extracted from different sources have significant therapeutic potential and represent a rich base for future discovery and development of novel compounds of medical value. The nonspecific nature of polysaccharide immunomodulatory activity makes them attractive because they can be used to treat a broad spectrum of infections and are not susceptible to antibiotic resistance. In general, immunomodulators mimic the natural mechanism used by pathogens to stimulate innate immunity and thus are potentially beneficial in preventing infection. Additionally, many plant polysaccharides exhibit antioxidant activity adding to the beneficial effects for the host. Thus, the balance between therapeutic and proinflammatory properties is an important consideration when evaluating polysaccharide immunomodulators and the goal is to enhance local defence without inducing excessive or systemic inflammation. This balance is dependent on pharmacodynamic and pharmacokinetic properties of the polysaccharides and will be important to evaluate as further research focuses on development of such compounds as novel therapeutics to treat diseases.

Cyclodextrins are of great value when preparing pharmaceutical forms because of their entrapping capacity. This is why cyclodextrins are considered very useful tools in modern drug delivery systems as carriers capable of specific targeting in the organism. The interest for cyclodextrins in pharmaceutical uses is determined by reduction or elimination of bitter taste drugs, change of state (stabilisation of many volatile or aromatic substances, converting liquids and oil into free-flowing powders, formulation to maintain the higher dissolution characteristics and oral bioavailability of drugs), improvement in stability to maintain the efficiency of drugs, increase in solubility, improvement in bioavailability and decrease in side effects. For these reasons, cyclodextrins seem to be major tools in the formulation of a number of therapeutic substances. Cyclodextrins can be used for formulation improvements or drug delivery with protein, peptide and oligonucleotide dosage forms, the development of new derivatives and also new uses for existing derivatives to formulate dispersed systems.

Chondroitin sulfate is a constituent of tissue extracellular matrix, cell surface receptors and intracellular organelles such as lysosomes. The different variants of chondroitin sulfate can exhibit differences in biological activity towards the same type of cell. Chondroitin sulfate is a complex polysaccharide composed of the disaccharide repeat unit glucuronic acid and *N*-acetyl-D-galactosamine. The complexity of the molecule is determined by the presence of sulfate groups on the four hydroxyl groups of the disaccharide repeat unit. The number and location of sulfate groups per disaccharide unit can vary, making the encoding of information possible. This information can be used to guide cell behaviour by changes in intracellular signalling which could alter viability, proliferation, differentiation, extracellular matrix synthesis and matrix degradation. Chondroitin sulfate has shown beneficial effects for treating arthritis and was able to improve the biological activity of biomaterials for application such as heart valve engineering, wound healing and cartilage regeneration.

At present, many polysaccharides are used in nonmodified or modified forms to serve pharmaceutical needs as carriers for controlled drug release. In the form of micro- and nanoparticles, polysaccharides have been found to be useful in many diseases in different parts of the human body. There are a number of examples of micro- and nanoparticles based on polysaccharides applied in the fields of ophthalmic, respiratory, renal, cardiovascular, digestive, immunologic diseases, cancer therapy, neurologic and endocrine pathology.

Carbohydrates containing dendrimers in biomedical applications are reviewed to underline their implications in drug and gene deliveries, synthetic vaccines and prevention of pathological processes caused by bacteria and viruses. Two main approaches of glycodendrimer syntheses - convergent and divergent - are described. Since glycopeptide dendrimers suppress lectin-carbohydrate interactions, glycodendrimeric libraries provide potent inhibitors of bacterial adhesion and biofilm formation.

I want to sincerely thank all the authors who contributed to this volume for their dedicated effort and their excellent contributions. I am very grateful to these scientists for their willingness to contribute to this reference work and for their engagement. Without them and without their commitment and enthusiasm, it would not have been possible to compile such a book. I am also grateful to the publisher for recognising the demand for such a book, for taking the risk to bring out such a book and for realising the excellent quality of the publication. Special thanks are due to Helene Chavaroche and many of her colleagues, especially production and marketing, for their constant effort, their helpful suggestions, constructive criticism and wonderful ideas. Last but not least, I would like to thank my family for their patience, and I sincerely apologise for the many hours I spent in the preparation of this book, which kept me away from them.

This book will be useful to scientists and postgraduate students working with polysaccharides and their possible uses; and the book should also be of interest to people working in the areas of chemistry, biology, pharmacy and medicine. It may not only help in research and development but may be also suitable in the line of teaching.

I hope that you as a reader will enjoy the volume.

Valentin I. Popa

4HH, UK

Jan Jezek

Institute of Organic Chemistry and Biochemistry, Flemingovo nam. 2, 166 10 Prague, Czech Republic

Didier Letourneur

Inserm, U698, Bio-ingénierie de Polymères Cardiovasculaires; and Institut Galilée, Université Paris 13, 99 Av. J.B. Clément, 93430 Villetaneuse, France

Anne Meddahi-Pellé

Inserm, U698, Bio-ingénierie de Polymères Cardiovasculaires; and Institut Galilée, Université Paris 13, 99 Av. J.B. Clément, 93430 Villetaneuse, France

Cătălina Anişoara Peptu

Technical University Gheorghe Asachi, Faculty of Chemical Engineering and Protection of the Environment, D. Mangeron Street, 71A, 700050, Iasi, Romania

Marcel Popa

Technical University Gheorghe Asachi, Faculty of Chemical Engineering and Protection of the Environment, D. Mangeron Street, 71A, 700050, Iasi, Romania

Valentin I. Popa

Gheorghe Asachi Technical University of Iasi, Faculty of Chemical Engineering and Environmental Protection, Blvd. Mangeron No.71, Iasi, 700050, Romania

Mark T. Quinn

Immunology and Infectious Diseases, Montana State University, Bozeman, MT Inserm,

1 Configurations, Structures and Morphologies of Cellulose

Stephanie Schubert, Kerstin Schlufter and Thomas Heinze

1.1 Introduction

Cellulose is the world's most abundant natural polymer with an estimated annual global production and decomposition of about 1.5×10^{12} t, which is of the same order as the finite reserves of the most important fossil and mineral resources [1]. Because of the increasing environmental consciousness, the use of cellulose as a renewable and biodegradable raw material in various application fields is becoming increasingly attractive to successfully meet the environmental and recycling problems [2]. A versatile structuring of cellulose by mechanical and chemical modifications has led to its use in a number of applications, for example, as a filler, in building and coating materials, and for laminates, papers, textiles, optical films, sorption media, viscosity regulators and advanced functional materials for biomedical devices and enantioselective chromatography [1].

In ancient times, people used cellulose not only for generating fire but also for manufacturing clothes and papyrus [3]. In particular, hemp is supposed to be the oldest cultivated fibre plant used for the production of rope and cordage. Later on, the industrial applications of cellulose had a profound influence on the economy and the interrelated science and technology.

However, the structure and morphology of cellulose were unknown for centuries. The first systematic clarification began in 1837 with investigations of the French agricultural chemist Anselme Payen [4]. He showed that various plant materials yielded a fibrous substance after purification with acid-ammonia treatment and extraction with water, alcohol and ether. The French Academy finally named the resulting carbohydrate 'cellulose' [5]. Nowadays, there are various processes for isolating cellulose, for example, the alkaline, bisulfite and sulfate (kraft) process, in combination with thermal and mechanical treatments [6]. The different processes result in varying fibre strengths of the pulp.

1.2 Structure

1.2.1 Molecular Structure

The cellulose molecule consists of glucose units linked by β-1,4-glycosidic bonds. The glucose structure is a D-glucopyranose ring in $^4C_1$-chair configuration, which exhibits the lowest energy conformation [7]. The β-linkage results in a turning around of the cellulose chain axis by 180° at each glucose unit. Consequently, the actual repeating unit of cellulose is cellobiose with a length of 1.3 nm [8]. An anhydroglucose unit (AGU) within the cellulose chain has three reactive hydroxyl groups, one primary (C_6) and two secondary ones (C_2, C_3), which are able to undergo the typical reactions of hydroxyl groups including esterification, etherification and oxidation (**Figure 1.1**). The hydroxyl groups are positioned in the plane of the ring. A prolonged hydrolysis of cellulose under acidic conditions leads to depolymerisation and, finally, glucose.

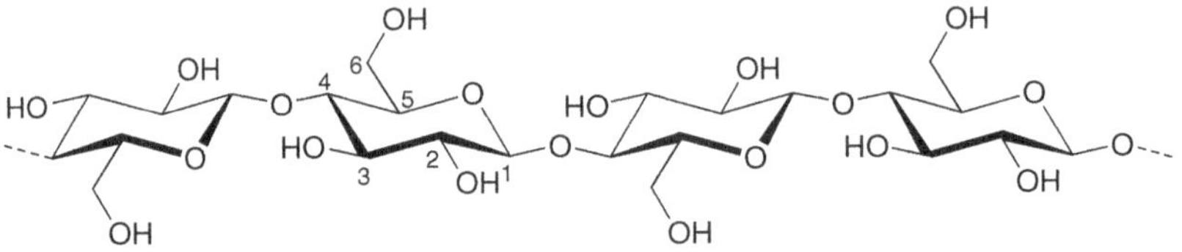

Figure 1.1 Schematic representation of a cellulose molecule

The chain ends of the cellulose molecule are chemically different [9]. One end has the anomeric carbon atom involved in the glycosidic linkage, whereas the other end consists of the unbound D-glucopyranose unit, which is in equilibrium with the aldehyde function and its reducing properties.

Native cellulose may also contain small amounts (<1%) of other constituents like lignin that were already built into or onto the cellulose fibres during photosynthesis [8]. However, most changes in the molecular structure originate from isolating reactions leading to hydrolysis or oxidation of the cellulose chain. Such reactions mainly occur on the surface of the fibrils or in amorphous regions.

The degree of polymerisation (DP) of native cellulose of various origins is in the range of 1000–30,000, which corresponds to chain lengths from 500 to 15,000 nm. The cellulose samples that are obtained after isolation methods have DP values ranging between 800 and 3000 [8]. However, the DP values must be considered as average values due to the fact that cellulose samples are always polydisperse. There are several techniques for obtaining information about the molar mass and its distribution, including viscosity and osmotic pressure measurements, size-exclusion chromatography (SEC) and light scattering.

1.2.2 Hydrogen Bonding

The formation of hydrogen bonds in cellulose is known to have a significant influence on the properties of cellulose [10]. The limited solubility in most solvents, the reactivity of the hydroxyl groups and the crystallinity of various cellulose samples all have their origin in the strong hydrogen bonding systems. Due to the fact that cellulose also contains hydrophobic areas (around the C atoms), there is an ongoing discussion about the influence of hydrophobic interactions on the overall properties including insolubility.

The three hydroxyl groups of the AGU and the oxygens of the ring and the glycosidic linkage are able to interact with each other within the chain or with the related groups of another cellulose chain by forming secondary valence bonds, namely intramolecular and intermolecular hydrogen bonds. The versatile possibilities for the formation of hydrogen bonds give rise to various three-dimensional structures. Solid-state carbon-13 nuclear magnetic resonance (^{13}C-NMR) measurements [11] and infrared spectroscopy [12, 13] reveal intramolecular bonds between the OH of C_3 and the adjacent ether oxygen of the AGU as well as a second one between the hydroxyl oxygen at C_6 and the adjacent OH at position C_2. The intramolecular hydrogen bonds are responsible for the relative stiffness and rigidity of the cellulose molecules in combination with the β-glycosidic linkage [8]. The chain stiffness results in highly viscous solutions (compared to the β-glycosidically linked polysaccharides like starch or dextran), a high tendency to crystallise and the formation of fibrillar assemblies.

In the crystal lattice, the cellulose molecules are connected to each other by intermolecular hydrogen bonds, in particular between the OH of C_6 and the oxygen of C_3 of an adjacent chain along the (002) plane of native cellulose (cellulose I) [14]. As a consequence, the cellulose molecules are linked together in a layer, and the layers are held together by hydrophobic interactions and weak C–H–O bonds, which could be confirmed by synchrotron X-ray and neutron diffraction data [15].

Cellulose II obtained by regeneration of the dissolved polymer shows, for example, a different hydrogen bonding system. Due to the existence of an intermolecular hydrogen bond between the OH of C_6 and C_2 of another chain, the intramolecular bonding of OH in C_2 is avoided and an intermolecular hydrogen bond of OH-C_2 to OH-C_2 of the next chain is formed [16]. This extra intermolecular hydrogen bond explains that, in comparison to cellulose I, the cellulose II molecules are more densely packed and strongly interbonded and, therefore, less reactive as commonly observed [8]. **Figure 1.2** is a schematic representation of the hydrogen bonding system in cellulose I and II.

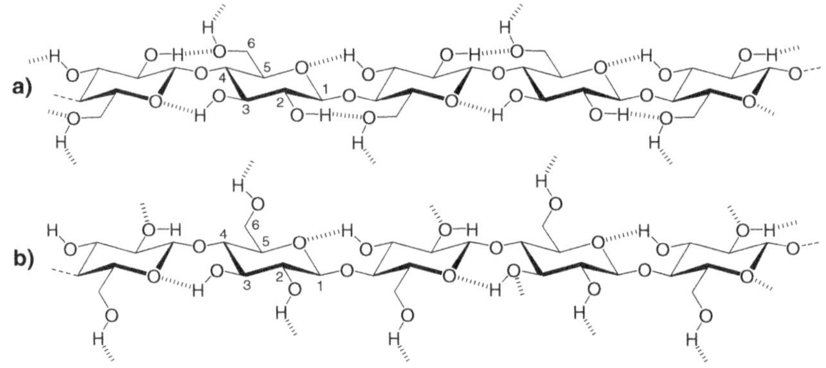

Figure 1.2 Hydrogen bonding system of (a) cellulose I and (b) cellulose II. Reproduced with permission from K. Tashiro and M. Kobayashi, *Polymer*, 1991, **32**, 1516. ©1991, Elsevier [17]

1.2.3 Crystal Modifications

The structure of cellulose is sufficiently regular that cellulose exhibits a crystalline X-ray diffraction pattern. In the past, there were a number of inconsistencies in the description of the crystalline structure of cellulose I and other cellulose modifications (cellulose II to IV), which were obtained after certain treatments [18, 19]. X-ray and NMR experiments later confirmed the dimorphism [20]. Therefore, the X-ray diffraction patterns (**Figure 1.3**) and the solid-state ^{13}C-NMR spectra (**Figure 1.4**)

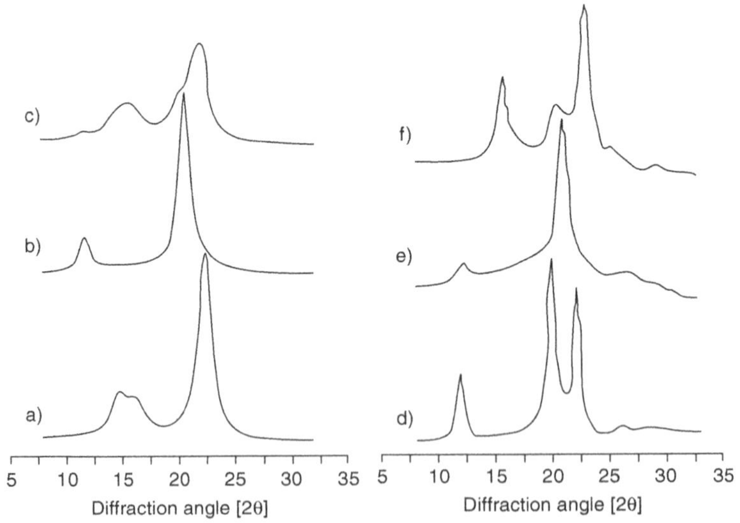

Figure 1.3 X-ray diffraction patterns of (a) cellulose I_{β}, (b) cellulose III_{I}, (c) cellulose IV_{I}, (d) cellulose II, (e) cellulose III_{II} and (f) cellulose IV_{II}. Reproduced with permission from A. Isogai, M. Usuda, T. Kato, T. Uryu and R.H. Atalla, *Macromolecules*, 1989, **22**, 3168. ©1989, ACS [21]

of cellulose conformations turned out to be valuable tools to elucidate the detailed crystalline structure and the mechanisms of the transformations of the different allomorphs [21].

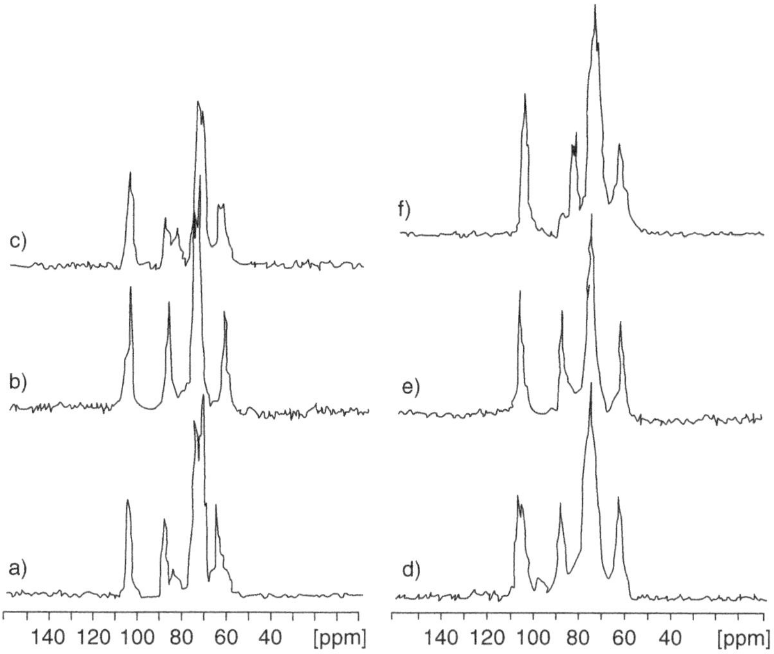

Figure 1.4 Solid-state ^{13}C-NMR spectra of (a) cellulose I_β, (b) cellulose III_I, (c) cellulose IV_I, (d) cellulose II, (e) cellulose III_{II} and (f) cellulose IV_{II}. Reproduced with permission from A. Isogai, M. Usuda, T. Kato, T. Uryu and R.H. Atalla, *Macromolecules*, 1989, **22**, 3168. ©1989, ACS [21]

Celluloses of different sources possess a comparable crystallinity, that is, the modification of cellulose I. However, based on solid-state ^{13}C-NMR studies, cellulose has been shown to crystallise in different phases namely cellulose I_α and I_β with varying proportions. Plant cellulose mainly consists of cellulose I_β, whereas cellulose produced by primitive organisms crystallises in the I_α phase. The monoclinic unit cell of cellulose I_α with the space group $P2_1$ consists of two cellulose molecules, each containing a cellobiose unit, in the 002 corner plane and 002 centre plane in a parallel fashion [14]. Cellulose I_α corresponds to a triclinic symmetry with space group P_1 containing one chain in the unit cell as schematically displayed in **Figure 1.5**.

Cellulose I can be transformed into a more stable crystalline form (cellulose II) by regeneration or mercerisation. The unmodified cellulose sample or a cellulose derivative can be dissolved following regeneration by precipitation and crystallisation. Mercerised cellulose II can easily be achieved by alkali treatment of native cellulose

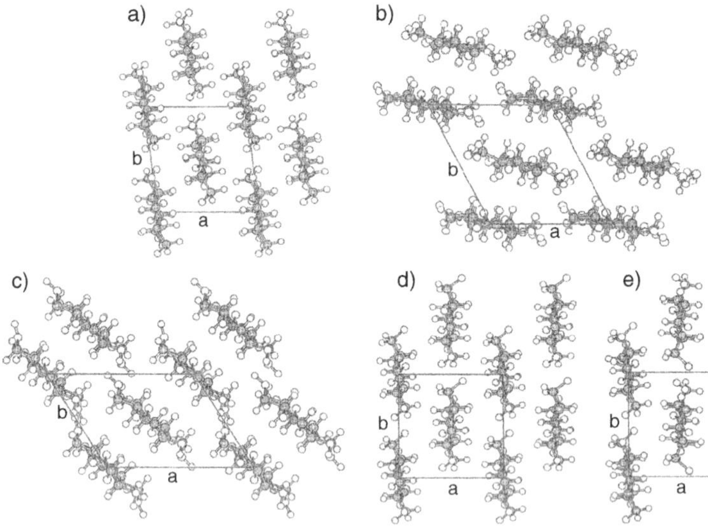

Figure 1.5 Representation of the model of (a) cellulose I_{β}, (b) cellulose II, (c) cellulose III_{I}, (d) cellulose IV_{I} and (e) cellulose VI_{II} on the *a-b* plane. Reproduced with permission from P. Zugenmaier, *Progress in Polymer Science*, 2001, **26**, 1341. ©2001, Elsevier [21]

at more than 18 wt% and by subsequent thorough washing procedures. Cellulose II can also be produced by a mutant strain of *Gluconacetobacter xylinum* [22]. The irreversible transition to cellulose II is used for improving the qualities of natural fibres and yarns. The structure of cellulose II could be revised by neutron fibre diffraction analysis [23]. Two chains of cellulose are located antiparallel on the 2_1 axis of the monoclinic cell (**Figure 1.5**), while the chains are displaced relative to each other by about one-fourth of the AGU.

The treatment of cellulose I and II with liquid ammonia and certain amines results in the formation of cellulose III_{I} and III_{II}, respectively [24]. The structures can easily be reconverted by mild heating into cellulose I or II. The unit cells for both crystalline structures are the same, and only the meridional reflexions differ in X-ray measurements. The crystalline structure of cellulose III_{I} can be described with a one-chain unit cell and a $P2_1$ space group with the cellulose chain axis on one of the 2_1 screw axes of the cells [25]. A single chain of cellulose III_{I} is similar to one of the two chains existing in the crystal of cellulose II.

Cellulose III_{I} and III_{II} can be transformed in glycerol at high temperatures into cellulose IV_{I} and IV_{II}, respectively, corresponding to the starting materials used. However, the conversion is never quantitative, which complicates the complete analysis of crystallinity [26]. The space group P_1 is assumed for both structures.

In addition to the crystalline domains in cellulose samples, there are also amorphous or noncrystalline regions, which influence the physical and chemical properties of celluloses intensely [27]. Interactions between solid cellulose and water, enzymes, or reactive or adsorptive substances occur first at the noncrystalline, amorphous domains or at the surface of cellulose crystals. Amorphous cellulose samples can be prepared by ball-milling of cellulose [28], deacetylation of cellulose acetate under nonaqueous alkaline conditions [29] or precipitation from nonaqueous cellulose solutions into nonaqueous media avoiding stress [30]. However, the amorphous structures are usually unstable in the presence of water and form partly crystalline cellulose II [29]. Interestingly, it was found that Raman and solid-state ^{13}C-NMR spectra of the amorphous and the highly crystalline cellulose IV_{II} are almost identical, which confirms that the secondary structures of the two cellulose types are similar [31].

1.2.4 Morphology

The ability of hydroxyl groups to form hydrogen bonds between the cellulose molecules is the basic requirement for their organisation into parallel arrangements of elementary crystallites, and, hence, into fibres [8]. The ordered crystallites are packages of chains folded in a longitudinal direction, whereas the areas containing turns between adjacent chain packages are the less-ordered, noncrystalline regions. The cellulose chains located in these regions are randomly oriented in a spaghetti-like arrangement leading to a lower density [32]. These together form the basic fibrillar units, the so-called elementary fibrils, with lateral dimensions between 1.5 and 3.5 nm and lengths of 100 nm [1]. The elementary fibrils are arranged into fibrillar bundles, namely microfibrils, and microfibrillar bands thereof with diameters in the range of 10–30 nm (microfibril) and 100 nm (microfibrillar band) and lengths of several hundred nanometres to some micrometres.

This architecture can be found in both native and man-made fibres, whereas the main difference lies in the position of the fibrillar elements to the fibre axis [8]. **Figure 1.6** illustrates the composition of a plant cell wall with the microfibrillar structures of cellulose. The microfibril of a natural fibre is further covered with a sheath of amorphous cellulose surrounded by hemicelluloses [34].

During the period of growth, the fibre possesses only a very thin cell wall, the primary cell wall. After completing growth in length, the secondary cell wall is formed on the inner side of the primary wall. The central channel of the cell is a hollow capillary, the lumen, covered with the tertiary cell wall. The width of a typical plant fibre is 15–30 µm with a cell wall thickness of about 4–6 µm. The primary and tertiary cell walls are disordered in nets with dimensions of about 100 nm. The secondary wall consists of three layers (S1, S2 and S3) and is about

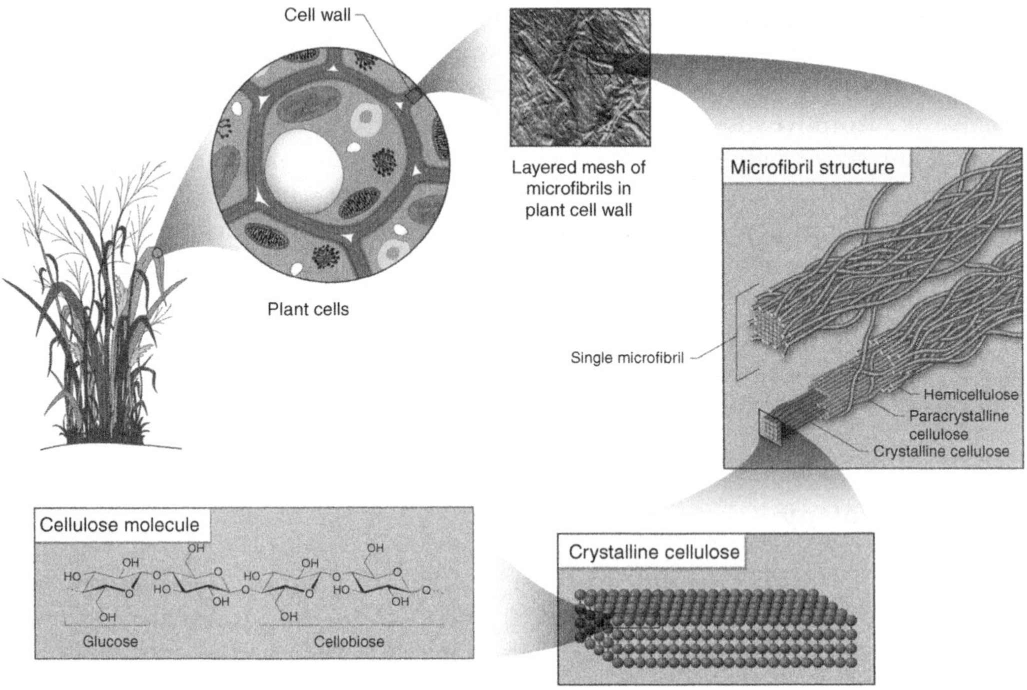

Figure 1.6 Association of cellulose molecules in the plant cell wall. Adapted from the US Department of Energy Office of Science. *http://genomics.energy.gov/gallery/gtl/detail.np/detail-36.html* [33]

3–5 µm thick; the dominant S2 layer is 2–4 µm. The microfibrils in the S2 layer are located parallel to each other and are oriented under a specific angle towards the fibre axis, the so-called 'microfibril angle', which correlates with the tensile strength of the fibres [35]. The build-up of the primary wall, in which the microfibrils are wound in a network-like and criss-cross helical fashion around the more extended and better oriented microfibrils of the secondary wall, has a significant influence on the swelling behaviour and, thus, on the fibre's physical and chemical properties. The fibres also contain various imperfections, in particular pores, cracks, nodes, compression failures, thin places and other sides of damages, which are weak points for chemical attack and mechanical forces. Fibres of various origins have different shapes and dimension, for example, cotton fibres are twisted (**Figure 1.7a**), whereas the fibres of spruce wood (**Figure 1.7b**) are generally untwisted [36]. Fibres of bast plants are straight and round (**Figure 1.7c**).

The fibrillar arrangement of regenerated cellulose is quite different from that of native cellulose fibres. Man-made fibres also consist of elementary fibrils but with a random location in the supramolecular structure [37]. In a precipitation process without any shear forces, the crystallites are randomly distributed in a semiamorphous matrix,

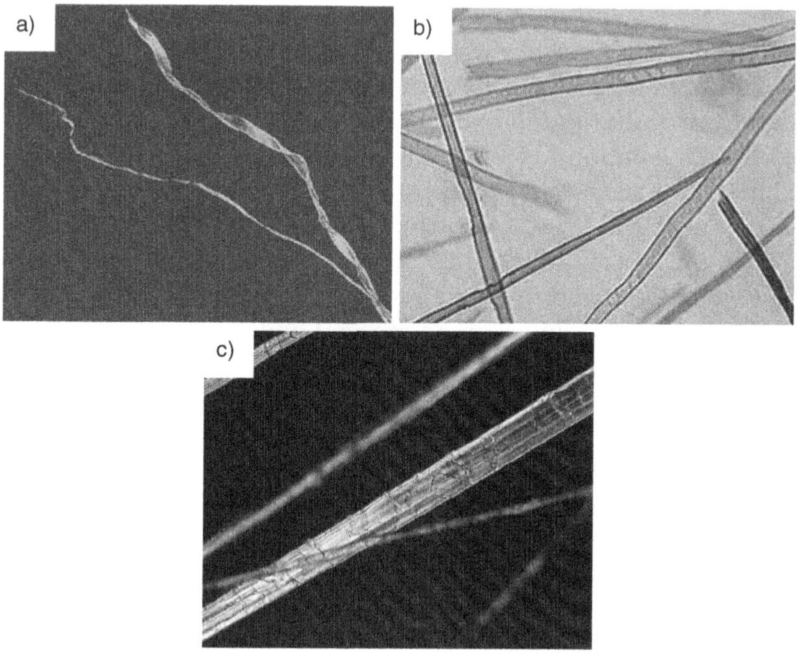

Figure 1.7 Micrographs of (a) twisted cotton fibres, (b) tracheids of spruce wood and (c) straight fibres of ramie. Adapted from M. Ioelovich, *Bioresources*, 2008, **3**, 1403. ©2008, [36]

whereas in a film-forming procedure the crystallites are positioned parallel to the film surface with an orientation in the direction of the draw [38]. In regenerated cellulose fibres, the crystallites are aligned with the longitudinal axis in the direction of stretch but with a certain transverse nonuniformity depending on the spinning conditions applied.

1.3 Properties

1.3.1 Solubility

Due to the extended hydrogen bonding system between the cellulose chains, there are only limited ways available to dissolve cellulose and functionalise it homogeneously or form it into the desired shape, which is necessary for the considerable production of cellulose fibres. The industrial-scale production was first conducted using cellulose nitrate as a soluble and, thus, formable cellulose derivative [1]. Later on, a process using a mixture of copper(II)hydroxide and aqueous ammonia with subsequent precipitation in dilute sulfuric acid was developed followed by probably the most important large-scale technical process in fibre production, the viscose process. For this process, cellulose is transformed into cellulose xanthogenate with subsequent spinning of the solution in aqueous sodium hydroxide. The Lyocell process is an environmentally friendly alternative to the viscose process, whereby cellulose is regenerated from a solution in *N*-methylmorpholine-*N*-oxide monohydrate. Other efficient cellulose solvents that are, up to now, used only in the laboratory scale are *N,N*-dimethylacetamide (DMAc)/LiCl, dimethyl sulfoxide (DMSO)/tetrabutylammonium fluoride (TBAF) and ionic liquids (e.g., 1,3-substituted imidazolium halides) [39, 40]. Among these, ionic liquids possess an enormous potential for the dissolution and chemical modification of cellulose in an efficient and environmentally friendly way even for industrial purposes [41].

1.3.2 Chemical Reactivity

The preparation of commercial cellulose derivatives is exclusively carried out under heterogeneous reaction conditions at least at the beginning of the conversion. However, the dissolution of cellulose prior to the chemical reaction offers a great opportunity to design novel and unconventional cellulose derivatives by homogeneous phase chemistry. Thus, homogeneous phase chemistry with cellulose stands for the dissolution of the biopolymer prior to the chemical reaction in either nonderivatising or derivatising solvents. In the case of derivatising solvents, not only the conversion of the soluble intermediate formed during dissolution but also the modification of the isolated intermediate, which is redissolved in an organic solvent (DMSO, *N,N*-dimethylformamide (DMF)), is considered as homogeneous reaction. On the contrary, the chemical modification of soluble but 'stable' cellulose derivatives like cellulose acetate in DMSO as well as the chemical modification of cellulose under dissolution of the cellulose derivative formed (as a result of the conversion) is not included in the context of homogeneous phase chemistry. In the following discussion, typical conversions will be briefly discussed in order to give the reader an idea about the importance of cellulose as a renewable resource for product design by chemical modification, which is important to take full advantage of this biopolymer.

1.3.2.1 Esterification

Conventional acylation of cellulose is carried out heterogeneously using anhydrides or chlorides of the carboxylic acids. The most common method, which is also realised in the technical scale, is the conversion of cellulose in a mixture of acetic acid and acetic acid anhydride (10–40% in excess of the amount needed for the formation of cellulose triacetate) in the presence of a strong mineral acid (H_2SO_4, $HClO_4$) as catalyst. For a better control of the reaction temperature and to reduce the amount of catalyst, the acetylation is carried out in methylene chloride, which leads to dissolution of the product in the final state of the reaction. Thus, the reaction starts heterogeneously and the cellulose product dissolves at a certain degree of substitution (DS) in the reaction medium (this process is not considered as homogeneous reaction). The reaction occurs as a peeling process, that is, at intermediate DS values cellulose chains with different amounts of functional groups appear, and, consequently, the cellulose acetate is only partly soluble or even insoluble [42, 43].

Due to the even accessibility of the hydroxyl groups under homogeneous reaction conditions, an even distribution of functional groups within the polymer chain can be realised. It is possible to prepare cellulose acetates soluble in acetone or water starting from the dissolved biopolymer. It is a general phenomenon of homogeneous conversions that well-soluble products can be synthesised due to the even distribution of the ester moieties within and between the polymer chains [44–46].

The homogeneous acylation of cellulose dissolved in nonderivatising solvents proceeds usually faster at the primary hydroxyl group at position 6 [47]. The evaluation of the functionalisation pattern of cellulose esters that are homogeneously prepared revealed a distribution in the order of $C_6 > C_2 > C_3$ (**Table 1.1**). However, conversion in ionic liquids may lead to products with a higher partial DS at position 3 compared to position 2 [53, 54].

The tosylation of cellulose with tosyl chloride can be carried out either heterogeneously in pyridine or homogeneously in DMAc/LiCl. While the heterogeneous conversion leads to products that contain nitrogen and chlorine and may also be insoluble due to a variety of side reactions, the homogeneous functionalisation yields well-soluble products with controlled DS (up to 2.3) when different molar ratios and a reaction temperature of about 10 °C are used [48, 55]. The increase of DS (up to a certain value) may be almost linear with the molar ratio of reagent to AGU as exemplified for the sulfation of cellulose in ionic liquids (**Figure 1.8**).

1.3.2.2 Etherification

Cellulose ethers are prepared in laboratory-scale synthesis mainly and in large-scale production exclusively under heterogeneous conditions using a slurry process, that is, conversion of alkali cellulose swollen in aqueous NaOH and an organic liquid containing the etherifying agent (e.g., methyl chloride, monochloroacetic acid, ethylene oxide and propylene oxide). The differences and similarities of heterogeneous *versus* homogeneous processes will be exemplified by discussing the most important ionic cellulose ether carboxymethyl cellulose (CMC, as it is usually called). The heterogeneous slurry process for the preparation of CMC yields products with DS values in the range of 0.5–1.3 (**Figure 1.9**) [57–59]. It was revealed that the carboxymethyl functions are distributed according to $C_2 \geq C_6$ C_3 within the AGU, by means of ^{13}C- and ^{1}H-NMR

Table 1.1 Functionalisation pattern of cellulose esters prepared from different cellulose materials (bacterial cellulose (BC), cotton linters (CL), sisal, bagasse, dissolving pulp, pulp, microcrystalline cellulose (MC)) dissolved in nonderivatising solvents applying acetic anhydride (Ac_2O), acetyl chloride (AcCl), vinyl acetate (VA) and 2-furoyl chloride (FurCl)

Conditions						Degree of substitution				Reference
Solvent	Cellulose	Reagent	Molar ratio[a]	Time (hours)	Temperature (°C)	O-6	O-2	O-3	Total	
[AllylMeIm][Cl]	Pulp	Ac_2O	1:5	0.25	80	0.71	0.01	0.14	0.94	[47]
[AllylMeIm][Cl]	Pulp	Ac_2O	1:5	2	80	0.87	0.46	0.56	1.80	[47]
[C_4MeIm][Cl]	BC	Ac_2O	1:1	2	80	0.41	0.11	0.17	0.69	[47]
[C_4MeIm][Cl]	BC	Ac_2O	1:2	2	80	1.00	0.29	0.37	1.66	[47]
[C_4MeIm][Cl]	BC	Ac_2O	1:3	2	80	1.00	0.57	0.68	2.25	[47]
[C_4MeIm][Cl]	BC	Ac_2O	1:5	2	80	1.00	0.71	0.79	2.50	[47]
[C_4MeIm][Cl]	MC	FurCl	1:1	3	65	0.38	0.08[b]		0.46	[48]
[C_4MeIm][Cl]	MC	FurCl	1:1	3	65	0.47	0.15[b]		0.62	[48]
[C_4MeIm][Cl]	CL	FurCl	1:3	3	65	0.48	0.19[b]		0.67	[48]
DMAc/LiCl	Bagasse	Ac_2O	1:3	18	60	0.80	0.58	0.72	2.10	[49]
DMAc/LiCl	CL	Ac_2O	1:3	18	60	0.80	0.42	0.68	1.90	[49]
DMAc/LiCl	Sisal	Ac_2O	1:3	18	60	0.92	0.52	0.66	2.10	[49]
DMAc/LiCl	MC	AcCl	1:1	2	80	0.63	0.37[b]		1.00	[44–46]
DMSO/TBAF	MC	AcCl	1:2.3	70	40	0.35	0.48[b]		0.83	[50]
DMSO/TBAF	MC	VA	1:2.3	70	40	0.49	0.55[b]		1.04	[50]
NaSCN/LiSCN*$2H_2O$	Pulp	Ac_2O	1:100	3	130	0.91	1.57[b]		2.41	[51]
NaSCN/LiSCN*$2H_2O$	Pulp	Ac_2O	1:100	1	130	0.86	1.12[b]		1.98	[51]
NaSCN/LiSCN*$2H_2O$	Pulp	Ac_2O	1:100	0.5	130	0.39	0.85[b]		1.23	[51]

[a]*Molar ratio cellulose:reagent*

[b]*Degree of substitution in total at O-2 and O-3*

Table 1.2 Degree of substitution (DS) of carboxymethyl cellulose depending on the concentration of aqueous NaOH (reaction of spruce sulfite pulp, degree of polymerisation = 650, in 2-propanol and aqueous NaOH with monochloroacetic acid for 5 hours at 55 °C)

NaOH concentration (%, *w/v*)	5	8	10	15	20	30
DS (carboxymethyl cellulose)	0.59	0.93	1.00	1.24	1.03	0.95

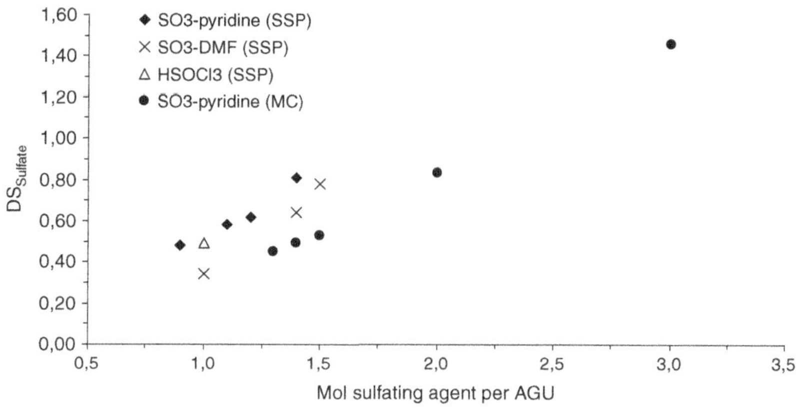

Figure 1.8 Degree of substitution (DS) of cellulose sulfates prepared from spruce sulfite pulp (SSP) and microcrystalline cellulose (MC) depending on the molar ratio of sulfating reagent to anhydroglucose unit (AGU). Reproduced with permission from M. Gericke, T. Liebert and T. Heinze, *Macromolecular Bioscience*, 2009, **9**, 343. ©2009, Wiley InterScience [56]

spectroscopy starting with the pioneering studies of Reuben and Conner [60], as well as by means of high-pH anion-exchange chromatography with pulsed amperometric detection of hydrolytically degraded samples [61].

A mathematical processing of the ^{13}C-NMR spectra reveals the monomer composition, that is, the molar ratio of all differently modified units that build up the polymer chain in the case of partially functionalised samples (2-, 3- and 6-mono-O-carboxymethyl unit; 2,3-, 2,6- and 3,6-di-O-carboxymethyl unit as well as 2,3,6-tri-O-carboxymethyl unit) (**Figure 1.10**) [60]. Another efficient method to gain information about the content of the repeating units is based on high-performance liquid chromatography (HPLC) analysis after chain degradation with perchloric acid [62, 63]. The mole fractions of non-, mono-, di- and tricarboxymethylated units and of glucose can be analysed to determine the different positions of both mono- and dicarboxymethylated units. For a broad variety of CMC samples prepared by the conventional slurry process, no significant deviation from a statistical pattern of functionalisation is achievable [62, 64]. In this context, statistical functionalisation

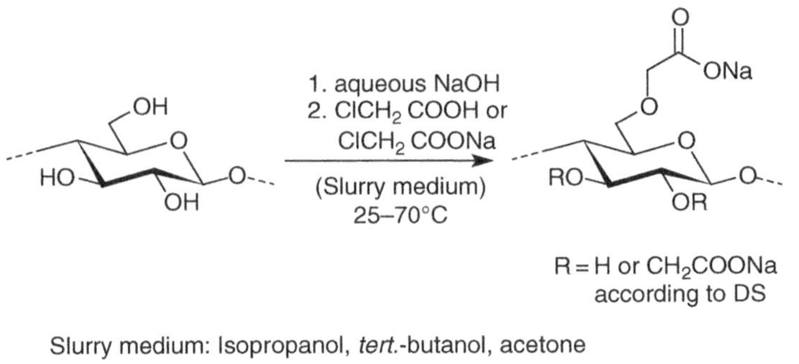

Figure 1.9 Commercial synthesis of carboxymethyl cellulose by Williamson ether synthesis

means a composition of the repeating units according to the binomial distribution (**Equation (1.1)**). Representative results are demonstrated in **Figure 1.11** for CMC samples within a wide DS range prepared in a completely heterogeneous manner using aqueous sodium hydroxide of mercerisation concentration.

$$c_i = \binom{3}{k}\left(\frac{DS}{3}\right)^k\left(1-\frac{DS}{3}\right)^{3-k} \tag{1.1}$$

Cellulose dissolved in an aqueous solution of $Ni(tren)(OH)_2$ offers the possibility for a completely homogeneous carboxymethylation of the biopolymer [65]. In the case of a careful addition of an aqueous NaOH up to concentrations of 31% (*w*/*v*), neither

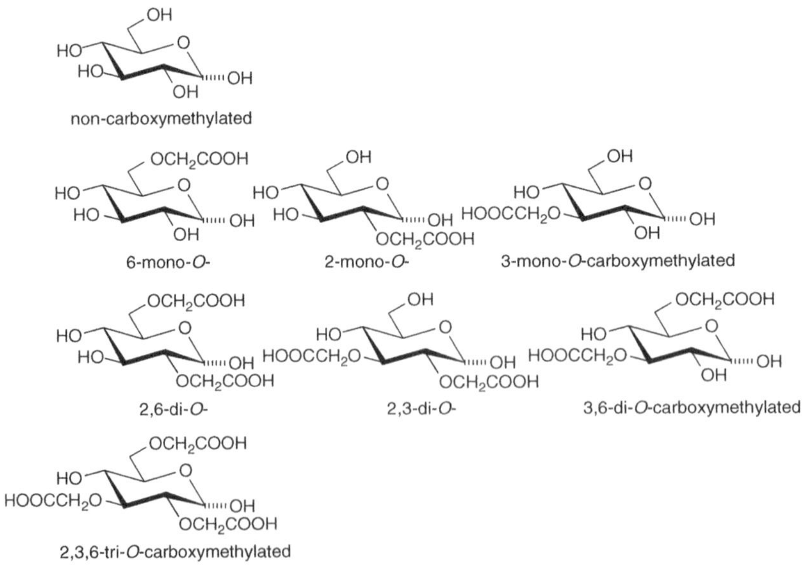

Figure 1.10 Different repeating units appearing in carboxymethyl cellulose at a degree of substitution <3 (after hydrolytic chain degradation)

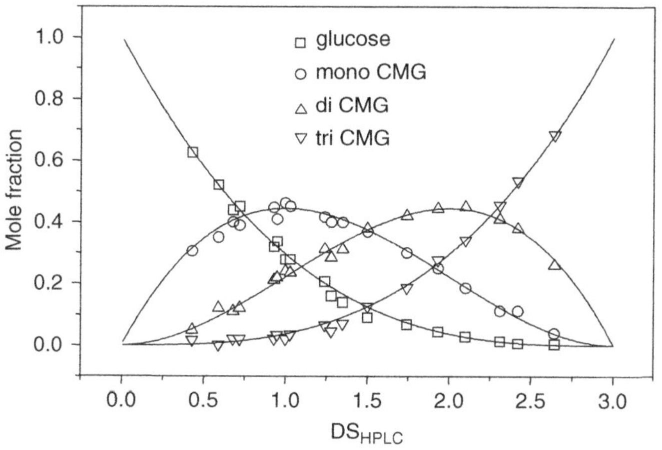

Figure 1.11 The mole fractions of glucose, mono-*O*-carboxymethyl glucose (mono-CMG), di-*O*-carboxymethyl glucose (di-CMG) and 2,3,6-tri-*O*-carboxymethyl glucose (tri-CMG) in hydrolysed carboxymethyl cellulose samples (heterogeneous synthesis) plotted as a function of degree of substitution determined by means of high-performance liquid chromatography (DS_{HPLC}). The curves are calculated as described in the text. Reproduced with permission from T. Liebert and T. Heinze, *ACS Symposium Series*, 1998, **688**, 61. ©1998, ACS [63]

gelation nor precipitation of the polymer occurs. No regeneration was observed during the subsequent addition of 36% (*w/v*) aqueous sodium monochloroacetate. This totally homogeneous procedure gives a maximum DS_{CM} of 0.54 by application of a molar ratio of AGU:NaOH:sodium monochloroacetate of 1:20:10 (**Table 1.3**). The stepwise addition of reagents leads to a slight increase of the DS to 0.71 due to the reduced hydrolysis of sodium monochloroacetate (an important side reaction) during the etherification. Nevertheless, regarding the total DS *versus* the molar excess of etherifying agent, the carboxymethylation in the aqueous solution is less effective compared to the heterogeneous conversion [66].

Structure determination of CMC samples obtained by homogeneous carboxymethylation (solvent $Ni(tren)(OH)_2$) was carried out by means of ^{1}H-NMR spectroscopy and HPLC (both after complete hydrolytic depolymerisation) indicating a distribution of substituents at the level of the AGU in the order O-2 > O-6 > O-3 (e.g., for a sample with DS 0.44, the partial DS values at positions 2 = 0.17, 3 = 0.08 and 6 = 0.19) [67]. An analogous distribution of functional groups within the AGU was evaluated for CMC prepared in totally heterogeneous reactions (slurry process), using different analytical methods [68]. The amounts of functionalised units (2-, 3- and 6-mono-*O*-carboxymethyl glucose; 2,3-, 2,6- and 3,6-di-*O*-carboxymethyl glucose; and 2,3,6-tri-*O*-carboxymethyl glucose) as well as unmodified glucose, which build up the polymer

Table 1.3 Conditions for and results of the homogeneous carboxymethylation of cellulose in $Ni(tren)(OH)_2$

Reaction conditions			Carboxymethyl cellulose	
Molar ratio[a]	Time (hours)	Temperature (°C)	DS	Solubility in water
1:5:2.5	3	80	0.11	No
1:10:5	3	80	0.25	No
1:20:10	3	80	0.54	Yes
1:40:20	3	80	0.50	Yes
1:20:10	24	80	0.44	Yes
1:40:20[b]	4	80	0.71	Yes

[a]Molar ratio of anhydroglucose unit:NaOH (20% in water):sodium monochloroacetate
[b]Addition in four steps
Reproduced with permission from T. Heinz, T. Liebert, P. Klüfers and F. Meister, Cellulose, 1999, **6***, 153. ©1999, SpringerLink [61]*

chains, are given in **Table 1.4**. Comparison of these values with statistically calculated values [60], given in parentheses in **Table 1.4**, shows a good fit of the data.

Thus, the homogeneously prepared CMC samples possess a statistical content of different repeating units, that is, both the simple activation of cellulose with aqueous NaOH in a wide range of concentrations [58, 66] and the dissolution and complete homogeneous conversion of the polysaccharide yield a reactive polymer with an almost even accessibility of the reactive sites along the polymer chain. Hence, the reaction is not diffusion controlled.

The treatment of cellulose dissolved in DMAc/LiCl with NaOH powder in order to initiate the etherification reactions leads to gelation of the systems. In the case of the water-free system, the gelation is due to the regeneration of cellulose (crystal modification of cellulose type II) on the interface of solid particle-solution as evaluated by Fourier transform infrared (FTIR) spectroscopy and polarising light microscopy [62, 68, 69]. This process is called induced phase separation, leading to reactive microstructures. The FTIR spectra possess signals of free OH groups (i.e., not included or less included in hydrogen bonds) as observed for low-molar-mass alcohols in the gas phase indicating a very reactive polymer. The reaction of cellulose with sodium monochloroacetate yields CMC with DS values as high as 2.2 in a one-step procedure, which is significantly higher compared to the heterogeneous reaction. ^{1}H- and ^{13}C-NMR spectroscopy revealed that the CMC prepared via induced phase

Table 1.4 Results of high-performance liquid chromatography analyses of homogeneously prepared carboxymethyl cellulose samples (solvent: Ni(tren)$(OH)_2$) after hydrolytic chain degradation in comparison to statistically calculated data (in parentheses) according to the Spurlin statistics

DS	Mole fractions of...			
	Glucose	Mono-*O*-carboxymethyl glucose	Di-*O*-carboxymethyl glucose	Tri-*O*-carboxymethyl glucose
0.22	0.789	0.199	0.012	0
	(0.7957)	(0.1889)	(0.015)	(0.0004)
0.11	0.890	0.107	0	0
	(0.8940)	(0.1021)	(0.0032)	(0.0000)
0.25	0.765	0.216	0.020	0
	(0.7702)	(0.2101)	(0.0191)	(0.0006)
0.54	0.574	0.333	0.072	0.020
	(0.5514	(0.3631)	(0.0797)	(0.0058)
0.50	0.564	0.367	0.070	0
	(0.5787)	(0.3472)	(0.0694)	(0.0046)
0.44	0.634	0.310	0.082	0.047
	(0.6214)	(0.3204)	(0.0551)	(0.0031)
0.71	0.436	0.435	0.128	0.005
	(0.4448)	(0.4137)	(0.1283)	(0.0132)

separation has a much higher degree of etherification at positions 6 and 3 compared to conventionally prepared samples of similar total DS.

A comparison of the mole fractions of the different repeating units, determined by HPLC after chain degradation, with values calculated by statistics [60], which simulate the conversion along the polymer chain without preference for any of the OH groups and without the influence of the DS already reached, leads to very interesting results. As mentioned before, the statistical data meet the mole fractions of conventionally obtained CMC completely in the case of polymers synthesised via induced phase separation, although significant differences between the data sets exist (**Figure 1.12**). The comparably high amount of glucose and 2,3,6-tri-*O*-carboxymethyl glucose units is an evidence for a gradient-like distribution of ether functions along the backbone. After the induced phase separation, the carboxymethylation is more or less limited to the area directly at the interface between polymer and solid NaOH particle.

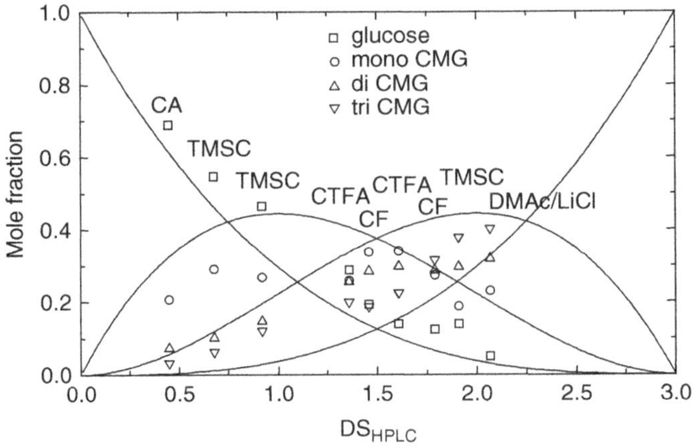

Figure 1.12 The mole fractions of glucose, mono-*O*-carboxymethyl glucose (mono-CMG), di-*O*-carboxymethyl glucose (di-CMG) and 2,3,6-tri-*O*-carboxymethyl glucose (tri-CMG) in hydrolysed carboxymethyl cellulose samples (polymers were synthesised via induced phase separation starting from cellulose acetate (CA), trimethylsilyl cellulose (TMSC), cellulose trifluoroacetate (CTFA), cellulose formate (CF) and cellulose dissolved in *N,N*-dimethylacetamide/LiCl) plotted as a function of degree of substitution determined by means of high-performance liquid chromatography (DS_{HPLC}). Reproduced with permission from T. Liebert and T. Heinze, *ACS Symposium Series*, 1998, **688**, 61. ©1998, ACS [63]

This new synthesis concept is not limited to cellulose dissolved in DMAc/LiCl. The reaction of cellulose dissolved in DMSO/TBAF trihydrate via activation with NaOH powder gives CMC of high DS of up to 2 possessing a nonstatistical content of the different repeating units [70]. Using *N*-methylmorpholine-*N*-oxide (NMNO) as solvent, comparable results regarding the functionalisation pattern are obtained (**Table 1.5**) [66].

It should be noted that various cellulose intermediates (derivatising solvents) and even cellulose derivatives of different hydrolytic stability that reacted under water-free conditions may give products with a nonstatistical content of the repeating units. Solutions of cellulose trifluoroacetate (DS_{CTFA} 1.5, DP 460), cellulose formate (DS_{CF} 2.2, DP 260), commercial cellulose acetate (DS_{CA} 1.8, DP 220) and trimethylsilyl cellulose (DS_{TMSC} 1.1, DP 220) in DMSO (5.7%, *w/v*, polymer) treated with solid NaOH particles suspended in DMSO show phase separation and formation of a reactive microstructure (see **Figure 1.12**). Examples of the conditions and the results of the described synthesis method are given in **Table 1.5**. Overall DS values of up to 2.2 can be obtained [71].

The addition of solid NaOH to the dissolved intermediates and derivatives of cellulose results in the cleavage of the primary substituents, for example, trifluoroacetate, and the formation of cellulose II, which is regenerated on

Table 1.5 Conditions for and results of the carboxymethylation of cellulose in NMNO/DMSO and DMSO/TBAF and the results of the high-performance liquid chromatography analysis (after complete depolymerisation) in comparison to statistical calculations

Reaction conditions				Mole fractions				DS_{CMC}
Medium	Molar ratio[a]	Time (hours)	Temperature (°C)	Glucose	Mono-*O*-CMG	Di-*O*-CMG	Tri-*O*-CMG	
NMNO	1:10:20	2	80	0.772 (0.6815)[b]	0.130 (0.2788)	0.066 (0.0380)	0.031 (0.0017)	0.36
NMNO + 2 ml DMSO per gram cellulose	1:20:10	2	80	0.307 (0.1951)	0.282 (0.4238)	0.227 (0.3069)	0.183 (0.0741)	1.26
NMNO + 2.5 ml DMSO per g cellulose	1:20:10	2	80	0.419 (0.3191)	0.295 (0.4436)	0.198 (0.2056)	0.087 (0.0317)	0.95
NMNO + 3 ml DMSO per g cellulose	1:20:10	2	8	0.267 (0.4625)	0.159 (0.4067)	0.120 (0.1192)	0.093 (0.0116)	0.68
DMSO/TBAF	1:5:10	0.5	70	0.106 (0.016)	0.267 (0.282)	0.327 (0.434)	0.299 (0.223)	1.82
DMSO/TBAF	1:5:10	2	70	0.081 (0.035)	0.215 (0.198)	0.444 (0.305)	0.305 (0.428)	2.02
DMSO/TBAF	1:5:10	4	70	0.068 (0.028)	0.198 (0.192)	0.305 (0.442)	0.428 (0.338)	2.09
DMSO/TBAF	1:5:10	16	70	0.084 (0.048)	0.253 (0.252)	0.331 (0.442)	0.332 (0.258)	1.91
DMSO/TBAF	1:10:20	48	70	-	-	-	-	1.89

[a]*Molar ratio of anhydroglucose unit:NaOH:sodium monochloroacetate*

[b]*Calculated according to the Spurlin statistics*

DS_{CMC}, degree of substitution of carboxymethyl cellulose

the solid particles. The particle size of the NaOH powder affects the overall DS but does not influence the distribution of the functional groups. A decrease in the size of NaOH particles from 1.00 to <0.25 mm increases the DS from 0.64 to 1.12 if the reaction is performed with cellulose trifluoroacetates (**Table 1.6**).

Endoglucanase fragmentation of the CMC followed by analytical and preparative SEC proved that samples with a DS of up to 1.9 are intensively degraded, supporting a block-like pattern of functionalisation. The detailed analysis of the

Table 1.6 Conditions for and results of the carboxymethylation of cellulose via induced phase separation starting from dissolved cellulose and cellulose derivatives (reaction temperature: 70 °C)

Reaction conditions			Carboxymethyl cellulose	
Starting material	**Molar ratio[a]**	**Time (hours)**	**DS**	**Solubility in water**
Cellulose in DMAc/LiCl	1:2:4	48	1.13	–
Cellulose in DMAc/LiCl	1:4:8	48	1.88	+
Cellulose in DMAc/LiCl	1:5:10	48	2.07	+
Cellulose trifluoroacetate	1:5:10	2	0.11	–
Cellulose trifluoroacetate	1:10:20	4	1.86	+
Cellulose trifluoroacetate	1:10:20	16	1.54	+
Cellulose trifluoroacetate	1:10:20[b]	4	0.62	–
Cellulose trifluoroacetate	1:10:20[c]	16	0.97	–
Cellulose formate	1:10:20	2	1.46	+
Cellulose formate	1:10:20	4	1.91	+
Cellulose formate	1:20:40	2	2.21	+
Trimethylsilyl cellulose	1:10:20	0.5	2.04	+
Trimethylsilyl cellulose	1:10:20	1	1.91	+
Trimethylsilyl cellulose	1:10:20	2	1.97	+
Cellulose acetate	1:10:20	2	0.36	–
Cellulose acetate	1:10:20	4	0.45	–

[a]*Molar ratio of anhydroglucose unit:NaOH:sodium monochloroacetate*
[b]*NaOH particle size: 0.63–1.00 mm*
[c]*NaOH particle size: 0.25–0.63 mm*

fragments obtained was carried out by preparative SEC, hydrolysis and anion-exchange chromatography with pulsed amperometric detection. From these studies, a structure can be proposed with chains containing segments of very high DS alternating with areas of limited substitution. It was clearly shown that the highly carboxymethylated fragments were dominated by 2,3,6-tri-O-carboxymethyl glucose units [72].

Based on the reaction starting from the dissolved polymer in either nonderivatising or derivatising solvents and induced phase separation, a novel synthesis concept was developed that yields cellulose derivatives with a nonstatistical content of the differently substituted repeating units and a block-like structure [64].

1.4 Sources of Cellulose

Cellulose is distributed throughout the nature in plants, animals, algae as well as minerals. Plant fibres are, however, the major source of cellulose. In fact, cellulose contributes approximately 40% to the carbon fraction in plants. In plant materials, the cellulose macromolecules serve as structural elements within the complex architecture of plant cell walls. Cellulose occurs in the pure form in plants but it is usually found in a mixture with hemicelluloses, lignins and comparably small amounts of extractives. Wood contains about 40–50 wt% cellulose. Comparable amounts can be found in bagasse (35–45 wt%), bamboo (40–55 wt%) and straw (40–50 wt%), and even higher amounts can be found in flax (70–80 wt%), hemp (75–80 wt%), jute (60–65 wt%), kapok (70–75 wt%) and ramie (70–75 wt%); cotton represents the purest cellulose source containing more than 90 wt% [73]. It is an impressive amount of cellulose that is produced in 1 year not only by wood fibres from trees with 1,750,000 kt world production but also in bamboo 10,000 kt, cotton linters 18,450 kt, jute 2300 kt, flax 830 kt, sisal 378 kt, hemp 214 kt and ramie 100 kt, just to mention some of the most important cellulose sources [34]. The abundance of such a raw material is of great benefit for the manufacturing industry.

In addition, several fungi as well as green algae (e.g., *Valonia ventricosa*, *Chaetomorpha melagonium*, *Glaucocystis*) contain cellulose in their cell walls, and some animals such as ascidians, which are marine animals, contain cellulose in the outer membrane. Several bacteria of the genera *Gluconacetobacter*, *Agrobacterium*, *Pseudomonas*, *Rhizobium* and *Sarcina* can synthesise bacterial cellulose directly from glucose [74, 75]. However, not all bacterial species mentioned are able to secrete the cellulose extracellularly. Bacterial cellulose that is produced directly in the form of a fibrous network contains no lignin, pectin, hemicelluloses as well as other biogenic products, and is highly crystalline and has high DP values (see **Chapter 5**). A selection of important cellulose sources is given in **Figure 1.13**.

The crystal modification of the extracted cellulose biopolymer is always I_β such as cellulose present in the outer membrane (tunic) of marine animals [81]. Cellulose from algae (*Glaucocystis*) is uniquely composed of cellulose I_α [82].

A synthetic preparation of cellulose can be realised by ring-opening polymerisation of 3,6-di-O-benzyl-α-D-glucopyranose-1,2,4-orthopivalate [83] or by stepwise reactions of selectively protected β-D-glucose [84]. Products having comparatively low DP values are very helpful in the determination of structure-property relationships. Moreover, cellulose can be produced enzymatically by *in vitro* synthesis starting from cellobiosyl fluoride using cellulase as catalyst [85].

Figure 1.13 A selection of important cellulose sources: (a) hard wood (beech tree) [76]; (b) bamboo [77]; (c) cotton [78]; (d) sisal [79]; (e) tunicate [80] and (f) *Gluconacetobacter xylinum*. ©2006, Vera Chimscholli ihr sein TextBlog, Wordpress, *http://textblog.wordpress.com/2006/08/29/mein-freund-der-baum/* [76]; ©Bamboo Fiber and Bamboo Bedding, Linenplace, *http://www.linenplace.com/boutiques/product-ideas/bamboo-bedding.html* [77]; ©Reallynature, *www.reallynatural.com/archives/cat_business.php* [78]; ©San Marcos Growers, *http://www.smgrowers.com/products/plants/plantdisplay.asp?plant_id=2273* [79]; and ©Fisheries and Oceans Canada, *http://www.dfo-mpo.gc.ca/science/Publications/article/2007/17-09-2007-eng.htm* [80]

1.5 Micro- and Nano-scale Cellulose Materials

The high potential of cellulose as raw material in almost all applications derives from its structural diversity due to the supramolecular architecture and the glucose main chain. For example, the ease of adhesion on cellulose surfaces has contributed to its use as a high-value fibre-based material. Cellulose can also be considered as a source for micro- and nano-scale cellulose materials. Specific treatments of natural cellulose of various origins result in small, defined fibres like microfibrillar cellulose (**Figure 1.14**), microcrystalline cellulose (**Figure 1.15**), cellulose whiskers (**Figure 1.16**) and nanofibres; cellulose nanofibres prepared by electrospinning have unique characteristics depending on the preparation technique used and the source of natural cellulose. Basically, wood is used for the preparation of nanomaterials; in

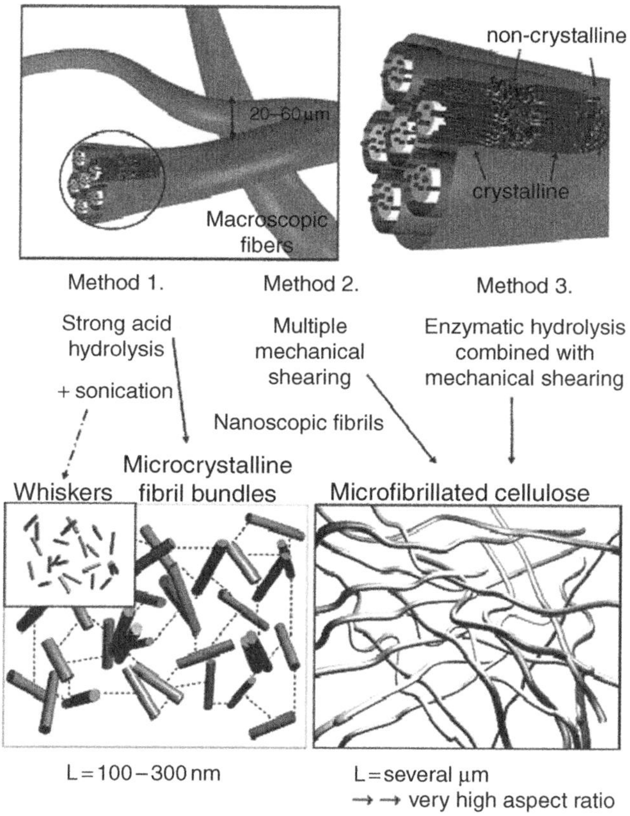

Figure 1.14 Methods for preparing nanoscale cellulosic materials. Reproduced with permission from M. Pääkkö, M. Ankerfors, H. Kosonen, A. Nykänen, S. Ahola, M. Österberg, J. Ruokolainen, J. Laine, P.T. Larsson, O. Ikkala and T. Lindström, *Biomacromolecules*, 2007, 8, 1934. ©2001, American Chemical Society [86]

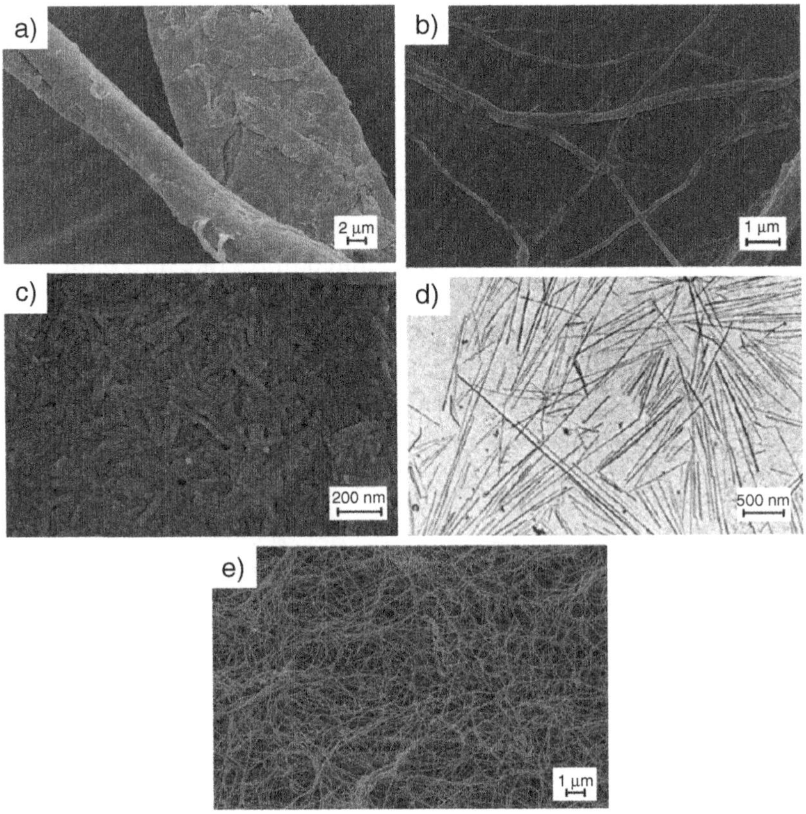

Figure 1.15 Scanning electron micrographs of (a) fibres of cotton linters, (b) microfibrillar cellulose, (c) microcrystalline cellulose, (d) tunicate whiskers [34] and (e) bacterial cellulose. Part (d) is reproduced with permission from S.J. Eichhorn, C.A. Baillie, N. Zafeiropoulos, L.Y. Mwaikambo, M.P. Ansell, A. Dufresne, K.M. Entwistle, P.J. Herrera-Franco, G.C. Escamilla, L. Groom, M. Hughes, C. Hill, T.G. Rials and P.M. Wild, *Journal of Materials Science*, 2001, **36**, 2107. ©2001, SpringerLink [87]

addition, agricultural by-products such as wheat straw, potato tubers, sugar beet pulp and banana rachis and some animals such as tunicates are also the source of nanomaterials [89]. **Figure 1.14** gives an overview of the commonly applied processes for the preparation of micro- and nano-structured cellulose. Bacterial cellulose has a special position (**Figure 1.17**) because it is produced by microorganisms. This cellulose conformation is of importance particularly for medical applications since its proven use as a scaffold to direct the growth of tissue or bones.

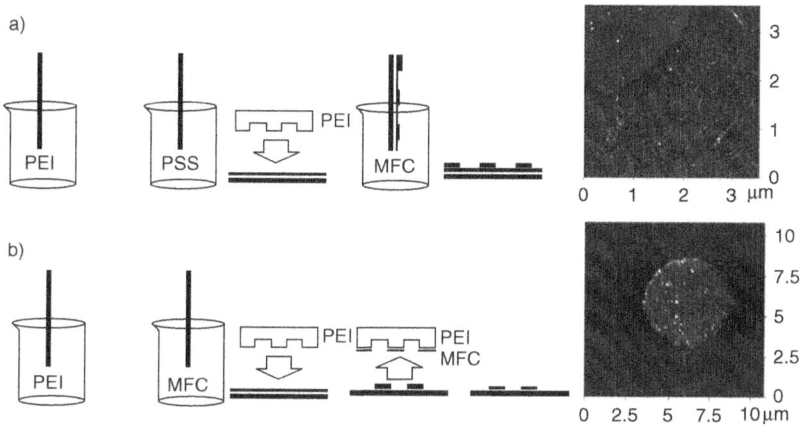

Figure 1.16 Schematic illustration of the preparation and a representative atomic force microscopy image of (a) the selective adhesion technique using poly(ethylene imine) (PEI) and poly(styrene sulfate) (PSS) to pattern microfibrillated cellulose (MFC) and (b) the lift-off technique, where MFC is partially removed by a PEI-modified stamp. Reproduced with permission from O. Werner, L. Persson, M. Nolte, A. Fery and L. Wagberg, *Soft Matter*, 2008, **4**, 1158. ©2008, RSC Publishing [88]

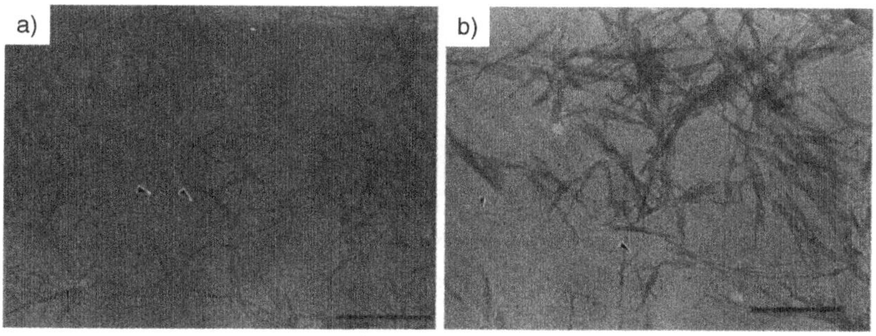

Figure 1.17 Transmission electron micrographs of microcrystalline cellulose prepared by (a) H_2SO_4 and (b) HCl treatment with typical single microcrystals highlighted by arrowheads. Scale bars = 500 nm. Reproduced with permission from J. Araki, M. Wada, S. Kuga and T. Okano, *Colloids and Surfaces A: Physicochemical and Engineering Aspects*, 1998, **142**, 75. ©1998, Elsevier [90]

The various configurations of cellulose mainly differ in their DP and crystallinity according to the disintegration technique used and, consequently, influence the shape of the particles. **Figure 1.15** shows selected examples of native fibres and nano- and microdispersed cellulose samples. In addition, **Table 1.7** gives an overview of the dimensions of several nanocelluloses from different sources.

Table 1.7 Length (*L*) and cross section (*D*) of cellulose nanocrystals obtained from various sources				
Source	*L* (nm)	*D* (nm)	Remarks	Reference
Bacterial cellulose	100 to >1000	5–50		[101]
Cotton	100–300	5–10		[91]
Cellulose whiskers from sisal	5	215		[92]
Microcrystalline cellulose	3.5	180	HCl treatment	[93]
Microfibrillar cellulose	5–15	Up to 1000	From carboxymethyl cellulose	[94]
Microfibrillar cellulose			By enzymatic hydrolysis	[86]
Sisal	100–500	3–5		[95]
Sugar beet pulp	210	5		[97]
Tunicate	100 to >1000	10–20		[98]
Wheat straw	150–300	5		[99]
Wood	100–300	3–5		[100]
Valonia ventricosa	>1000	10–20		[101]

Due to their high strength, the different fibrillar structures are interesting basic materials not only in paper industries but also in various other fields. The almost ideal crystalline structure of the cellulose molecules in the fibrils in combination with the enhanced degree of hydrogen bonds between the cellulose fibrils results in fibrillar cellulose of unique strength. Therefore, the elastic modulus is theoretically estimated to be between 130 and 170 GPa, which is indeed two to three times higher compared to glass fibres [102]. Moreover, micro- and nano-scale cellulose particles in suspension possess a remarkably higher reactivity, accessibility, solubility and solvent sorption due to their increased surface-to-volume ratio. Small, pure cellulose crystals possess a lower melting point [103]. Suspensions containing nano-scale cellulose also have unusual rheological properties. Due to the sorption of a high amount of water, the suspensions are gel-like aqueous systems with a viscosity much higher than that of comparable microscale cellulose suspensions.

Due to its biocompatibility, the natural polysaccharide cellulose finds use in medicine, pharmacy, cosmetics and food industry [104]. Cellulose is generally considered recognised as safe for use as a food additive by the US Food and Drug Administration.

There are no limitations concerning the acceptable daily intake levels when it is used in accordance with good manufacturing practices. However, the highly fibrous form has a poor mouthfeel. This drawback can be overcome by using nano- and small microdispersed cellulose such as microfibrillated cellulose, microcrystalline cellulose and cellulose whiskers. These cellulose modifications are still harmless and biocompatible. Therefore, they can be used for hygiene products, cosmetics, medicine, pharmaceutics and nutrition, mainly as stabilising agents against phase separation and sedimentation of heavy ingredients but also as peeling agents in cosmetics [36]. Chemically modified nanocellulose can also be used for immobilisation of enzymes or drugs.

Nanoscale cellulose may also be utilised in large-scale applications as composite material. For example, the increased surface-to-volume ratio implies interaction with other components. The mechanical performance of such composite materials is highly depending on the degree of dispersion of cellulose particles in the polymer matrix and the nature and intensity of cellulose-polymer adhesion interactions. The nanoscale cellulose fillers have a great impact on the permeability to gases and liquids, thermal stability, surface appearance and optical transmittance of the resulting nanocomposites [105]. This could create innovative breakthroughs for the use of these nanocomposites in areas like packaging, coating, high-modulus films and dispersion technology [96]. For example, very dense papers could be produced using dispersions of microfibrillated cellulose showing excellent tensile strengths [86]. Strong but transparent composite films may be developed as organic display systems, and a colour change may be effected by applying polyelectrolyte multilayer films to nanocrystalline cellulose [106]. Very recently, the concept of paper as a substrate for memory and advanced electrical storage devices has been realised. However, there is still a need for research and development investments in science as well as engineering to produce advanced and cost-competitive cellulose nanoscale products. It is necessary to obtain a better understanding of the adhesion interactions beyond hydrogen bonding, including mechanical interlocking and interpenetrating networks, on a fundamental level to improve the interfacial properties of cellulose composite material [107].

1.5.1 Microfibrillated Cellulose

For the manufacture of microfibrillated cellulose, wood pulp is disintegrated by applying high shear forces giving a material where the fibres are moderately degraded and where their substructural fibrils and microfibrils are exposed [108]. The fibrils and fibril aggregates are highly entangled and form mechanically strong networks and gels. The interactions are inherent resulting in much stronger gels than formed only by weak hydrogen bonds between water and fibrils [86]. However, the preparation of microfibrillated cellulose requires extensive and multiple treatments in the homogeniser, which consequently can easily be blocked. A less energy consuming procedure for the manufacture of microfibrillated cellulose is the dissolution of carboxymethylated pulp when exposed to a very high shear stress [94]. Subsequent ultrasound treatment results in smaller, highly charged and more heterogeneously microfibrillated cellulose. The combination of high-pressure shear forces and mild enzymatic hydrolysis constitutes an additional method to prepare microfibrillated cellulose with a well-controlled diameter in the nanoscale range and, thus, a tunable storage modulus useful for multicompartment mixtures. The cellulase enzymes are expected to favour the attack on the amorphous regions of the cellulose substrate.

By using lithographic methods, patterned surfaces of microfibrillar cellulose can be prepared [88]. The geometric features can be patterned either by microcontact printing of oppositely charged poly(ethylene imine) (PEI) on a PEI/poly(styrene sulfate) surface with subsequent treatment with microfibrillated cellulose (**Figure 1.16a**) or by using a PEI-coated poly(dimethyl siloxane) stamp to remove homogeneously deposited cellulose (**Figure 1.16b**). These methods are described as selective adhesion technique and lift-off technique, respectively. Such modified surfaces may be used to create filters or membranes, where both the pore openings and open areas can be controlled by varying the pattern of the microstamp.

The addition of microfibrillated cellulose to a variety of suspensions improves their homogeneity and stability significantly [108]. Microfibrillated cellulose is therefore used as an additive in food, cosmetic and pharmaceutical formulations, and colours. The tailoring of the flow behaviour and the reduction of droplets as well as the improved coverage of surfaces are the positive effects of the usage of microfibrillated cellulose in colours. Oil-in-water emulsions can be stabilised by the addition of only 1 wt% microfibrillated cellulose as a completely biocompatible component. The stability can be further improved in combination with hydrophilic polymers like cellulose ethers or starch. Besides water, other solvents that swell the cellulose matrix can also be applied, for instance ethanol, glycerine and propylene glycol.

Microfibrillated cellulose cannot be dispersed in dichloromethane, probably due to residual hemicelluloses at the surface [109]. This limitation can be overcome by chemical modification of the surface, for example, with *N*-octadecyl isocyanate, leading to materials that can be used for film casting processes in combination with synthetic polymers (for composites, see also **Section 1.5.3**) [92]. Moreover, polymer brushes, vinyl groups and charge can be introduced to the surface of the microfibrils by using glycidyl methacrylate, succinic anhydride and maleic anhydride, respectively [96]. Acetylation [110], silanisation [111] and carboxymethylation [112] can also be used for surface functionalisation as well as corona or plasma treatment [113] leading to cellulose-based materials possessing defined characteristics. For example, a layer-by-layer self-assembly of polycations such as PEI and anionic poly(thiophenes) was used to construct multilayer nanofilms on wood microfibres to develop papers for monitoring electrical and optical signals [114].

1.5.2 Microcrystalline Cellulose

For the production of microcrystalline cellulose, the biomass is generally treated with aqueous NaOH in order to purify the cellulose by removing other constituents [105]. The bleached material is then submitted to an acidic treatment. The natural cellulose fibrils are not uniform because they consist of crystalline and amorphous regions. The acidic hydrolysis of the cellulose substrate results in the dissolution of regions of low lateral order with the consequence that the water-insoluble, highly crystalline residues are converted into a stable suspension after vigorous stirring. The initially fast decrease of the DP slows down with time and finally reaches a nearly constant DP value (from 25 to 300) [115]. Subsequent dialysis of the suspension is performed to remove the acid. The spray-drying process that is commonly applied for microcrystalline products generally results in reagglomeration of smaller crystalline domains [116]. Therefore, microcrystalline cellulose may appear from stubby to fibrillar. These crystallites can be larger in dimension than the original microfibrils due to higher freedom of motion after hydrolysis [32].

Sulfuric acid is commonly used for hydrolysis [90]. During the preparation, a certain amount of sulfate groups (sulfur content 0.5–2%) is introduced to the surface of microcrystals. With increasing hydrolysis time, the length of the microcrystalline cellulose particles decreases while the surface charge increases [117]. The resulting negative charge contributes to the stability of the suspensions. However, the negative surface charge can prevent the substrate from being recognised by enzymes, and therefore preparation in HCl leading to noncharged microcrystalline cellulose is encouraged. The preparation in H_2SO_4 and HCl results in similar particle sizes and shapes as observed by electron microscopy (**Figure 1.17**).

The viscosity behaviour of the microcrystalline cellulose suspensions strongly depends on the surface change. In particular, HCl-treated suspensions show thixotropic ($c > 0.5\%$) and antithixotropic ($c < 0.3\%$) viscosity behaviour. The defined introduction of sulfur groups by subsequent H_2SO_4 treatment reduces the viscosity significantly [118]. The flow behaviour of the suspensions including accessibility to enzymatic degradation can thus be tailored in a very simple way.

Cellulose crystallites show self-assembly into chiral nematic phases, whose crystal pitches are in the order of the wavelength of visible light and reflect circularly polarised light of the same handedness as the chiral nematic phase [119]. Above a critical concentration, the cellulose suspension changes spontaneously into a chiral nematic liquid crystalline phase and forms regularly twisted fibrillar layers after drying that mimic the structural organisation of helicoids in nature [120]. This phenomenon will also be discussed for cellulose nanocrystallites, namely whiskers, in **Section 1.5.3**.

Microcrystalline cellulose is a fine, white, odorless crystalline powder, and is available commercially under the trade names Avicel®, Heweten®, Microcel®, Nilyn® or Novagel®. It has a high potential for application in pharmaceuticals (tablet binder), food (rheology control), and paper and composite manufacturing [121]. The free-flowing, nonfibrous particles can be compressed into self-binding tablets for compounding pharmaceuticals with subsequent rapid disintegration in water [104]. Aqueous suspensions of fine cellulose particles have a smooth texture comparable to butter. In addition, the suspensions behave like pseudoplastics, including a stable viscosity over a wide range of temperature. Therefore, they can be used as excipients to trigger the flow, lubrication and bonding properties of the ingredients to be tableted, to improve the stability of the drugs and to provide for rapid disintegration in the stomach. The ability to form stable colloid gels, suspensions and foams is further applied in cosmetics (hair conditioners, dyes, shampoos, toothpastes) and pharmaceutics for creams as well as solid suspensions. In food industries, microcrystalline cellulose is used as a noncaloric bulking agent in dietary foods, for example, to prepare reduced-lipid ice cream, mayonnaise-like products and low-oil pourable salad dressings.

Microcrystalline cellulose represents a convenient starting material for the development of new cellulose derivatives in the laboratory scale due to its very high purity and a sufficiently low viscosity; for example, it is used to acquire well-resolved liquid-state NMR spectra for structural analysis.

1.5.3 Cellulose Whiskers

The intense hydrolysis of microcrystalline cellulose results in pure cellulose crystallites. By subsequent treatment with ultrasound, the cellulose crystallites assemble into rigid rod-like cellulose particles called cellulose whiskers [105]. The preparation is also possible by mechanical treatments whereby the amorphous parts are cleaved by mechanical disintegration of a cellulose suspension. This high energy consuming process could be optimised by an enzymatically produced precursor resulting in a more efficient two-step process [86]. The milder enzymatic hydrolysis in comparison to the more aggressive acid hydrolysis yields longer and highly entangled nanoscale fibrils with a drastically enhanced strength of the resulting gel network.

The nanocrystals occur as rod-like particles with dimensions typically on the order of a few hundred nanometres in length and a few nanometres in diameter [105]. Using an electron microscope, the small but long crystals resemble a cat's whiskers in terms of straightness and the length-to-width ratio. The cellulose whiskers show no chain folding and contain only a small number of defects. Therefore, it is not surprising that the whiskers have a large modulus of elasticity (~150 GPa), strength (~7 GPa) and a very low coefficient of thermal expansion (~10^{-7} K^{-1}) [122, 123]. The precise dimensions that can be determined by microscopy and scattering techniques depend on the amount of amorphous regions, and thus on the origin of the substrate, the exact hydrolysis conditions and the ionic strength. The particles can be separated, for example, in isotropic suspensions. With increasing concentration, the smaller particles are located in the isotropic phase (top) whereas the larger particles are in the anisotropic phase [124]. A further separation could be achieved for tunicate whiskers by ultracentrifugation using a saccharose gradient [125].

The stability of the cellulose whiskers strongly depends on the dimensions of the particles, the size polydispersity and their surface charge. Suspensions prepared in H_2SO_4 possess a negative surface charge due to sulfate groups introduced to the surface [126]. In contrast, hydrolysis with HCl results in neutral particles that are less stable because of the absence of electrostatic repulsion and, thus, exhibit less interesting properties than the H_2SO_4-treated cellulose microfibrils.

Due to the rigid rod-like character of the cellulose whiskers, a macroscopic birefringence can be directly observed through crossed polarisers [127]. At low concentrations, the particles are randomly oriented and appear as spherical or oval droplets [128]. With increasing concentration, the whiskers self-align along a vector director resulting in a typical cholesteric liquid crystalline state. The chiral nematic orders can even be retained after evaporation of the solvents resulting in iridescent films of cellulose I. The colour of the films can easily be tuned by varying the ionic strength of the suspension [129]. At higher ionic strength, for example, after the addition of HCl, NaCl and KCl, the electrical double layer effect is screened out and the chiral interactions become stronger. The counterion also affects the interactions between the particles. In the presence of protons, the cellulose suspensions form ordered phases at the lowest critical concentration, whereas with increasing ordering number of the counterion a higher critical concentration can be observed. The submission to a magnetic field during the drying process of cellulose films results in perfect orientation of the whiskers leading to coloured films that can be used as security paper for making banknotes, passports and certificates. The different colour effect that results depending on the viewing angle is useful to produce novel pigments for optically variable coatings and inks. **Figure 1.18** shows different coloured domains suggesting an ordered phase of the cellulose rods (**Figure 1.18a**) and the well-defined cholesteric phase (**Figure 1.18b**) [32]. Small-angle neutron scattering experiments further point

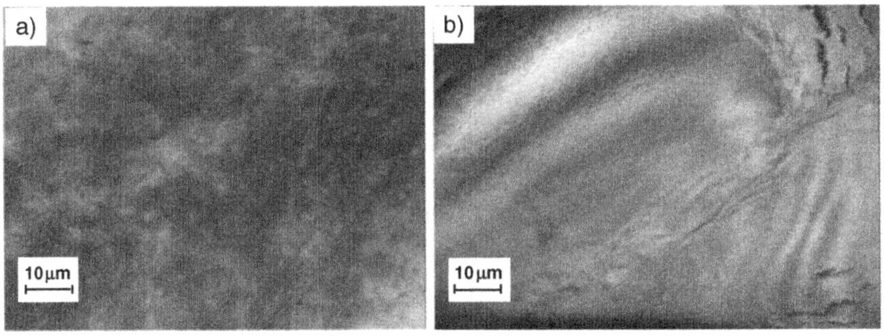

Figure 1.18 Crosspolarised optical microscopy images of tunicate whiskers (a) at the initial ordered phase and (b) at the cholesteric phase. Reproduced with permission from M.M. de Souza Lima and R. Borsali, *Macromolecular Rapid Communications*, 2004, **25**, 771. ©2004, Wiley InterScience [32]

out that the cholesteric axis of the chiral nematic phase aligns along an introduced magnetic field [130]. The distance between the cellulose particles is shorter along the cholesteric axis than when perpendicular to it, which evidences the suggestion that cellulose whiskers are helically twisted.

The alignment of the cellulose whiskers is strongly influenced by the shear rate. From another point of view, the hydrodynamic properties of cellulose whiskers are directly correlated to their size and their length distribution as well as their orientation in suspensions [131]. The typical rheological behaviour of the cellulose suspensions shows three distinct regions [132, 133]. The first region was observed at low shear rates corresponding to the shear thinning, which indicates the initial alignment of the domains formed by the particles. With increasing shear rate, the domains are broken up, which results in a plateau in the flow curve. At even higher shear rates, the viscosity shows a constant decrease due to the alignment of individual rods, which is characteristic of liquid crystals. The rheological behaviour of the cellulose suspensions also depends on the particle charge. The H_2SO_4-treated suspensions show no time dependence in viscosity, whereas the HCl-treated suspensions are thixotropic at higher (>0.5 wt%) and antithixotropic at lower concentrations (>0.3 wt%) [90, 118].

Besides several advantages such as low cost, biocompatibility, nontoxicity and biodegradability, there are also some disadvantages of cellulose whiskers. For instance, the high moisture absorption, poor wetting, incompatibility with most polymeric materials and the limited processing temperature must be considered as limitations for the use of cellulose whiskers [92]. Furthermore, cellulose is only a poor material to produce films with the required thermomechanical properties. However, one attempt to improve these properties while remaining the biodegradability is the preparation of cellulose-based nanocomposites. Here, cellulose whiskers show excellent properties, because of their regular and precise rigid rod shape, to improve the mechanical characteristics of a variety of materials produced from natural or synthetic materials. The nanocomposites show significantly enhanced mechanical properties due to the formation of a rigid whiskers network, even when the whiskers network is only a few percent [105].

Because of the high stability of cellulose in aqueous media, water is the preferred medium for processing composites of cellulose with other materials. The polar surfaces of the particles result in a poor dispersibility with nonpolar solvents or resins. This limits the choice of the matrix to water-soluble or water-dispersible polymers. However, one benefit of cellulose whiskers is that they can be dispersed in polar aprotic solvents such as DMF and DMSO, for example, for the preparation of birefringent cellulose films [134]. Moreover, dichloromethane can be used as dispersing media, which allows film casting with poly(ε-caprolactone) [92]. Such materials possess an increased glass transition, crystallisation and melting temperature compared to the pure poly(ε-caprolactone) matrix. Latex [135], poly(β-hydroxyalkanoate) [136, 137], starch [138], cellulose acetate butyrate [139], poly(vinyl chloride) [140], poly(vinyl alcohol) [32] and several other natural and synthetic polymers can be blended with cellulose whiskers resulting in an enhanced reinforcement of the material. The reinforcing effect is probably due to the strong interactions existing between the cellulose whiskers and the polar surfaces of the composite material and the creation of a percolating network linked by hydrogen bonds between the cellulose particles [107]. Cellulose whiskers can increase the crystallinity of the matrix with cellulose particles probably acting as a nucleating agent [141]. The nucleating effect is mainly governed by the surface characteristics, and unmodified whiskers have the largest nucleation effect [142]. X-ray diffraction and differential scanning calorimetry studies further show that whiskers with a more hydrophilic surface tend to crystallise in β-phase while modified particles appear to favour the α-phase [105].

Attempts were made to disperse the whiskers in nonpolar solvents by coating with surfactants [143] or after a chemical modification for example by grafting with poly(ethylene glycol) [144]. The surfactant or polymer layer covering the particles results in a steric rejection between the particles, while the rod shape of the whiskers is maintained as well as the chiral nematic structure. However, the large amount of surfactant that is required to coat the high surface area of the particles (150 m^2/g) limits the use of this technique for composite applications [145, 146]. Another mechanism to stabilise the suspensions is silylation [147]. The silyl groups are randomly distributed on the surface and allow dispersion in organic solvents with low polarity. However, by applying silylation with a DS higher than 1, the core of the whiskers becomes silylated, which leads to a loss of the whisker character. As a consequence, it is no longer possible to obtain any birefringent suspension.

One application of cellulose whiskers is related to their use as mechanical reinforcing agents of low-thickness polymer electrolytes for lithium batteries [148, 149]. A cellulosic nanocomposite was produced with poly(ethylene oxide) and a lithium imide salt for conducting ions. Cellulose whiskers incorporated in a mixture for sol-gel mineralisation can be incinerated during annealing to produce ceramic [150]. The resulting mesoporous silicas have unique narrow and uniform pores. The cellulose nanocrystals also find application in medicine. The functionalisation of the surfaces with fluorescein gives labelled cellulosic nanocrystals that can be used for imaging in biological applications [151].

1.5.4 Bacterial Cellulose

Bacterial cellulose, which was first described by Brown in 1886, is of particular interest in research and development [152]. There are various bacterial strains that produce cellulose extracellularly, including *Acetobacter*, *Agrobacterium*, *Aerobacter*, *Achromobacter*, *Azotobacter*, *Rhizobium*, *Sarcina* and *Alcaligenes* [153]. In particular, the bacterial strain *Gluconacetobacter xylinum* is used for the production of bacterial cellulose in a larger scale. The nonpathogenic bacteria produce cellulose within several days starting from D-glucose. Other carbon sources can also be metabolised, for example, sucrose, fructose, glycerol and maltose, but glucose is the preferred source [154]. The resulting molecular structure of bacterial cellulose is identical to the cellulose originating from plants. However, the detailed structure differs significantly resulting in outstanding properties. Bacterial cellulose is very pure and possesses high crystallinity, excellent water absorption capacity and extraordinary mechanical strength particularly in the wet state. An important advantage is the *in situ* mouldability, that is, shaping up during biosynthesis [155]. Furthermore, the biopolymer contains no hemicelluloses and lignin as well as no or only a very small amount of carbonyl and carboxyl moieties. Bacterial cellulose shows a very high crystallinity of more than 80% compared to plant-based celluloses with about 50% crystalline parts. The biopolymer consists of a three-dimensional network of ultrafine cellulose fibrils with a diameter in the range of 80–150 nm and may contain up to 99% water in the initial never-dried state. In addition, the DP value of bacterial cellulose is significantly higher with up to 10,000 compared to plant cellulose [156].

1.5.4.1 Preparation of Bacterial Cellulose

The most commonly studied strain for cellulose formation in the laboratory and commercial scale is *G. xylinum* (**Figure 1.19**). These microorganisms are generally found in vinegar, fruits, vegetables and alcoholic beverages. The Gram-negative, rod-shaped and aerobe bacteria produce the biopolymer at the interface of liquid and air

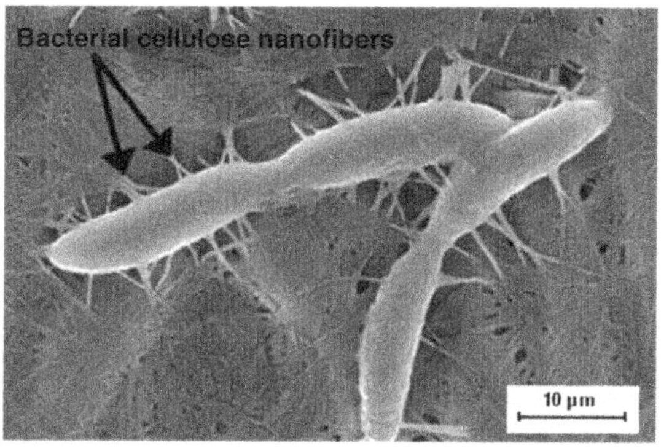

Figure 1.19 *Gluconacetobacter xylinum* bacteria embedded in the fibrillar network of cellulose. Reproduced with permission from Virginia Polytechnic Institute and State University. ©Virginia Polytechnic Institute and State University [157]

as a leather-like pellicle. The biosynthesis of cellulose is beneficial for the survival of the bacteria because the cellulose matrix on the surface of the nutrient solution protects them from ultraviolet light while preventing from drying and assuring an aerobic environment [158]. In addition, the cellulosic environment can serve as a protecting shield, for example, against enemies and heavy metal ions while nutrients can easily pass through [159]. The acetic acid bacteria are grown in a culture medium containing glucose and other ingredients like yeast extract, bactopeptone, disodium hydrogen phosphate and citric acid (known as Schramm-Hestrin medium) [160]. Apart from glucose, other carbon sources such as other types of monosaccharides, disaccharides, organic acids and products from beets (molasses, sugar syrup and saccharose), corn (starch, hydrolysed starch, glucose syrup) and potatoes (starch and starch hydrolysates) can also be used for producing bacterial cellulose [161]. The utilisation of inexpensive residues from agriculture or food processing is one interesting alternative to perform the fermentation process of bacterial cellulose in a more economical way.

Cellulose is produced between the outer membrane and the cytoplasmic membrane of the bacteria by a cellulose-synthesising complex, which is in association with the pores at the surface of the bacterium. The cellulose synthase is considered to be the crucial enzyme in this process. About 35–40% of the inserted glucose is converted into cellulose during 8–10 days with an increasing thickness of the fleeces [155]. The thickness of the layers may increase up to 4 cm within 4 weeks of cultivation. The kinetics and the resulting thickness, composition and microstructure of the cellulose gels strongly depend on the fermentation conditions such as the substrate constitution, the pH value, additives such as vitamins and water-soluble polymeric substances, and the particular bacterial strain used.

The formation of cellulose microfibrils proceeds by bundling of protofibrils of approximately 2–4 nm in diameter to fibrillar ribbons of around 4 nm in diameter and 80 nm in length. That means that the cellulose is produced in the pores as fibrils and this together with many synthesised fibrils forms a ribbon of crystalline cellulose [162, 163]. A self-assembly process is responsible for the crystallisation after polymerisation of the fibrils is completed [164, 165]. Various authors have suggested that the process of cellulose formation should occur at the medium/pellicle interface and the producing bacteria should be near this interface [166, 167]. However, it was shown that no cellulose formation occurs in the medium [168]. The real site for the formation of cellulose is at the upper side of the pellicle at the air/pellicle interface, which means that the producing cells as well as nutritive sources must be transported through the pellicle to its surface [169]. For purification, the cellulose fleeces are treated with water, boiled with a 0.1 N sodium hydroxide solution and washed again thoroughly with water to remove the cells and culture ingredients.

The bacterial cellulose fibres consist of pure crystalline cellulose I_α as investigated by synchrotron measurements. The orientational state of the cellulose fibrils is important with regard to the principles of structure formation and the mechanical properties of the cellulose [159]. The cellulose films exhibit basically a uniplanar orientation based on X-ray investigations [170]. The stretchability and, thus, the orientability of the microfibrils can be improved markedly by loosening or cutting interfibrillar hydrogen bonds by NaCl treatment and simultaneous drawing of the bacterial cellulose [171].

The bulk cellulose is deposited as a fleece in a highly swollen state containing about 99% water at the air-liquid interface while the bacterial cells become entrapped in the pellicles. Consequently, such cellulose fleeces exhibit an enormous water retention value in the range of 1000% due to the extensive surface area of the nanofibrils. However, after casual drying, the water retention is comparable to that of plant cellulose (60%) but can be significantly increased to ~600% using lyophilisation as the drying technique [155].

1.5.4.2 Methods for Production of Bacterial Cellulose

Bacterial cellulose has been conventionally produced by static and agitated culture methods. Under static culture conditions, bacterial cellulose layers up to several centimetres are formed on the surface of the culture medium. In the static culture, it is important to control the pH because the accumulation of gluconic, acetic or lactic acids in the culture broth decreases the pH far below the optimum for growth and polysaccharide production [172]. Additionally, a large surface area is important for good productivity. Relatively low glucose concentrations give better productivity and yields than higher concentrations. While stationary culture conditions have been investigated successfully and described very intensely, agitated culture of *Gluconacetobacter* strains causes many problems, among which strain instability, non-Newtonian behaviour during mixing of bacterial cellulose and proper oxygen supply are the most common ones [172, 174]. Despite these problems, some researchers have suggested that agitated culture might be the most suitable technique for economical scale production [175, 176]. Various reactors were developed to carry out bacterial cellulose production under agitated culture conditions preventing conversion of bacterial cellulose-producing strains into cellulose-negative mutants. This can be achieved for instance by a rotating disc fermenter. The rotating disc reactor is designed in a way that half of the surface of its disc is submerged in the medium broth while the other half is exposed to the atmosphere. As the discs rotate continuously, the surface of the discs alternates between the medium and the atmosphere. When this reactor is used for the production of bacterial cellulose, cells that are stuck to the disc surface take nutrients when they are immersed in the medium, and are exposed to oxygen when the disc surface is exposed to the atmosphere. Further reactors are described by Chawla and co-workers [177]. Morphological differences between cellulose produced by static and agitated cultures contribute to varying degrees of crystallinity, different crystalline sizes and cellulose I_α content. The difference in cellulose I_α content between cellulose produced under agitated and static conditions exceeds that of the crystallinity index. The crystallinity index is closely related to the I_α value [178].

The use of a novel aerosol bioreactor working on the fed-batch principle has been described previously [179]. This involved the generation of an aerosol spray of glucose and its even distribution to the living bacteria on the medium-air interface. This process guarantees optimal supply of oxygen and nutrients during the whole cultivation process. The apparatus operated up to 8 weeks with a constant rate of cellulose production. The aerosol system provided the basis for an economic production of bacterial cellulose in surface culture.

1.5.4.3 Applications of Bacterial Cellulose

The pure bacterial cellulose has long been used as the raw material of an indigenous dessert in the Philippines. The gel sheets fermented with coconut water are cut into cubes and immersed in sugar syrup giving *nata-de-coco*, which is nowadays likely to be eaten in the eastern Asian area in puddings or fruit salads. The cellulose material is also used as a thickener and binder for food and as noncaloric supplement in dietary products. Bacterial cellulose was found to be an excellent ingredient in ice cream to resist meltdown and heat stock [180].

Bacterial cellulose is further used as artificial skin for the covering of wounds while creating optimal moist conditions and providing permeability for gases and liquids

[155]. Cases of burns and ulcers could be treated successfully by bacterial cellulose, and it provides benefits such as immediate pain relief, close adhesion to the wound bed, diminished postsurgery discomfort and faster healing. Moreover, improved exudate retention, spontaneous detachment following re-epithelisation, reduced infection rate and facile wound observation due to the transparency of the cellulose cover promote its use in wound healing [181].

As early as in the 1980s, Johnson & Johnson started to commercialise bacterial cellulose on a large scale for medical applications. The main issue was a product for the treatment of different types of wounds [182, 183]. The Brazilian company BioFill Produtos Biotecnologicos (Curitiba, PR Brazil) independently created a new wound healing system based on bacterial cellulose produced by *G. xylinum* [181, 184, 185]. At present, commercial products are available (Suprasorb X®) and are distributed by Lohmann & Rauscher GmbH & Co. KG, Neuwied.

One advantage of bacterial cellulose is its mouldability during the cultivation process for the manufacture of biocompatible materials for microsurgical applications [155]. The shaping process proceeds in a static culture by applying a template matrix. In this way, tubes based on BASYC® (bacterial synthesised cellulose) with different length, wall thickness and inner diameter can be designed (**Figure 1.20**).

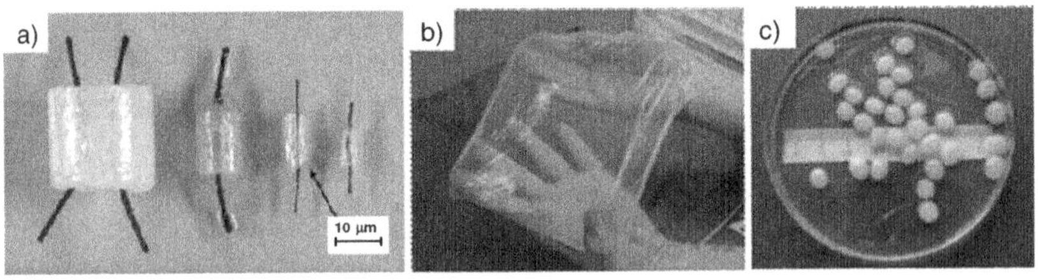

Figure 1.20 Bacterial cellulose of different shapes and the potential uses of the various forms: (a) BASYC®-tubes of different diameter and length for microsurgical application; (b) cellulose foils for masks in cosmetic use and (c) spheres as absorber. Reproduced with permission from a) D. Klemm, B. Heublein, H-P. Fink and A. Bohn, *Angewandte Chemie International Edition*, 2005, **44**, 3358, ©2005, Wiley InterScience [1] and b) and c) D. Klemm, D. Schumann, U. Udhardt and S. Marsch, *Progress in Polymer Science*, 2001, **26**, 1561. ©2001, Elsevier [155]

The roughness of the BASYC®-tubes in the wet state resembles that of blood vessels and ranges between 7 and 14 nm. The tremendous mechanical strength provides a stability that is necessary to resist mechanical strain during microsurgical preparation and also the blood pressure of the living body. Such cellulose tubes are actually used as artificial blood vessels for microsurgery, as protective cover for micronerve sutures and as a vessel-like model for the training of microsurgical suture techniques. Postoperative observation of a microvessel endoprosthesis shows that the cellulose matrix is completely incorporated in the body, wrapped up with connective tissue and pervaded with small vessels, without any rejection reaction. Similar results were achieved in micronerve surgery, as demonstrated in **Figure 1.21**. Consequently, the bacterial cellulose-based tubes are excellent prosthesis materials for microsurgery as they provide an easy handling, blood and tissue compatibility, consistency comparable to those of blood vessels and vitalisation in the body. A systematic evaluation of the *in vivo* biocompatibility evidences the absence of microscopic signs of inflammation, fibrotic capsules or giant cells in implanted bacterial cellulose scaffolds, while fibroblasts infiltrate the cellulose matrix, and new blood vessels around and inside the scaffold are formed [186]. Bacterial cellulose is further used as a carrier for animal cell cultures and as an adhesive in epidermal cell grafts.

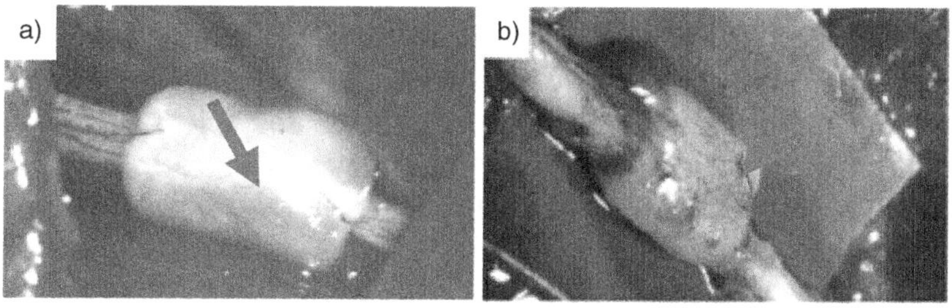

Figure 1.21 Sciatic nerve of a rat with a BASYC®-tube as protective cover (a) immediately and (b) 10 weeks after the operation. Reproduced with permission from D. Klemm, D. Schumann, U. Udhardt and S. Marsch, *Progress in Polymer Science*, 2001, **26**, 1561. ©Elsevier, 2001 [155]

Because of its outstanding properties, bacterial cellulose is applicable in cosmetics as a moistening mask as well as an ingredient of moistening creams [187]. At present, NanoMasque®, a commercial product, is distributed by fzmb GmbH, Bad Langensalza, Germany. Due to its high water retention value, NanoMasque® is an ideal base for cosmetic active substances. It has an extremely smooth and elastic surface, adapts perfectly to the shape of the face and feels like a second skin (**Figure 1.22**) [188].

Further attempts are made for the immobilisation of biologically active substances in the cellulose structure, for example, for improving the healing process in insured tissue or regeneration and moisturisation of the skin in both medicine and cosmetics. For example, plant extracts, extracts from algae, essential oils and panthenol can be impregnated in the cellulose matrix by physical interactions and then allowed to migrate into the skin during application [187].

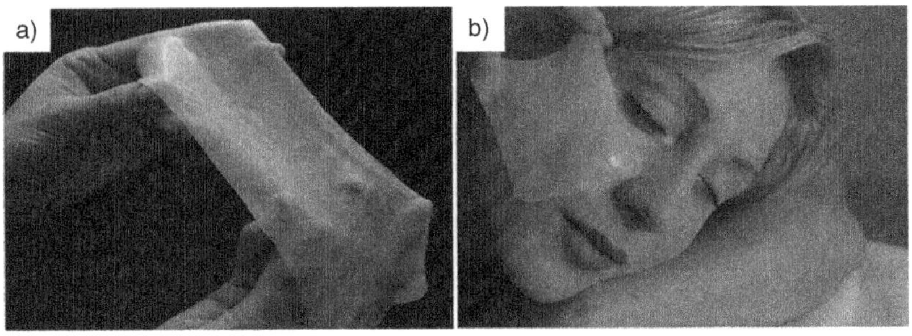

Figure 1.22 (a) Thin membrane of bacterial cellulose and (b) cosmetic application of NanoMasque® for face care. Reproduced with permission from Die Eigenschaften von NanoMasque, Forschungszentrum für Medizintechnik und Biotechnologie GmbH, [187]. *http://nanomasque.fzmb.de/seiten/deutsch/produkt2.html* [188]

The separation of DNA is enhanced by the addition of bacterial cellulose fibrils to an electrophoretic separation medium [189]. The high-resolution separation of a wide range of DNA fragments (10–15 kbp) including the single-nucleotide polymorphisms is probably due to a double-mesh concept combined with a stereo effect. An interesting method comprises a microchannel containing an optical compact disc (CD) grating that confines fragments of bacterial cellulose fibrils [190]. The lab-on-a-chip technology provides up to six times higher sensitivity to detect DNA and other biomolecules based on the multiple combination of diffraction on the CD grating, reflection on the CD mirror and refraction on the cellulose fibrils, leading to an efficient light-confining effect useful for various biological applications.

Sheets of bacterial cellulose have outstanding mechanical properties as indicated by the high Young's modulus from 16 GPa up to 30 GPa [191]. This can be attributed to the high crystallinity of bacterial cellulose, the good planar orientation and the ultrafine network structure of the ribbons, which allows the formation of extensive hydrogen bonds [192]. Recently, it was found that nanofibres produced by bacteria have an extraordinary potential as reinforcement materials for fibre-based composites. For instance, they can be used to reinforce transparent plastics with less than 10% loss of transmittance even at fibre contents as high as 70 wt% (see **Figure 1.23**). The nanofibre-reinforced composite showed incredibly low thermal expansion (3×10^{-6} K^{-1}) and high tensile strength (325 MPa), while maintaining the flexibility and ductility of many plastics [194].

Bacterial cellulose was also found to be effective for reinforcement of paper pulp. Fragmented bacterial cellulose has promising prospects in papermaking; test pieces of flexure-durable papers and high filler-content papers can be prepared, which are ideal for special paper employed in the making of banknotes [195]. Additionally, it was

Figure 1.23 Flexibility and transparency of a 64 μm thick bacterial cellulose sheet with acrylic resin. Reproduced with permission from H. Yano, J. Sugiyama, A.N. Nakagaito, M. Nogi, T. Matsuura, M. Hikita and K. Handa, *Advanced Materials*, 2005, **17**, 153. ©2005, Wiley InterScience [193]

found that a surface treatment with fibrous bacterial cellulose during the process of papermaking leads to excellent surface properties suitable for high-quality paper [196].

The high sonic velocity and low dynamic loss properties enable the cellulose membranes to be used as acoustic diaphragms for high-fidelity loudspeakers and headphones [197]. Sony Corporation (Japan), in conjunction with Ajinomoto (Japan), developed the first audio speaker diaphragms using bacterial cellulose. However, the production of the speaker membrane is not justifiable for the market because of the high costs [198].

1.6 Some Comments about Further Applications of Plant Cellulose

Besides the commonly known use of cellulose in paper and textile industry, cellulose is also applied in medical and pharmaceutical fields. Oxidised regenerated cellulose originating from plants is, for example, used in the form of woven cotton gauze dressings for wounds. Its haemostatic effect and action as an adhesion barrier are exploited for many years and continue to be exploited [199, 200]. Cellulose is recognised as an excellent wound dressing material because of the high water-holding capacity, elasticity, conformability, mechanical strength and porosity. In addition, it provides a barrier against external bacteria, and is easy to sterilise and mouldable *in situ* [201].

Furthermore, cellulose is applied over decades as a material for haemodialysis [202]. Haemodialysis is a rather complex topic since more mechanisms than diffusion are included such as adsorption of biomolecules on the surface. In order to minimise the responses with other biological materials existing in blood, the active surfaces are coated with poly(ethylene glycol). Another approach to minimise the activation of polymorphonuclear leucocytes is functionalisation of the cellulose membrane with vitamin E, which reduces oxidative stress [203].

Cellulose has also found an application as culture material and postoperative adhesion barrier, and in scaffolds for tissue engineering and skin substitution. Despite the absence of hydrolases that attack the β-1,4-glycosidic linkage in animal and human tissues, cellulose, used in the form of cellulose sponges, can be regarded as a slowly degradable implantation material [204]. **Figure 1.24** shows the changes in a cellulose sponge implant over time. In the beginning, the contours of the pore walls are sharp and the structure of the cellulose matrix is intact (**Figure 1.24a**). After 16 weeks, the pore walls are softened, cracks have appeared and the size of the micropores in the pore walls has increased (**Figure 1.24b**). However, the time needed for gradual degradation of the cellulose sponge is longer than 60 weeks.

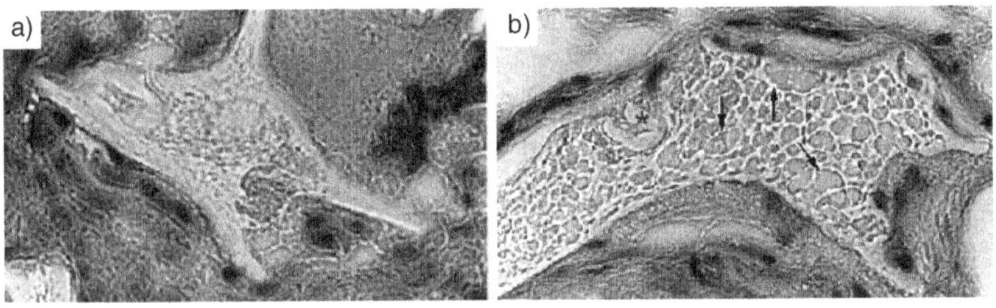

Figure 1.24 Changes in the structure of the cellulose matrix (a) 1 week and (b) 16 weeks after the implantation. Reproduced with permission from M. Märtson, J. Viljanto, T. Hurme, P. Laippala and P. Saukko, *Biomaterials*, 1999, 20, 1989. ©1989, Elsevier [204]

The structure of the cellulose matrix strongly affects its medical application. For example, wound dressings are nanoporous materials to keep the moisture during the healing process, whereas cellulose as artificial skin substitute should possess a high porosity in order to facilitate cell growth in the cellulose scaffold [201]. The minimal pore size that allows tissue growing is 50 μm.

1.7 Concluding Remarks

Cellulose is the most important renewable resource. It is used as a starting material for various modification reactions yielding cellulose ethers and esters of commercial importance. These technical modification reactions are exclusively carried out under heterogeneous reaction conditions at least at the beginning of the conversions. Moreover, research and development in the field of homogeneous phase chemistry with cellulose using various solvents including molten salts (in particular ionic liquids) and molten salt hydrates open new avenues for the product design with modern organic chemistry. It may be expected that homogeneous phase chemistry will enter the technical scale also: not only chemical modification of the bulk but also surface modification will provide important novel products with low DS. Last but not least, as discussed in this review paper, depending on its nano- and micro-structured architectures, versatile characteristics of the biopolymer cellulose can be achieved and addressed to a specific function. Consequently, cellulose is a promising and broadly applicable material not only (as commonly known) in paper and textile industry but also for medical and pharmaceutical devices. It finds application as an additive for pharmaceutical formulations, as biocompatible hydrogels or in tissue engineering as scaffold material. The application fields of nano- and micro-structured cellulose will be further broadened by chemical and physical surface treatments.

References

1. D. Klemm, B. Heublein, H-P. Fink and A. Bohn, *Angewandte Chemie International Edition*, 2005, **44**, 3358.

2. A.K. Mohanty, M. Misra and G. Hinrichsen, *Macromolecular Materials and Engineering*, 2000, **276/277**, 1.

3. D.N-S. Hon, *Cellulose*, 1994, **1**, 1.

4. A. Payen, *Comptes Rendus*, 1839, **8**, 51.

5. T.J. Brongniard, A.B.D. Pelouze and C.R. Hedb, *Academie des Sciences*, 1837, **8**, 51.

6. R.A. Young, *Cellulose*, 1994, **1**, 108.

7. V.S.N. Rao, P.R. Sundararajan, C. Ramakrishnan and G.N. Ramachandran in *Conformation of Biopolymers*, Ed., G.N. Ramachandran, Academic Press, London, UK, p.1957.

8. H.A. Krässig, *Cellulose: Structure, Accessibility and Reactivity*, Gordon and Breach Science Publishers, Yverdon, Switzerland, 1993.

9. S. Perez and K. Mazeau in *Polysaccharides: Structural Diversity and Functional Versatility*, Ed., S. Dumitriu, Marcel Dekker, New York, NY, USA, 2005, p.41.

10. T. Kondo, *Journal of Polymer Science, Part A: Polymer Chemistry Edition*, 1998, **35**, 717.

11. K. Kamide, K. Okajima, K. Kowsaka and T. Matsui, *Polymer Journal*, 1985, **17**, 701.

12. C.Y. Liang and R.H. Marchessault, *Journal of Polymer Science*, 1959, **37**, 385.

13. A.J. Michell, *Carbohydrate Research*, 1988, **173**, 185.

14. K.H. Gardner and J. Blackwell, *Biopolymers*, 1974, **13**, 1975.

15. Y. Nishiyama, P. Langan and H. Chanzy, *Journal of the American Chemical Society*, 2002, **124**, 9074.

16. T. Kondo in *Polysaccharides: Structural Diversity and Functional Versatility*, Ed., S. Dumitriu, Marcel Dekker, New York, NY, USA, 2005, p.69.

17. K. Tashiro and M. Kobayashi, *Polymer*, 1991, **32**, 1516.

18. M. Polanyi, *Naturwissenschaften*, 1921, **9**, 337.

19. A. Sarko and R. Muggli, *Macromolecules*, 1974, **7**, 486.

20. R.H. Atalla and D.L. VanderHart, *Science*, 1984, **223**, 283.

21. A. Isogai, M. Usuda, T. Kato, T. Uryu and R.H. Atalla, *Macromolecules*, 1989, **22**, 3168.

22. S. Kugas, S. Takag and M.R. Brown, *Polymer*, 1993, **34**, 3293.

23. P. Langan, Y. Nishiyama and H. Chanzy, *Journal of the American Chemical Society*, 1999, **121**, 9940.

24. P. Zugenmaier, *Progress in Polymer Science*, 2001, **26**, 1341.

25. M. Wada, L. Heux, A. Isogai, Y. Nishiyama, H. Chanzy and J. Sugiyama, *Macromolecules*, 2001, **34**, 1237.

26. E.S. Gardiner and A. Sarko, *Canadian Journal of Chemistry*, 1985, **63**, 173.

27. A. Isogai in *Cellulosic Polymers: Blends and Composites*, Ed., R.D. Gilbert, Hanser Publishers, Munich, Germany, 1994, p.1.

28. P.H. Hermans and A. Weidinger, *Journal of the American Chemical Society*, 1946, **68**, 1138.

29. I.L. Wadehra and R.S.J. Manley, *Journal of Applied Polymer Science*, 1965, **9**, 2627.

30. L.R. Schroeder, V.M. Gentile and R.H. Atalla, *Journal of Wood Chemistry and Technology*, 1986, **6**, 1.

31. R.H. Atalla, J.D. Ellis and L.R. Schroeder, *Journal of Wood Chemistry and Technology*, 1984, **4**, 465.

32. M.M. de Souza Lima and R. Borsali, *Macromolecular Rapid Communications*, 2004, **25**, 771.

33. US Department of Energy Office of Science. *http://genomics.energy.gov/gallery/gtl/detail.np/detail-36.html*

34. S.J. Eichhorn, C.A. Baillie, N. Zafeiropoulos, L.Y. Mwaikambo, M.P. Ansell, A. Dufresne, K.M. Entwistle, P.J. Herrera-Franco, G.C. Escamilla, L. Groom, M. Hughes, C. Hill, T.G. Rials and P.M. Wild, *Journal of Materials Science*, 2001, **36**, 2107.

35. M. Ioelovich and A. Leykin, *Bioresources*, 2008, **3**, 170.

36. M. Ioelovich, *Bioresources*, 2008, **3**, 1403.

37. L.M. Welch, W.E. Roseveare and H. Mark, *Industrial and Engineering Chemistry*, 1946, **38**, 580.

38. W.A. Sisson, *Journal of Physical Chemistry*, 1940, **44**, 513.

39. R.P. Swatloski, S.K. Spear, J.D. Holbrey and R.D. Rogers, *Journal of the American Chemical Society*, 2002, **124**, 4974.

40. T.F. Liebert and T. Heinze, *Biomacromolecules*, 2005, **6**, 333.

41. T. Liebert and T. Heinze, *Bioresources,* 2008, **3**, 576.

42. R.T. Bogan and R.J. Brewer in *Encyclopedia of Polymer Science and Engineering*, Eds., H.F. Mark, N.M. Bikales, C.G. Overberger, G. Menges and J.I. Kroshwitz, John Wiley & Sons, New York, NY, USA, 1985, p.158.

43. T. Miyamoto, Y. Sato, T. Shibata, M. Tanahashi and H. Inagaki, *Journal of Polymer Science, Part A: Polymer Chemistry Edition,* 1985, **23**, 1373.

44. B. Philipp, *Polymer News*, 1990, **15**, 170.

45. C. Deus, H. Friebolin and E. Siefert, *Makromolekulare Chemie*, 1991, **192**, 75.

46. T. Heinze and T. Liebert, *Macromolecular Symposia*, 2004, **208**, 167.

47. T. Heinze, T. Liebert, K.S. Pfeiffer and M.A. Hussain, *Cellulose*, 2003, **10**, 283.

48. K. Rahn, M. Diamantoglou, D. Klemm, H. Berghmans and T. Heinze, *Angewandte Makromolekulare Chemie*, 1996, **238**, 143.

49. S. Köhler and T. Heinze, *Cellulose*, 2007, **14**, 489.

50. O.A. El Seoud, G.A. Marson, G.T. Ciacco and E. Frollini, *Macromolecular Chemistry and Physics,* 2000, **201**, 882.

51. T. Heinze, R. Dicke, A. Koschella, A.H. Kull, E-A. Klohr and W. Koch, *Macromolecular Chemistry and Physics*, 2000, **201**, 627.

52. S. Fischer, H. Leipner, E. Brendler and W. Voigt in *Polysaccharide Applications, Cosmetics and Pharmaceuticals*, Eds., M.A. El-Nokaly and H.A. Soine, ACS Symposium Series 737, American Chemical Society, Washington, DC, USA, 1999, p.143.

53. J. Wu, J. Zhang, H. Zhang, J. He, Q. Ren and M. Guo, *Biomacromolecules*, 2004, **5**, 266.

54. K. Schlufter, H-P. Schmauder, S. Dorn and T. Heinze, *Macromolecular Rapid Communications*, 2006, **27**, 1670.

55. C.L. McCormick and P.A. Callais, *Polymer*, 1987, **28**, 2317.

56. M. Gericke, T. Liebert and T. Heinze, *Macromolecular Bioscience*, 2009, **9**, 343.

57. L. Dahlgren in *Wood and Cellulosics - Industrial Utilization, Biotechnology, Structure and Properties*, Eds., J.F. Kennedy, G.O. Phillips and P.A. Williams, Horwood Publishing, Chichester, UK, 1987, p.427.

58. M.D. Nicholson and F.M. Merritt in *Cellulose Chemistry: Its Application*, Eds., T.P. Nevell and S.H. Zeronian, Horwood Publishing, Chichester, UK, 1989, p.363.

59. R.L. Feddersen and S.N. Thorp in *Industrial Gums, Polysaccharides and Their Derivatives*, Eds., R.L. Whistler and J.N. BeMiller, Academic Press, Inc., San Diego, Boston, New York, USA, 1993, p.537.

60. J. Reuben and H.T. Conner, *Carbohydrate Research*, 1983, **115**, 1.

61. E.A. Kragten, J.P. Kamerling and J.F.G. Vliegenthart, *Journal of Chromatography*, 1992, **623**, 49.

62. T. Heinze, U. Erler, I. Nehls and D. Klemm, *Die Angewandte Makromolekulare Chemie*, 1994, **215**, 93.

63. T. Liebert and T. Heinze in Cellulose Derivatives, Eds., T.J. Heinze and W.G. Glasser, *ACS Symposium Series* No. 688, ACS, Washington, DC, USA, 1998, p.61.

64. T. Heinze in *Ionische Funktionspolymere aus Cellulose: Neue Synthesekonzepte, Strukturaufklärung und Eigenschaften*, Shaker Verlag, Aachen, Germany, 1998, p.227.

65. J. Burger, G. Kettenbach and P. Klüfers, *Macromolecular Symposia*, 1995, **99**, 113.

66. T. Heinze, T. Liebert, P. Klüfers and F. Meister, *Cellulose*, 1999, **6**, 153.

67. W-M. Baar, K. Kulicke, R. Szablikowski and R. Kiesewetter, *Macromolecular Chemistry Physics*, 1994, **195**, 1483.

68. P. Käuper, W-M. Kulicke, S. Horner, B. Saake, J. Puls, J. Kunze, H-P. Fink, U. Heinze, T. Heinze, E.A. Klohr, H. Thielking and W. Koch, *Angewandte Makromolekulare Chemie*, 1998, **260**, 53.

69. T. Liebert and T. Heinze, *Macromolecular Symposia*, 1998, **130**, 271.

70. T. Heinze and S. Köhler in Cellulose Solvents: For Analysis, Shaping and Chemical Modification, Eds., T.F. Liebert, T.J. Heinze and K.J. Edgar, *ACS Symposium Series* No.1033, Washington, DC, USA, 2010, p.103.

71. T. Liebert, D. Klemm and T. Heinze, *Journal of Macromolecular Science, Part A: Pure and Applied Chemistry*, 1996, **A33**, 613.

72. B. Saake, S. Horner, T. Kruse, J. Puls, T. Liebert and T. Heinze, *Macromolecular Chemistry and Physics*, 2000, **201**, 1996.

73. D.N-S. Hon in *Polysaccharides in Medical Applications*, Ed., S. Dumitriu, Marcel Dekker, New York, NY, USA, 1996, p.87.

74. E.J. Vandamme, S. De Baets, A. Vanbaelen, K. Joris and P. De Wulf, *Polymer Degradation and Stability,* 1998, **59**, 93.

75. R. Jonas and L.F. Farah, *Polymer Degradation and Stability*, 1998, **59**, 101.

76. Vera Chimscholli ihr sein TextBlog, Wordpress *http://textblog.wordpress.com/2006/08/29/mein-freund-der-baum/*

77. Bamboo Fiber and Bamboo Bedding, Linenplace. *http://www.linenplace.com/boutiques/product-ideas/bamboo-bedding.html*

78. Reallynature, *www.reallynatural.com/archives/cat_business.php*

79. San Marcos Growers, *http:// www.smgrowers.com/products/plants/plantdisplay.asp?plant_id=2273*

80. Fisheries and Oceans Canada, *http://www.dfo-mpo.gc.ca/science/Publications/article/2007/17-09-2007-eng.htm*

81. P.S. Belton, S.F. Tanner, N. Cartier and H. Chanzy, *Macromolecules*, 1989, **22**, 1615.

82. J. Sugiyama, J. Persson and H. Chanzy, *Macromolecules*, 1991, **24**, 2461.

83. F. Nakatsubo, H. Kamitakahara and M. Hori, *Journal of the American Chemical Society*, 1996, **118**, 1677.

84. T. Nishimura and F. Nakatsubo, *Cellulose*, 1997, **4**, 109.

85. S. Kobayashi, K. Kashiwa, T. Kawasaki and S. Shoda, *Journal of the American Chemical Society*, 1991, **113**, 3079.

86. M. Pääkkö, M. Ankerfors, H. Kosonen, A. Nykänen, S. Ahola, M. Österberg, J. Ruokolainen, J. Laine, P.T. Larsson, O. Ikkala and T. Lindström, *Biomacromolecules*, 2007, **8**, 1934.

87. S.J. Eichhorn, C.A. Baillie, N. Zafeiropoulos, L.Y. Mwaikambo, M.P. Ansell, A. Dufresne, K.M. Entwistle, P.J. Herrera-Franco, G.C. Escamilla, L. Groom, M. Hughes, C. Hill, T.G. Rials and P.M. Wild, *Journal of Materials Science*, 2001, **36**, 2107.

88. O. Werner, L. Persson, M. Nolte, A. Fery and L. Wagberg, *Soft Matter*, 2008, **4**, 1158.

89. M.A. Hubbe, O.J. Rojas, L.A. Lucia and M. Sain, *Bioresources*, 2008, **3**, 929.

90. J. Araki, M. Wada, S. Kuga and T. Okano, *Colloids and Surfaces A: Physicochemical and Engineering Aspects*, 1998, **142**, 75.

91. T. Ebeling, M. Paillet, R. Borsali, O. Diat, A. Dufresne, J.Y. Cavaillé and H. Chanzy, *Langmuir*, 1999, **15**, 6123.

92. G. Siqueira, J. Bras and A. Dufresne, *Biomacromolecules*, 2009, **10**, 425.

93. J. Araki, M. Wada, S. Kuga and T. Okano, *Journal of Wood Science*, 1999, **45**, 258.

94. L. Wagberg, G. Decher, M. Norgren, T. Lindström, M. Ankerfors and K. Axnäs, *Langmuir*, 2008, **24**, 784.

95. N.L. Garcia de Rodrigez, W. Thilemans and A. Dufresne, *Cellulose*, 2006, **13**, 261.

96. P. Stenstad, M. Andresen, B.S. Tanem and P. Stenius, *Cellulose*, 2008, **15**, 35.

97. M.A.S.A. Samir, F. Alloin, M. Paillet and A. Dufresne, *Macromolecules*, 2004, **37**, 4313.

98. V. Favier, G.R. Canova, J.Y. Cavaillé, H. Chanzy, A. Dufresne and C. Gauthier, *Polymers for Advanced Technologies*, 1995, **6**, 351.

99. W. Helbert, J.Y. Cavaillé and A. Dufresne, *Polymer Composites*, 1996, **17**, 604.

100. S. Beck-Candanedo, M. Roman and D.G. Gray, *Biomacromolecules*, 2005, **6**, 1048.

101. J.F. Revol, *Carbohydrate Polymers*, 1982, **2**, 123.

102. K. Tashiro and M. Kobayashi, *Polymer*, 1991, **32**, 1516.

103. M. Ioelovich and R. Luksa, *Wood Chemistry*, 1990, **3**, 18.

104. D.N-S. Hon in *Polysaccharides in Medical Applications*, Ed., S. Dumitriu, Marcel Dekker, New York, NY, USA, 1996, p.87.

105. A. Dufresne, *Canadian Journal of Chemistry*, 2008, **86**, 484.

106. E.D. Cranston and D. Gray, *Biomacromolecules*, 2006, **7**, 2522.

107. D.J. Gardner, G.S. Oporto, R. Mills and M.A.S.A. Samir, *Journal of Adhesion Science and Technology*, 2008, **22**, 545.

108. A.F. Turbak, F.W. Snyder and K.R. Sandberg, inventors; ITT Industries Inc., assignee; EP 19,810108847, 1982.

109. E. Dinand and M.R. Vignon, *Carbohydrate Research*, 2001, **330**, 285.

110. J.Y. Cavaille, H. Chanzy, E. Fleury and J-F. Sassi, inventors; Rhodia Chimie, assignee; US 6117545, 1997.

111. C. Gousse, H. Chanzy, G. Excoffier, L. Soubeyrad and E. Fleury, *Polymer*, 2002, **43**, 2645.

112. M.J. Cash, A.N. Chan, H.T. Conner, P.J. Cowan, R.A. Gelman, K.M. Lusvardi, S.A. Thompson and F.P. Tise, inventors; Hercules Incorporated, assignee; US 6602994, 1999.

113. S. Dong, S. Sapieha and H.P. Schreiber, *Polymer Engineering and Science*, 1993, **33**, 343.

114. M. Agarwal, Y. Lvov and K. Varahramyan, *Nanotechnology*, 2006, **17**, 5319.

115. H-H. Steege and B. Philipp, *Zellstoff und Papier*, 1974, **23**, 68.

116. D. Bondeson, A. Mathew and K. Oksman, *Cellulose*, 2006, **13**, 171.

117. X.M. Dong, J.F. Revol and D.G. Gray, *Cellulose*, 1998, **5**, 19.

118. J. Araki, M. Wada and S. Kuga, *Journal of Wood Science*, 1999, **45**, 258.

119. H.I. De Vries, *Acta Crystallographica*, 1951, **4**, 219.

120. J-F. Revol, H. Bradford, J. Giasson, R.H. Marchessault and D.G. Gray, *International Journal of Biological Macromolecules*, 1992, **14**, 170.

121. M.A.S.A. Samir, F. Alloin and A. Dufresne, *Biomacromolecules*, 2005, **6**, 612.

122. L.M.J. Kroonbatenburg, J. Kroon and M.G. Northolt, *Polymer Communications*, 1986, **27**, 290.

123. T. Nishino, I. Matsuda and K. Hirao, *Macromolecules*, 2004, **37**, 7683.

124. T. Odijk and H.N.W. Lekkerkerker, *Journal of Physical Chemistry*, 1985, **89**, 2090.

125. M.M. de Souza Lima and R. Borsali, *Langmuir*, 2002, **18**, 992.

126. H. Angellier, J.L. Putaux, S. Molina-Boisseau, D. Dupeyre and A. Dufresne, *Macromolecular Symposia*, 2005, **221**, 95.

127. R.H. Marchessault, F.F. Morehead and N.M. Walter, *Nature*, 1959, **184**, 632.

128. J.F. Revol, L. Godbout, G.X.M. Dong, D.G. Gray, H. Chanzy and G. Maret, *Liquid Crystals*, 1994, **16**, 127.

129. J.F. Revol, L. Godbout and D.G. Gray, *Journal of Pulp and Paper Science*, 1998, **24**, 146.

130. W.J. Orts, L. Godbout, R.H. Marchessault and J-F. Revol, *Macromolecules*, 1998, **31**, 5717.

131. R.H. Marchessault, F.F. Morehead and M.J. Koch, *Journal of Colloid Sciences*, 1961, **16**, 327.

132. S. Onogi and T. Asada, *Rheology*, 1980, **1**, 127.

133. W.J. Orts, L. Godbout, R.H. Marchessault and J.F. Revol, *ACS Symposium Series*, 1995, **597**, 335.

134. D. Viet, S. Beck-Candanedo and D.G. Gray, *Cellulose*, 2007, **14**, 109.

135. V. Favier, H. Chanzy and J-Y. Cavaillé, *Macromolecules*, 1995, **28**, 6365.

136. D. Dubief, E. Samain and A. Dufresne, *Macromolecules*, 1999, **32**, 5765.

137. A. Dufresne, M.B. Kellerhals and B. Witholt, *Macromolecules*, 1999, **32**, 7396.

138. M.N. Anglès and A. Dufresne, *Macromolecules*, 2000, **33**, 8344.

139. M. Grunert and W.T. Winter, *Journal of Polymers and the Environment*, 2002, **10**, 27.

140. L. Chazeau, J.Y. Cavaillé and J. Perez, *Journal of Polymer Science, Part B: Polymer Physics Edition*, 2000, **38**, 383.

141. A.P. Mathew and A. Dufresne, *Biomacromolecules*, 2002, **3**, 609.

142. N. Ljungberg, J.Y. Cavaillé and L. Heux, *Polymer*, 2006, **47**, 6285.

143. L. Heux, G. Chauve and C. Bonini, *Langmuir*, 2000, **16**, 8210.

144. J. Araki, M. Wada and S. Kuga, *Langmuir*, 2001, **17**, 21.

145. L. Chazeau, P. Terech and J.Y. Cavaillé, *Macromolecules*, 1999, **32**, 1872.
146. L. Heux, G. Chauve and C. Bonini, *Langmuir*, 2000, **16**, 8210.
147. C. Goussé, H. Chanzy, G. Excoffier, L. Soubeyrand and E. Fleury, *Polymer*, 2002, **43**, 2645.
148. M.A.S.A. Samir, F. Alloin, W. Gorecki, J-Y. Sanchez and A. Dufresne, *Journal of Physical Chemistry B*, 2004, **108**, 10845.
149. M. Schroers, A. Kokil and C. Weder, *Journal of Applied Polymer Science*, 2004, **93**, 2883.
150. E. Dujardin, M. Blaseby and S. Mann, *Journal of Materials Chemistry*, 2003, **13**, 696.
151. S. Dong and M. Roman, *Journal of the American Chemical Society*, 2007, **129**, 13810.
152. A.J. Brown, *Journal of the Chemical Society: Transactions*, 1886, **49**, 432.
153. M. Shoda and Y. Sugano, *Biotechnology and Bioprocess Engineering*, 2005, **10**, 1.
154. R. Jonas and L.F. Farah, *Polymer Degradation and Stability,* 1998, **59**, 101.
155. D. Klemm, D. Schumann, U. Udhardt and S. Marsch, *Progress in Polymer Science*, 2001, **26**, 1561.
156. N. Yoshinaga, T. Aboshi and K. Watanabe, *Bioscience, Biotechnoloy and Biochemistry*, 1997, **61**, 219.
157. Virginia Polytechnic Institut and State University, *http://www.vtnews.vt.edu/images/08680gatenholm.jpg*
158. M. Schramm and S. Hestrin, *Journal of General Microbiology*, 1954, **11**, 123.
159. M. Iguchi, S. Yamanaka and A. Budhiono, *Journal of Materials Science*, 2000, **35**, 261.
160. M. Schramm and S. Hestrin, *Biochemical Journal*, 1954, **58**, 345.
161. S. Kongruang, *Applied Biochemistry and Biotechnology*, 2008, **148**, 245.
162. S. Fiedler, M. Füssel and K. Sattler, *Zentralblatt für Microbiologie*, 1989, **144**, 473.

163. R.M. Brown, Jr., and D. Montezinos, *Proceedings of the National Academy of Sciences of the United States of America*, 1976, **73**, 143.

164. M. Benziman, C.H. Haigler, R.M. Brown, Jr., A.R. White and K.M. Cooper, *Proceedings of the National Academy of Sciences of the United States of America*, 1980, **77**, 6678.

165. C.H. Haigler and M. Benziman in *Cellulose and Other Material Polymer Systems*, Ed., R.M. Brown, Jr., Plenum Press, New York, NY, USA, 1982, p.273.

166. R.E. Cannon and S.M. Anderson, *Critical Reviews in Microbiology*, 1991, **17**, 435.

167. R.M. Brown, Jr., in the *Proceedings of the 9th International Biotechnology Symposium and Expos*, Washington, USA, 1992, p.76.

168. S. Masaoka, T. Ohe and N. Sakota, *Journal of Fermentation and Bioengineering*, 1993, **75**, 18.

169. W. Borzani and S.J. de Souza, *Biotechnology Letters*, 1995, **17**, 1271.

170. M. Iguchi, S. Mitsuhashi, K. Ichimura, Y. Nishi, M. Uryu, S. Yamanaka and K. Watanabe, inventors; Agency of Industrial Science and Technology, Sony Corporation, assignee; US 4742164, 1988.

171. A. Bohn, H-P. Fink, J. Ganster and M. Pinnow, *Macromolecular Chemistry and Physics*, 2000, **201**, 1913.

172. S. Kongruang, *Applied Biochemistry and Biotechnology*, 2008, **148**, 245.

173. T. Kouda, H. Yano, F. Yoshinaga, M. Kaminoyama and M. Kamiwano, *Journal of Fermentation and Bioengineering*, 1996, **82**, 382.

174. T. Kouda, T. Naritomi, H. Yano and F. Yoshinaga, *Journal of Fermentation and Bioengineering*, 1998, **85**, 318.

175. P. Ross, R. Mayer and M. Benziman, *Microbiological Reviews*, 1991, **55**, 35.

176. F. Yoshinaga, N. Tonouchi and K. Watanabe, *Bioscience, Biotechnology, and Biochemistry*, 1997, **61**, 219.

177. P.R. Chawla, I.B. Bajaj, S.A. Survase and R.S. Singhal, *Food Technology and Biotechnology*, 2009, **47**, 107.

178. K. Watanabe, M. Tabuchi, Y. Morinaga and F. Yoshinaga, *Cellulose*, 1998, **5**, 187.

179. M. Hornung, M. Ludwig, A.M. Gerrard and H-P. Schmauder, *Engineering in Life Sciences*, 2006, **6**, 537.

180. D. Klemm, D. Schumann, F. Kramer, N. Hessler, M. Hornung, H-P. Schmauder and S. Marsch, *Advances in Polymer Science*, 2006, **205**, 49.

181. J.D. Fontana, A.M. De Souza, C.K. Fontana, I.L. Torriani, J.C. Moresch and B.J. Gallotti, *Applied Biochemistry and Biotechnology*, 1990, **24/45**, 253.

182. D.F. Ring, W. Nashed and T. Dow, inventors; Johnson & Johnson Products, Inc., assignee; US 4655758, 1987.

183. D.F. Ring, W. Nashed and T. Dow, inventors; Johnson & Johnson Products, Inc., assignee; US 4588400, 1986.

184. L.F. Farah, inventor; Bio Fill Produtos Biotecnologicos SA, assignee; US 4912049, 1990.

185. W. Caja, A. Krystynowicz, S. Bielecki and R.M. Brown, Jr., *Biomaterials*, 2006, **27**, 145.

186. I. Helenius, H. Bäckdahl, A. Bodin, U. Nannmark, P. Gatenholm and B. Risberg, *Journal of Biomedical Materials Research, Part B: Applied Biomaterials*, 2006, **76A**, 431.

187. D. Klemm, D. Schumann, F. Kramer, N. Hessler, M. Hornung, H-P. Schmauder and S. Marsch, *Advances in Polymer Science*, 2006, **205**, 49.

188. Die Eigenschaften von NanoMasque, Forschungszentrum für Medizintechnik und Biotechnologie GmbH, *http://nanomasque.fzmb.de/seiten/deutsch/produkt2.html*

189. M. Tabuchi and Y. Baba, *Analytical Chemistry*, 2005, **77**, 7090.

190. M. Tabuchi, K. Kobayashi, M. Fujimoto and Y. Baba, *Lab on a Chip*, 2005, **5**, 1412.

191. S. Yamanaka, K. Watanabe, N. Kitamura, M. Iguchi, S. Mitsuhashi, Y. Nishi and M. Uryu, *Journal of Materials Science*, 1989, **24**, 3141.

192. S. Yamanaka and K. Watanabe, *Cellulosic Polymers, Blends and Composites*, Eds., R,D. Gilbert, Hanser Publishers, Cincinnati, O.H, USA, 1994, p.207.

193. H. Yano, J. Sugiyama, A.N. Nakagaito, M. Nogi, T. Matsuura, M. Hikita and K. Handa, *Advanced Materials*, 2005, **17**, 153.

194. M. Nogi, S. Ifuku, K. Abe, K. Handa, A.N. Nakagaito and H. Yano, *Applied Physics Letters*, 2006, **88**, 133124.

195. M. Iguchi, S. Yamanaka and A. Budhiono, *Journal of Materials Science*, 2000, **35**, 261.

196. D.C. Johnson, A.N. Neogi and H.A. Leblanc, inventors; Weyerhaeuser Company, assignee; WO 8808899, 1988.

197. Y. Nishi, M. Uryu, S. Yamanaka, K. Watanabe, N. Kitamura, M. Iguchi and S. Mitsuhashi, *Journal of Materials Science*, 1990, **25**, 2997.

198. M. Iguchi, S. Mitsuhashi, K. Ichimura, Y. Nishi, M. Uryu, S. Yamanaka and K. Watanabe, inventors; Agency of Industrial Science and Technology, Sony Corporation, assignee; US 4742164, 1988.

199. V. Frantz, *Annals of Surgery*, 1943, **118**, 116.

200. D.M. Wiseman, L. Gottlick-Iarkowski and L. Kamp, *Journal of Investigative Surgery*, 1999, **12**, 141.

201. W.K. Czaja, D.J. Young, M. Kawecki and R.M. Brown, Jr., *Biomacromolecules*, 2007, **8**, 1.

202. N. Hoenich, *Bioresources*, 2006, **1**, 270.

203. F. Galli, S. Rovidati, S. Chiarantini, G. Campus, F. Canestrari and U. Buoncristiani, *Kidney International*, 1998, **54**, 580.

204. M. Märtson, J. Viljanto, T. Hurme, P. Laippala and P. Saukko, *Biomaterials*, 1999, **20**, 1989.

2 Hemicelluloses in Pharmacy and Medicine

Valentin I. Popa

2.1 Sources and Structure of Hemicelluloses

The use of polysaccharides for pharmaceutical application is attractive because they are economical, readily available, nontoxic, capable of chemical modifications, potentially biodegradable and, with few exceptions, also biocompatible. Of increasing importance is the fact that polysaccharides are obtained from plant resources, which are renewable and if cultivated or harvested in a sustainable manner can provide a constant supply of raw material [1].

Polysaccharide-based biomaterials are an emerging class in several biomedical fields, such as tissue regeneration, particularly for cartilage, drug delivery devices and gel entrapment systems for the immobilisation of cells, implants, films, microparticles, nanoparticles, inhalable and injectable systems as well as viscous liquid formulations. Important properties of the polysaccharides include their controllable biological activity, biodegradability and ability to form hydrogels. Non-starch, linear polysaccharides are resistant to the digestive action of the gastrointestinal enzymes and retain their integrity in the upper gastrointestinal tract. Matrices manufactured from these polysaccharides therefore remain intact in the stomach and the small intestine, but once they reach the colon they are degraded by the bacterial polysaccharidases. This property makes these polysaccharides exceptionally suitable for the formulation of colon-targeted drug delivery systems [2, 3].

Most of the polysaccharides used derive from plants and they include hemicelluloses. Hemicelluloses are the second most abundant polysaccharides in nature after cellulose. They occur in close association with cellulose and lignin and contribute to the rigidity of plant cell walls in lignified tissue. Hemicelluloses represent about 20–30% of total mass of annual and perennial plants and have a heterogeneous composition of various sugar units, depending on the type of plant and extraction procedure. The most common constituent of polysaccharides is D-glucose, but D-fructose, D-galactose, L-galactose, D-mannose, L-arabinose, D-xylose and uronic acids are also common. Polysaccharides differ not only in the nature of their component monosaccharides but also in the length of their chains and in the amount of chain branching that occurs. Although a given sugar residue has only one anomeric carbon and thus can form only one glycosidic linkage with hydroxyl groups on other molecules, each sugar residue carries several hydroxyls, one or more of which may be an acceptor of glycosyl substituent. This ability to form branched structures distinguishes polysaccharides from proteins and nucleic acids, which occur only as linear polymers.

Xylan and glucomannan form the basic backbone polymers of wood hemicelluloses. In hardwood, the main hemicellulose is *O*-acetyl-4-*O*-methylglucuronoxylan, whereas in softwood arabino-4-*O*-methylglucuronoxylan constitutes about one-third of the total hemicelluloses. Native hardwood and softwood xylans differ in their degree of polymerisation, which in hardwood xylan is about 150–200 and in softwood xylan about 70–130. Hardwood xylan contains 4-*O*-methylglucuronic acid and acetyl side groups. Methylglucuronic acid is linked to the backbone by α-(1,2)-glycosidic bonds. Acetic acid is esterified at the hydroxyl group of carbon 2 and/or 3. Softwood xylan contains 4-*O*-methylglucuronic acid and L-arabinofuranoside side groups linked to the backbone by α-(1,2)- and α-(1,3)-glycosidic bonds, respectively. The average molar ratio of xylose:4-*O*-methylglucuronic acid:acetic acid in hardwood xylan is 10:1:7 and that of xylose:4-*O*-methylglucuronic acid:arabinose sugar units in softwood xylan is 8:1.6:1 [4].

In annual plants, the commonest hemicelluloses included in farm crop consist of (1,4)-β-D-xylopyranosyl units in the main chain, with side chains of one to several α-L-arabinofuranosyl, D-galactopyranosyl and β-D-glucuronopyranosyl units. Also, L-rhamnosyl, L-galactosyl and L-fucosyl units and units of various methylated sugars are presented. These hemicelluloses could be partially acetylated, with the amount of acetyl groups varying up to 12%. For xylans from *Gramineae*, every seventh xylose unit from the main chain of internodes is substituted by an L-arabinofuranose residue. For leaves, the xylose:arabinose ratio is 1:2. Nodes also contain the same amount of arabinose. The acetyl content of xylans from *Gramineae* represents 1–2% of dry wall cell; also, a small quantity of phenolic acids (pherulic and *p*-coumaric acids) is linked to C_6 of heteroxylans (**Figure 2.1**) [5].

Hemicelluloses from bran cell wall contain (over 80% by mass) acidic arabinoxylan (different from the endospermic arabinoxylan structure) as the major component and xyloglucan as the minor component.

The main group of mannans in the cell walls of higher plants is represented by galactomannans. The backbone is composed of 1,4-β-linked D-glucose and D-mannose units, which are distributed randomly within the molecule. D-Galactose side groups are attached to mannose or glucose units via α-1,6 bonds. The mannose or glucose units are also partially substituted by acetyl groups; on average 20–30% of the hexose units are acetylated at the 2- or 3-position. *O*-Acetyl-galactoglucomannans

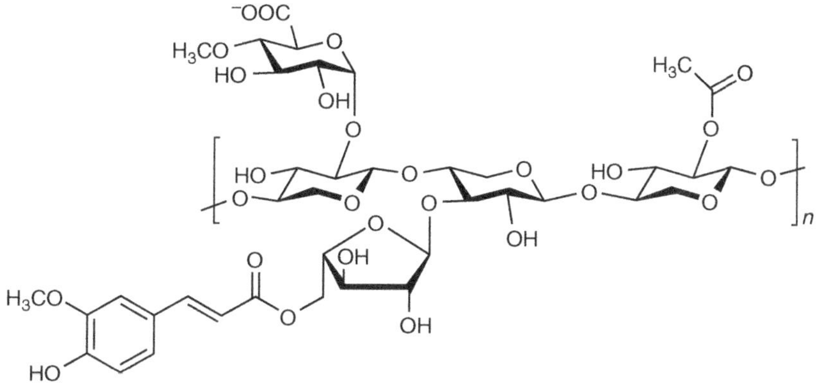

Figure 2.1 Structure proposed for arabinoxylan from annual plants

are the principal hemicelluloses in softwoods (about 20%) and can be divided into two main fractions, with ratios of galactose:glucose:mannose of about 1:1:3 and 0.1:1:4. The latter fraction with low galactose content is also often referred to as glucomannan. Hardwoods contain only some glucomannan (2–5%) with glucose and mannose residues in the ratio of 1:1–2. No galactose or acetic acid side groups are present.

As reserve substances, hemicelluloses from monocotyledonous seeds are galactoglucomannans. They are present in the seeds of *Liliaceae* and *Iridiaceae* and consist of glucosyl and mannosyl units in approximately equal amounts, together with 3% galactosyl units. The glucosyl and mannosyl units are (1,4)-linked with the galactosyl units forming the polysaccharide end groups. The reserve materials in the cell wall of leguminous seeds are mostly galactomannans. They are deposited outside the P-wall and S-wall and fill the endosperm cells completely during seed maturation. In the *Caesalpinaceae* family, galactosyl units represent 20–25%. From this family, in the seeds of *Ceratonia*, a galactomannan with galactosyl-to-mannosyl unit ratio of 20:80 has been demonstrated.

The seeds of *Cassia laevigata* contain a water-soluble galactomannan used in indigenous medicine in India. The main chain of the galactoglucomannan is composed of (1,4)-glycosidic linkages of D-galactose units. Because the ratio between D-galactose and D-mannose is 5:2, every second repeating unit is composed of three hexoses with two D-galactose units in the main chain (**Figure 2.2**) [5].

The mannan (80%) of date seeds (*Phoenix dactylifera*) is associated with about 10% galactose and 10% glucose. It is interesting that five exoenzymes from the germinated seeds (α-mannosidase, β-mannosidase, α-galactosidase, β-galactosidase and β-glucosidase) are isolated.

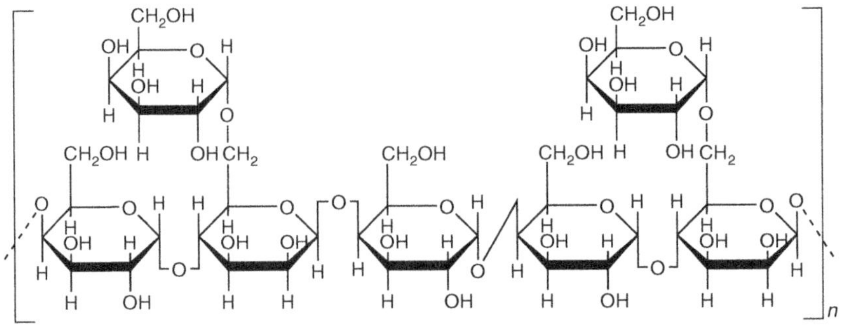

Figure 2.2 Structure proposed for galactoglucomannan of *Cassia laevigata* seeds

Arabinogalactans are a class of polysaccharides found in a wide range of plants; however, they are most abundant in plants of the genus *Larix* [6]. Arabinogalactan represents 5–40% by mass of *Larix* wood. Larch arabinogalactan is composed of galactose and arabinose units in a 6:1 ratio, with a trace amount of uronic acid. The molecular mass of the major fractions of arabinogalactan in larch is 16,000 and 100,000. Glycosyl linkage analysis is consistent with a highly branched structure consisting of a backbone of (1,3)-linked β-D-galactopyranosyl units, each of them bearing a substituent at C_6 α position. The side chain contains (1,3)-linked β-D-galactopyranosyl or 3-*O*-β-arabinopyranosyl-α-L-arabinofuranosyl units.

β-Glucans were identified as the major component in the bran of *Gramineae* (such as barley, oats, rye and wheat), consisting of linear unbranched polysaccharides of linked β-(1,3)- and β-(1,4)-D-glucopyranose units [7]. β-Glucans form 'worm'-like cylindrical molecules containing up to about 250,000 glucose residues that may produce crosslinks between regular areas containing consecutive cellotriose units. They form thermoreversible infinite network gels. Ninety per cent of the β-(1,4) links are in cellotriosyl and cellotetraosyl units joined by single β-(1,3) links, with no single β-(1,4) or double β-(1,3) links. The ratio of cellotriosyl/cellotetraosyl is between 2.0 and 2.4 in oats, about 3.0 in barley and about 3.5 in wheat. The high-molecular-mass β-glucans from barley and oats are viscous due to labile cooperative associations, whereas lower molecular mass β-glucans can form soft gels as the chains are easier to rearrange to maximise linkages. Barley β-glucan is highly viscous and pseudoplastic, both the properties decreasing with increasing temperature.

A glucan of cellular origin has been isolated from Libyan dates (*P. dactylifera* L.). This polysaccharide contains two major fractions of different molecular masses (*M*), which have been separated by Sephadex column chromatography into fraction I (*M* ~ 200,000) and fraction II (*M* ~ 10,000). These fractions differ by the presence of (1,6) branched chains consisting of D-glucose and [D-Glcp-(Glcp-(1,3)-D-Glc] side chains for the lowest molecular mass polysaccharide or [D-Glcp-(1,3)-D-Glcp] for the highest molecular mass polysaccharide. These polysaccharides were characterised as a mixture of linear (1,3)-β-D-glucan with various (1,6)-linked mono-, di- and tri-saccharide branches having 0, 1 or 2 (1,3)-β-D-glucopyranosyl residues [8].

In many seeds, collenchym-like wall thickenings are present as xylogalactoglucans, consisting of a main chain of β-(1,4)-glucan, with xylose and galactose as side groups. This hemicellulose appears only later, when S-wall is formed in the case of species of *Primulates*, *Linaceae*, *Fabaceae*, *Ranunculaceae* and *Sapotaceae*.

A study on the cell wall polymers of rice grain tissues during differentiation demonstrated a major difference in the structure of hemicellulosic polysaccharides between the preparations obtained from the endosperm cell walls and those from the cell walls of the other parts of the grains [5]. Structural analysis has shown that the endosperm cell wall contains acidic, highly branched arabinoxylans and xyloglucans and β-(1,3)/(1,4)-glucan as minor components.

A hemicellulose less used as a matrix substance, but mainly as independent wall deposition without a cellulosic framework is callose. It is a β-polyglucan-like cellulose, consisting of (1,3)-linked glucose molecules and having a helical conformation. Another (1,3)-β-glucan is laminarin, which is found in the brown seaweeds and this might be identical to callose.

Polysaccharides of algal origin include agar, alginates and carrageenans. From the various species of red-purple seaweeds (the *Rhodophyceae* class), a hydrophilic colloid, called agar, has been separated. Commercially, agar is produced from the species of *Gelidium* and *Gracilaria*: Agar is insoluble in cold water but dissolves in boiling water to give random coils. Agar is unique among polysaccharides in that gelation occurs at a temperature relatively far below the gel-melting temperature, and its main applications depend upon this high hysteresis. Gelation is reported to follow a phase separation process and association on cooling (about 35 °C), forming gels with up to 99.5% water and remaining solid up to 85 °C. Agar contains two components: one forming a strong gel (agarose) and a nongelling fraction (agaropectin). Agarose is a linear chain consisting of sequences of (1,3)-linked β-D-galactopyranosyl units joined by (1,4) linkages to 3,6-anhydro-α-D-galactopyranosyl units. On cooling, followed by aggregation, agarose forms double-helical structures (**Figure 2.3**) [9, 10].

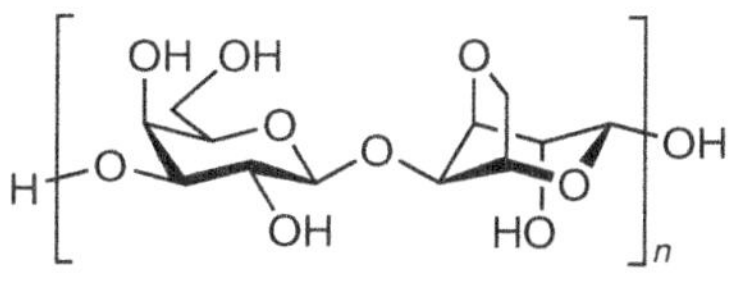

Figure 2.3 Structure of agarose

Agaropectin is a heterogeneous mixture of smaller molecules that occur in lesser amounts. Their structures are similar, but slightly branched and sulfated, and may have methyl and pyruvic acid ketal substituents. They gel poorly and may be easily removed from agarose molecules using their charge. The quality of agar is improved by an alkaline treatment that converts L-galactose-6-sulfate into 3,6-anhydro-L-galactose. The mechanical properties are directly related to the structure of the gel, especially the number of double helices forming the junction zones. The double helices of agarose are threefold and left-handed. The agarose polymer has 'kinks' in its structure, and Smith degradation results in short blocks or segments.

Another polysaccharide extracted from seaweeds is carrageenan. Carrageenans are linear, water-soluble polymers that form viscous solutions. The viscosity of the solution depends mainly on the concentration of the polymer, the temperature of the solution and the type of carrageenan. Chemically carrageenans are highly sulfated galactans, and because of their half-ester sulfate moieties they are strongly anionic polyelectrolytes. In this respect, they differ from agars and alginates, the other two classes of commercially exploited seaweed hydrocolloids. Carrageenans are susceptible to depolymerisation by acid-catalysed hydrolysis. High temperatures and low pH values lead to a complete loss of functionality. Their rate of hydrolysis at a given pH and temperature is markedly lower if they are in the gel rather than the sol state. This can be achieved by ensuring that gel-promoting cations are present in a sufficient concentration to raise the gel-melting temperature above the temperature at which they will be treated. Carrageenans can be used in acidic media, provided that they are not subjected to prolonged heating.

Carrageenan is frequently preferred over other thickening, suspending and binding ingredients, because it is natural, economical, readily available and functional in an extremely broad application base. Carrageenans can be produced via a variety of process techniques such as alcohol extraction, potassium chloride gel press or extraction with various alkalis. The process technique is important, because it influences the gel characteristics. Likewise, different seaweeds also influence the gel characteristics.

The three preferred carrageenan types are lambda, iota and kappa (**Figure 2.4**).

The kappa and iota types dissolve only in a heated water medium, whereas the lambda type is soluble in cold water. Gels created from different carrageenan types may be fluid, elastic or rigid and are heat-reversible. Gelling temperature and gel strength are also influenced by added ingredients such as salts and proteins.

From brown seaweeds (*Laminaria hyperborea*, *Macrocystis pyrifera* and *Ascophyllum nodosum*), alginates are extracted. Alginate is a linear, anionic block copolymer

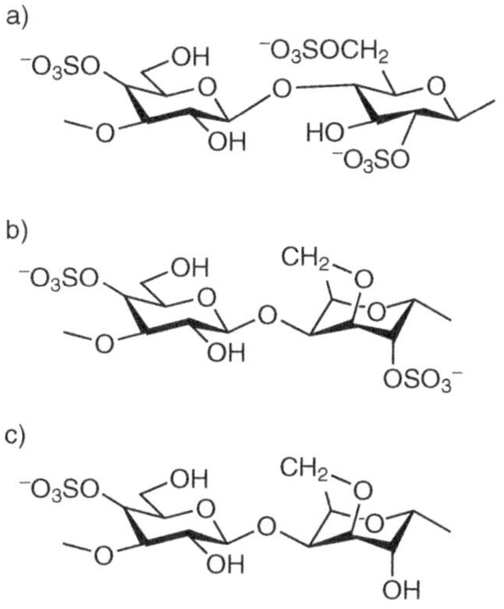

Figure 2.4 Structures of different carrageenan types: (a) lambda, (b) iota and (c) kappa

heteropolysaccharide consisting of β-D-mannuronic acid (M) and α-L-guluronic acid (G). The relative amount and sequential distribution of homogeneous M-M segments (M blocks), homogeneous G-G (G blocks) and alternating M-G segments (MG blocks), which represent the primary structure of alginate, depend on the producing species, and for marine sources, on seasonal and geographical variations (**Figure 2.5**).

The primary structure is generally defined by the F_G value, which is the fraction of overall guluronic acid residues in the polymer, and by N_G, the number-average of guluronic units in G blocks [11]. Poly β-(1,4)-linked D-mannuronic acid forms a threefold left-handed helix with (weak) intramolecular hydrogen bonding between the hydroxyl group in the 3-position and the subsequent ring oxygen, while poly

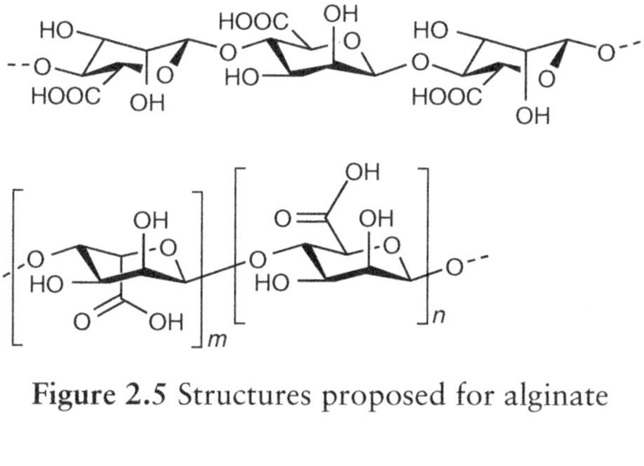

Figure 2.5 Structures proposed for alginate

α-(1,4)-linked guluronate forms stiffer (and more acid-stable) twofold screw helical chains, favouring intramolecular hydrogen bonding between the carboxyl group and the 2-OH group of the prior residues and the 3-OH group of the subsequent residues, the latter interaction being weaker.

Therefore, in practice, an alginate molecule can be regarded as a block copolymer containing M, G and MG blocks; the proportion of these blocks varies as a function of the seaweed source, and influences the physical properties of the corresponding alginates [11, 12]. Thus, gel formation by the addition of calcium ions involves G blocks so that the higher their proportion, the greater the gel strength, whereas the solubility of alginates in an acid medium depends on the proportion of MG blocks. The monovalent cation salts [Na^+, K^+, NH_4^+, $(CH_2(OH)_3NH^+)$] of alginic acid and its propylene glycol ester dissolve in water, but alginic acid and its calcium salt do not. Neutral alginate solutions of low-to-medium viscosity can be kept at 25 °C for several years, without any appreciable viscosity loss, as long as a suitable microbial preservative is added. Solutions of highly polymerised alginates will lose viscosity at room temperature within a year, and in order to achieve high, stable viscosities, it is better to add calcium ions to a solution of alginate of moderate molecular mass. In the presence of calcium ions, alginate forms irreversible gels. The ratio of calcium over which thixotropic gels are formed depends on the alginate type and pH and solid content of the system [13]. All alginate solutions depolymerise more rapidly as the temperature is raised. Alginates are most stable in the pH range of 5–9.

The alginate is known to form a physical gel by hydrogen bonding at low pH (acid gel) and by ionic interactions with divalent (Ca, Sr, Ba) or trivalent (Fe(III) and Al) ions, which act as crosslinkers between adjacent polymer chains. G blocks are the ones mainly responsible for such ionic interactions, as in the presence of multivalent cations they can associate to form aggregates of the 'egg-box' type [14].

The term 'gums' is applied to water-soluble substances refers to natural non-starch polysaccharides and their structurally modified derivatives. Mucilage is a term that is used to describe the slimy aqueous dispersions produced by plants, animals and microbes, which consist basically of water-soluble polysaccharides. Gums and mucilages are used in many pharmaceutical applications such as emulsifiers, suspending agents, binders and disintegrants as well as sustaining agents in tablets and as gelling agents [15].

Guar gum is also called guaran, clusterbean, Calcutta lucern, Gum cyamposis, Cyamposis gum, Guarina, Glucotard and Guyarem [16].

Guar gum is a galactomannan (**Figure 2.6**), which occurs as a storage polysaccharide in the seed endosperm of plants of the *Fabaceae* family. Galactomannans are linear

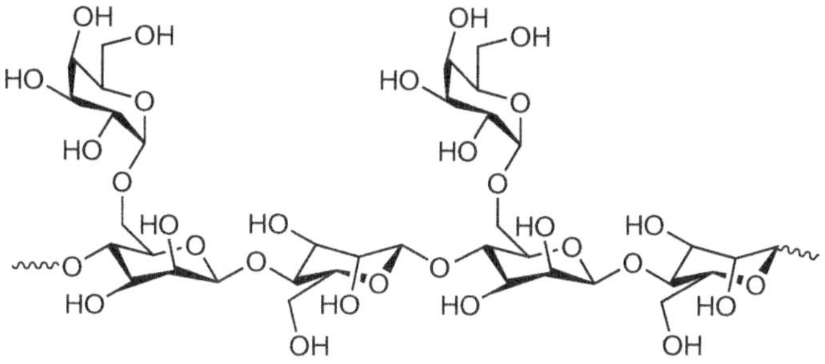

Figure 2.6 Structure proposed for guar gum

polysaccharides consisting of (1,4)-diequatorially linked β-D-mannose monomers, some of which are linked to single sugar side chains of α-D-galactose attached [17]. Galactomannan of guar gum has a backbone composed of β-1,4-linked D-mannopyranoses to which, on average, an α-D-galactose is linked at every alternate mannose by 1,6 bonds [18]. Guar gum, known for its thickening properties, is derived from the seed bean plant *Cyamopsis tetragonolobus*. It has an overall mannose-to-galactose ratio of around 2:1, the galactose substituents being regularly distributed along the mannose chain. The galactose units solubilise the polymer through steric effects, whereas galactose-poor regions, on the contrary, are less soluble and can associate both intramolecularly and intermolecularly to form partially crystalline complexes. Because of these associations, guar has remarkable rheological properties, which are used in food, personal care and oil recovery industries. To fully control the properties of guar solutions in applications, it is essential to characterise the structural changes induced by the different additives and to correlate them with the modification of rheological behaviour. Thus it was found that isopropyl alcohol promotes the formation of a network of large-scale structures via intermolecular associations, thus increasing dramatically the elastic response of guar solutions. On the other hand, salts (NaCl, Na_2CO_3 and NaSCN, tested at a concentration of 1 M) affect the guar on a local scale, leading to a more collapsed chain configuration, thus to a lower effective volume fraction and reduced viscosity [19]. The FDA has affirmed guar gum as generally safe [20].

Guar gum has been highlighted as an inexpensive and flexible carrier for oral extended-release drug delivery [21].

Guar gum is particularly useful for colon delivery because it can be degraded by specific enzymes in this region of the gastrointestinal tract. The gum protects the drug while in the environment of the stomach and small intestine and delivers the drug to the colon where it undergoes assimilation by specific microorganisms or degradation by the enzymes excreted by these microorganisms. As a hydrogel, guar gum was not found to be highly suitable for controlled release of water-soluble drugs because of their relatively fast delivery, but is useful for poorly water-soluble drugs. It was also used as a thickener for lotions and creams, as a tablet binder and as an emulsion stabiliser [18, 20]. Guar gum on its own showed high potential to serve as a carrier for oral controlled-release matrix systems. In addition, it was found that inclusion of excipients can be used as a tool to modulate drug release from these systems [22].

Locust bean gum, also known as carob bean gum, is derived from the seeds of the leguminous plant *Ceratonia siliqua* Linn.

This gum is widely cultivated in the Mediterranean region and to a smaller extent also in California. The brown pods or beans of the locust bean tree are processed by milling the endosperms to form locust bean gum and it is therefore not an extract of the native plant but a flour. Locust bean consists mainly of a neutral galactomannan polymer made up of 1,4-linked D-mannopyranosyl units and every fourth or fifth unit of the chain is substituted at C_6 by a D-galactopyranosyl unit (**Figure 2.7**). The ratio of D-galactose to D-mannose differs and this is believed to be due to the varying origins of the gum materials and growth conditions of the plant during production. Locust bean gum is a neutral polymer and its viscosity and solubility are therefore little affected by pH change within the range of 3–11 [23].

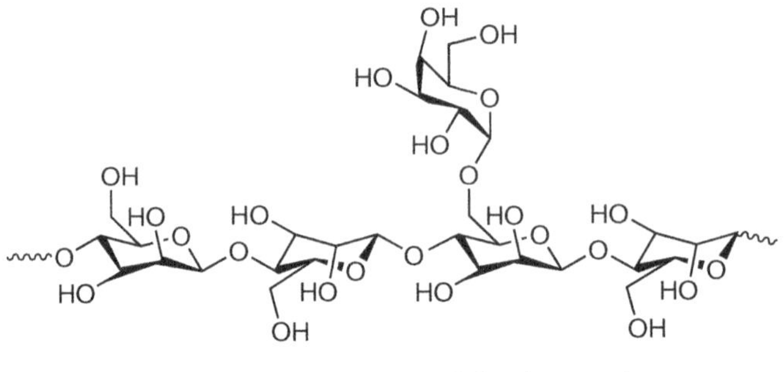

Figure 2.7 Structure proposed for locust bean gum

Locust bean gum was used to produce matrix tablets with and without the crosslinker glutaraldehyde, and the matrix tablets thus produced showed similar drug release profiles for different model drugs as guar gum and scleroglucan [18]. In another study, sustained release of diclofenac sodium could be obtained for mini-matrix systems made from locust bean gum [24]. A commercially available tablet system (TIMERx®) developed by Penwest Pharmaceuticals Company consisting of locust bean gum and xanthan gum showed both *in vitro* and *in vivo* controlled-release potential [25].

Gum arabic is a natural polysaccharide obtained from the exudates of Acacia trees [26]. Structurally, gum arabic is a branched molecule with the main chain consisting of 1,3-linked β-D-galactopyranosyl units, with other carbohydrates such as arabinose, glucuronic acid and rhamnose also present. Gum arabic was successfully used as a matrix microencapsulating agent for the enzyme endoglucanase, which resulted in slow release of the encapsulated enzyme and in addition an increase in the stability of the enzyme [27]. Gum arabic was used as an osmotic suspending and expanding agent to prepare a monolithic osmotic tablet system. The optimum system delivered the water-insoluble drug naproxen at a rate of approximately zero order for up to 12 hours at a pH of 6.8 [28].

Acidic polysaccharides (polyuronan) are an essential part of some of the cell wall plant hemicelluloses that are extractable with hot water. These substances present a high hydration capacity and, in contact with water, they form soluble colloidal glue or gel with very high water content; they are called mucilages (slimes) or gums. The main property of gums is their easy hydration to produce aqueous solutions with high viscosity at low gum concentrations. Gums can give thinner process viscosity, especially at higher temperatures. This can offer better heat transfer, easier pumping and more accurate filling. Gums also produce, under controlled conditions, gels of different strength and stability.

There are exudate gums, seaweed gums, seed gums and microbial gums. When gums are produced by pathogenic degradation of certain cells or whole tissue (cherry tree gum), certain cell groups derived from cambium are not differentiated into xylem cells but remain in the parenchymatous state. Gum production starts in these cells and then spreads to the adjacent differentiated xylem tissues where the walls are autolysed so that the pocket gum is formed inside the cambium. The swelling capacity of the gums may break the cortex of the cherry tree and the gum exudes through this split.

Industrial gums are available as powders with different degrees of fines. To be used they will be mixed in water, or with aqueous solutions of food components for hydration and dissolution. To prevent partial gelation of gums it is necessary to add the dry powdered gums to a highly turbulent stream of water. Another method for gum dissolution in water begins with mixing of the gum with a rapidly dissolving material (such as sucrose or alcohol), followed by separation of fine gum particles and maintenance of their separation for a short period of high shear necessary for their dissolution.

Chia gum of Salvia species from *Labiatae* family appears either in the seed coat or in the adjacent layer. The exudate is either partially crosslinked or bound to the seed surface, as it is not easily separated from the seeds. It is composed of α-D-xylopyranosyl, β-D-glucopyranosyl and 4-O-methyl-β-D-glucopyranosyluronic acid units in the ratio of 2:1:1.

Okra gum presents a rhamnogalacturonan as the major component, whereas the main constituent of acacia gum is arabic acid. Acacia gum contains D-galactose, L-arabinose, L-rhamnose and D-glucuronic acid in a 3:3:1:1 ratio.

Plantago ovata gum contains 63.6% D-xylose, 20.4% L-arabinose, 6.5% L-rhamnose and 9% D-glucuronic acid [29].

Analysis of corn fibre gum hydrolysates shows 48–54% D-xylose, 33–35% L-arabinose, 7–11% D,L-galactose and 3% D-glucuronic acid. The gum structure appears to consist of highly branched xylan, with attached branching containing one to three sugar units. Corn fibre gum solutions produce films of high strength. After conditioning at 65% relative humidity, the tensile strength of the unplasticised film is high, elongation is low and elastic modulus is high [30]. Corn fibre gum is an arabinoxylan and it can be extracted from the corn kernel pericarp and/or endosperm fibre fractions; it could also be used as a potential gum arabic replacer for beverage flavour emulsification.

Pectins are polycarboxylic acids. They function together with hemicelluloses as matrix substances for the fibrillar network. They exist in the middle lamella of the cell walls and in the pectic lamella of the colenchyma walls.

Pectins are negatively charged at neutral pH and approach zero charge at low pH. Galacturonic acid is the major constituent of all natural pectins. Pectins also contain varying quantities of neutral sugars, mainly arabinose, galactose and rhamnose. Galacturonic acid molecules polymerise by (1,4) bonds into linear chains (**Figure 2.8**). Because the position of –OH at C_4 is exchanged as compared with glucose, this bond is α-glycosidic. The α-(1,4)-polygalacturonic acid is termed pectic acid but its degree of dissociation is too small to make pectic acid water soluble. Its salts are called pectates. In the cell wall, a considerable part (up to 50%, of the carboxyl groups) is esterified with methanol. Such methylated pectic acids are pectins. Because a certain number of carboxyl groups are not esterified, pectins still behave as polyanions. In contrast to pectic acids they are water soluble and therefore extractable with hot water. Their solubility increases with the degree of esterification. Pectins dissolved in water can be precipitated with ethanol or calcium

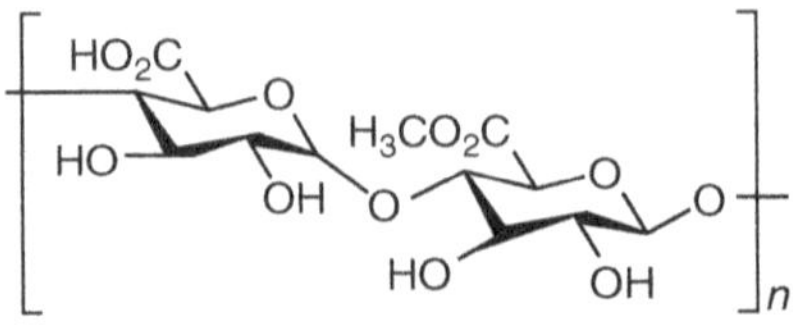

Figure 2.8 Structure proposed for pectins

ions. The pectin of the middle lamella in the cell walls is insoluble. Therefore, it is considered as a calcium pectate.

Biogenesis of pectins proceeds during deposition on the P-wall. At the same time, labelled representatives of this series (glucose, glucuronic acid and xylose) appear, indicating that the epimerase in the Golgi vesicles no longer quantitatively transforms the resorbed glucose into galactose. Later on, deposition of the S-wall follows and the activity of epimerase is stopped, so the pectic substances (indispensable for the extension growth of the P-wall) are replaced by hemicellulosic substances such as polyglucuronic acids and xylans. Conversion of hexose into uronic acid and pentose units of cell wall polysaccharides in higher plants may occur not only by the sugar nucleotide oxidation pathways, which leads to uridine diphosphate (UDP)-uronate, but also by the myoinositol oxidation pathway. In this case, the homocyclical inositol ring is split, oxidised at C_6 and concurrently replaced by a heterocycle with an –O-bridge between C_1 and C_5 as shown by label experiments. Glucuronic acid can be generated in this way.

The relative stability of pectin molecules depends on the temperature and pH. Pectins have optimal stability at pH 3.5–4. A change in pH may result in the loss of stability, especially at elevated temperatures. Highly esterified pectins are vulnerable to a high pH. Thus, even at pH 5, the depolymerisation process is considerable; this means that it is difficult to raise the pH of pectin solutions without causing a decline in the degree of polymerisation. If depolymerisation at low pH is due to hydrolysis of glycosidic bond, high-pH depolymerisation is due to β-elimination of methyl ester groups at the anhydrogalacturonic residue, which has its $C_{(4)}$ attached to the bond being split [31, 32].

Pectins are well known for their gelling, thickening and stabilising properties. For these reasons, new applications are found for pectins in the food industry and also in the pharmaceutical and cosmetic fields. Factors that affect gelation and gel characteristics include temperature, pectin concentration, pH, concentration of cosolutes (sugars) and concentration of ions such as Ca^{2+}. Thus, a pectin gel is prepared, in most cases, hot and then solidified by cooling, and a pH of about 3.0–3.1 is typical for high-sugar jams (with a concentration of approximately 65% soluble solids). Most divalent cations could be effective in pectin gelation, but only Ca^{2+} is used in food applications. An increase in the cation concentration results in an increase in both gel strength and gelling temperature. Also, it must be mentioned that if some of the galacturonic acid subunits in the galacturonan chain contain acetyl groups at O_2 or O_3, gelation will be hampered [5].

Inulin consists of a mixture of oligomers and polymers that belong to the glucofructans; it occurs in plants such as garlic, onion, artichoke and chicory. The inulin molecules

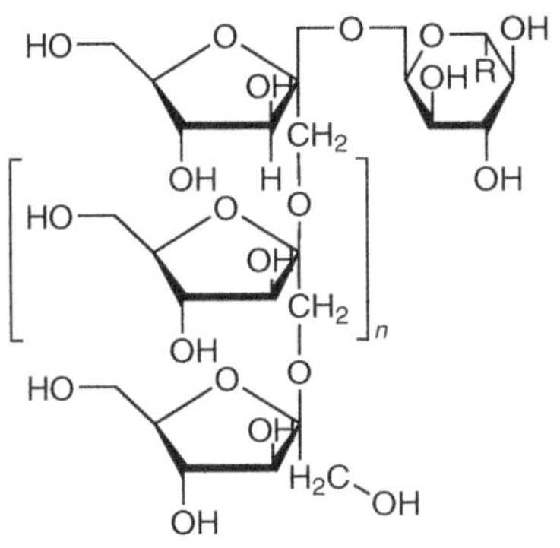

Figure 2.9 Structure proposed for inulin

contain from 2 to more than 60 fructose molecules linked by β-2,1 bonds (**Figure 2.9**). Inulin is resistant to digestion in the upper gastrointestinal tract, but is degraded by colonic microflora [33, 34].

2.2 Applications of Hemicelluloses

The beneficial health effects of hemicelluloses include improving lipid metabolism and mineral balance [35], improving colon function [36], protection against colon cancer [37], reducing the risk of heart disease and generally improving body health [38]. Hemicelluloses from various sources have potential health benefits as immunity enhancers [39, 40].

The chemical structure of hemicelluloses is the most important factor that influences their properties. This explains why polysaccharides present large differences in solubility in solution and gel properties. Their chemical structure determines the shape the molecules adopt both in the aqueous system and in the solid state.

Strong/concentrated alkalis are used to extract hemicelluloses from corn fibres, which are mainly composed of arabinoxylans, for possible uses as corn fibre gum, nontoxic adhesives, thickeners, emulsifiers, stabilisers, film formers and paper additives.

In industrial applications, hemicelluloses are used to control water content and the rheology of aqueous phases. Thus they may be used as food additives, thickeners, emulsifiers, gelling agents, adhesives and adsorbents. Thus, derivatives of xylans (acetate, butyrate and benzoate) are used as extruders for fatty acids. Carboxymethyl xylans are used as detergents, flocculants and adhesives in coating paper. Due to their antitumoral and hypocholesterolemic activities, xylan sulfate derivatives have applications in medicine. Arabinoxylans are used as emulsifiers, thickeners or stabilisers in the food, cosmetic or pharmaceutical industries. It was proved that a modified arabinoxylan from rice bran can be used as a safe alternative or as an adjuvant to existing immunotherapeutic modalities for human natural killer (NK) cell function [41]. 4-*O*-Methylglucuronoxylan is a water absorption agent and presents antitumoral activity.

Biopolymer films are excellent vehicles for incorporating a wide variety of additives, such as antioxidants, antifungal agents, antimicrobials, drugs, colours and other nutrients. Thus xyloglucan can be used to obtain a transparent film for various applications, especially in controlled-release drugs and cosmetics. The xyloglucan was extracted from tamarind seed powder by hot water extraction method. Streptomycin was encapsulated in a composite xyloglucan/chitosan (1:1) film and then the controlled-release efficiency of the film was studied, which will be a useful parameter for its usage in capsules, patches and other aroma and cosmetic products. The results showed that controlled release of the drug occurs at pH 7.4 and exhibits a small burst release in the initial stage and then a slow constant release. This initial burst effect could be attributed to the diffusion of the drug caused by rapid swelling of the membrane and also the faster release of the drug adsorbed towards the surface of the film matrix. The slow release of the drug in buffer of pH 7.4 has proved that the xyloglucan/chitosan blend film can be used as a good carrier of drugs. It is generally accepted that increased hydrophobicity, electrostatic attractive forces and swelling behaviour of the materials could improve the controlled drug release properties [42].

Arabinoxylan gels could have potential applications for colon-specific protein delivery due to their macroporous structure with mesh sizes varying from 200 to 400 nm, their aqueous environment and their dietary fibre nature. Ferulated arabinoxylan can be crosslinked by laccase (1.6 nkat/mg ferulated arabinoxylan), and stable gels are formed after 4 hours incubation at 25 °C. The crosslinking method used allowed the formation of maize bran arabinoxylan gel in the presence of insulin or β-lactoglobulin without modifying the rheological properties of the gel. The protein release rate and quantity are dependent on protein molecular mass. A low amount of entrapped protein is released by diffusion; in turn, most of the protein would be liberated only after gel degradation by colonic bacteria. These results indicate that maize bran arabinoxylan gels could be matrices suited for protein delivery in specific sites such as the colon [43, 44].

Arabinoxylan hemicelluloses showed positive oxidative burst activity in murine macrophages *in vitro*, and tended to increase body mass gain and reduced attachment of the pathogen *Salmonella* to ileal tissue in broiler chicks undergoing mild heat stress *in vivo*. These data implicate beneficial effects of arabinoxylans in cell culture or animal models and therefore, potentially, these would have positive effects on humans.

Several polysaccharides were previously tested in cell culture and animal models and subsequently displayed health-promoting effects in humans. An arabinoxylan hemicellulose was used for manufacturing a vaccine adjuvant. Arabinogalactan-proteins are thought to function as immune regulators in human health. Water-soluble hemicelluloses from soybean hull have also activated macrophages in rats. Although the activation mode of polysaccharides is not established, chemical structure, molecular mass and appropriate conformation are considered to be important in enhancing activity [45].

Glucomannans are used in the food industry (as caviar substituent), whereas arabinogalactans have applications in pharmaceutical industry (as a tablet binder or emulsifier). The most commonly used type of glucomannan is referred to as konjac glucomannan, which is extracted from the tubers of *Amorphophallus konjac* K. Koch and is a very promising polysaccharide for incorporation into drug delivery systems. Since konjac glucomannan by itself forms very weak gels, it has been investigated as an effective excipient in controlled-release drug delivery devices in combination with other polymers or by modifying its chemical structure [46, 47].

It was shown that konjac glucomannan gel systems were able to maintain integrity and control the release of theophylline and diltiazem for 8 hours. This was, however, dependent on the country of origin (i.e., Japan, Europe or the United States) due to the differences in the degree of acetylation of the konjac glucomannan [47]. Matrix tablets prepared from konjac glucomannan alone showed the ability to sustain the release of cimetidine in the physiological environments of the stomach and small intestines but the presence of β-mannanase (colon) accelerated the drug release substantially. Mixtures of konjac glucomannan and xanthan gum in matrix-type tablets showed high potential to sustain and control the release of the drug due to stabilisation of the gel phase of the tablets by a network of intermolecular hydrogen bonds between the two polymers, which may effectively retard drug diffusion. Wen and co-workers [48] used konjac glucomannan to form hydrophilic cylinders for controlled release of DNA. Konjac glucomannan crosslinked with trisodium trimetaphosphate formed hydrogel systems that could sustain hydrocortisone release depending on crosslinking density and enzymatic degradation [49].

Other hemicelluloses such as gums are stabilisation agents in medicine (e.g., stabilisation of barium sulfate suspensions for X-ray diagnostic preparations).

Purified arabinogalactans derived from western larch are known to bind *in vitro* to liver asialoglycoprotein receptors. *In vivo* experimental evidence has also demonstrated this strong binding property to liver asialoglycoprotein receptors. Larch arabinogalactan, reaching the liver through the portal circulation, is rapidly and specifically internalised by hepatocytes by receptor-mediated endocytosis. Because of the high percentage of larch arabinogalactan arriving at the liver, and its active uptake by hepatocytes,

Groman and co-workers [50] have suggested that arabinogalactan might be an ideal vehicle to deliver drugs to the liver.

Evidence also indicates that human consumption of larch arabinogalactan has a significant effect on enhancing gut microflora, specifically increasing anaerobes such as *Bifidobacteria* and *Lactobacillus* while decreasing *Clostridia*.

Experimental studies indicate larch arabinogalactan stimulates NK cell cytotoxicity, enhances other functional aspects of the immune system and inhibits the metastasis of tumour cells to the liver. Metastatic disease most commonly spreads to the liver, in preference to other organ sites. This has been theorised to be the result of a reaction between the galactose-based glycoconjugate on the metastatic cells and hepatic-specific lectin-like receptor (e.g., the D-galactose-specific hepatic binding protein) found in liver parenchyma [51].

Larch arabinogalactan pretreatment induced an increased release of interferon gamma (IFNγ), tumour necrosis factor alpha, interleukin-1 beta (IL-1β) and interleukin-6, but only IFNγ was involved in the enhancement of NK cell cytotoxicity [52].

Other studies have shown that larch arabinogalactan enhances vascular permeability [53]. At the same time, larch arabinogalactan improves contrast enhancement of the liver and has significant effects on hepatic lesion detection as assessed by contrast-to-noise ratio [54].

Also arabinogalactan can be used in pharmaceutical dispersions as a tablet binder and as an emulsifier for water-in-oil or oil-in-water emulsions.

Powdered endosperm called carrulin (from *Ceratonia* seeds) is used in the textile and pharmaceutical industries. From this plant a galactomannan can be extracted [5].

β-Glucans have important health benefits, especially reducing the risk of coronary heart disease, lowering cholesterol level and reducing the glycaemic response. These positive effects are linked to their high viscosity, although it may be that some of the effects arise from appetite suppression. Almost all carboxymethyl glucans possess properties useful in cosmetics and dermatological applications [55].

The antitumour activity of date glucan was tested on the allogeneic solid Sarcoma-180 in mice. This tumour model is known to be very useful for testing immunomodulating substances. In all experiments, $\sim 2 \times 10^6$ Sarcoma-180 tumour cells (ascites form) were subcutaneously transplanted into the right side of female CD1 mice. The test samples were dissolved in saline solution and sterilised for 20 minutes at 120 °C, and then injected intramuscularly every day for 10 days, starting 24 hours after tumour implantation. The evaluation of antitumour activity was performed by measuring the tumour diameter at 10-day intervals and by determining the mass of the excised tumours at day 30 after tumour inoculation. The antitumour effect of the date glucans was dose dependent, with an optimum activity at 1 mg/kg. The mechanisms of the influence of molecular mass and of the nature of branching chains on the antitumour activity are under investigation [8].

Agar is used in the food industry in icings, glazes, processed cheese, jelly sweets and marshmallow and sometimes as a substitute for gelatin. Another interesting application of agar comes from the fact that it constitutes a very good microbiological medium [56]. In its hydrate form, agar provides the smoother nonirritating bulk necessary for normal peristalsis. Sodium agar, agarose and sodium agarose are used in globulin electrophoresis, immunodiffusion diagnostic techniques and gel filtration. The hydrophilic properties of agarose make it a good moistening additive for bread. It binds to proteins and is used to clarify wines, juices and vinegars. It is used as a binder in medicinal tablets and capsules. Some red algal polysaccharides may control cold sore herpes. They are active compounds in Zovirax, for their ability to interfere with virus binding to human cell membranes.

Carrageenans and semisynthetic sulfated polysaccharides, like laminarin sulfate, were shown to be potent angiogenesis inhibitors [57]. Carrageenan extracted from seaweed is not assimilated by human body and provides only bulk but no nutrition. The carrageenans investigated were suitable tablet excipients for the manufacturing of controlled-release tablets [58]. In another study, matrices made of i-carrageenan and λ-carrageenan sustained the release of three different model drugs and showed release profiles that approached zero-order kinetics. It was found that factors such as tablet diameter, drug-to-carrageenan ratio and ionic strength of the dissolution medium may play a role in the release of drug from these matrices [59].

Hydrogel beads were prepared from a mixture of crosslinked κ-carrageenan with potassium and crosslinked alginate with calcium and they exhibited a smoother surface morphology than the one-polysaccharide network beads. The carrageenan part of the hydrogel pronouncedly enhanced the thermostability of the polymeric network. These beads were introduced as novel carriers for controlled drug delivery systems [60].

Alginates must be regarded as a family of polymers with a wide range of chemical composition, molecular size and, hence, functional properties. They are compatible with a wide variety of materials including other thickeners, lattices, oil sugars, fats, pigments, various surfactants and alkali metal solutions. The incompatibility occurs as a result of reaction with divalent cations (except magnesium) or other heavy metal ions, cationic quaternary amines or chemical agents that cause alkaline degradation or acid precipitation. Alginates have a haemostatic function and are able to absorb specific solutes, whereas calcium alginates are used as hydrogel dressing due to large pore size and high water absorbency. Such alginate gel capsules should ideally have high mechanical and chemical stability; controllable swelling properties; low content of toxic, pyrogenic and immunogenic materials; defined pore size; and narrow pore size distribution. These characteristics could be obtained by selection and purification of alginates from seaweeds followed by further refinement (by fractionation and chemical or enzymatic modification) and control of the gelling process, in combination with other biopolymers.

Alginates generally show high water absorption and may be used as low-viscosity emulsifiers and shear-thinning thickeners. They can be used to stabilise phase separation in low-fat fat substituents, for example as alginate/caseinate blends in starch three-phase systems. Alginate is used in a wide variety of foodstuff such as pet food chunks, onion rings, stuffed olives, low-fat spreads, sauces and pie fillings [61]. Propylene glycol alginate has widespread use as an acid-stable stabiliser, for example, preserving the head on beers [62, 63].

Alginates are suitable for various clinical applications, such as cell encapsulation, drug delivery and tissue engineering [64, 65].

Thus numerous studies [66–68] have analysed the biocompatibility of alginates/sodium alginate and drugs, to estabilsh which are the most important factors affecting the drug release from matrix tablets, and their effects on cell proliferation, cell migration and so on [69]. Because the properties of alginate hydrogels are readily controllable through modification of the chemical structure of the sugar residues and different crosslinked molecules, and these materials interact with cells [70], they are used to engineer bones [71], cartilage [72], muscles, blood vessels [73], liver and nerve tissue [74, 75]. Alginate was used as a scaffold to transplant subcultured dental pulp cells subcutaneously into the backs of nude mice, and it was proved that subcultured dental pulp cells actively differentiated into odontoblast-like cells and induced calcification in an alginate scaffold [76].

Alginates enhance efficient treatment of oesophageal reflux, and create multiquality calcium fibres for dermatology and wound healing. They are also used for high and low gel strength dental impression materials. Besides this, alginate is an effective natural disintegrant tablet binder and offers an attractive alternative for sustained-release systems. Alginates offer advantages over synthetic polymers as they form hydrogels under relatively mild pH and temperature and are generally regarded as nontoxic, biocompatible, biodegradable, less expensive and abundantly available in nature; in addition, alginates meet the important requirement of being amenable to sterilisation and storage. All these advantages make alginates very useful materials for biomedical applications especially for controlled delivery of drugs and other biologically active compounds and for the encapsulation of cells. Calcium alginate is a natural haemostat, so alginate-based dressings are indicated for bleeding wounds. The gel-forming property of alginates helps in removing the dressing without much trauma.

Alginates have been successfully used as a matrix for the entrapment and/or delivery of biological agents, such as drugs and proteins. In particular proteins can be loaded and released by alginate matrices without loss of their biological activity because of the relatively mild gelation process of alginate. In pharmaceutical formulations, the alginate gel can be prepared prior to use, or it can spontaneously form *in situ* in physiological fluids, by low pH and/or calcium ions naturally present at the site of administration. Alternatively, the gelling agent can be either added as a part of the formulation or separately administered. The microencapsulation technique has been specifically developed for the oral delivery of proteins, as they are quickly denatured and degraded in the hostile environment of the stomach. The protein is encapsulated in a core material, which, in turn, is coated with a biocompatible, semipermeable membrane that controls the release rate of the protein while protecting it from biodegradation. Several examples are reported, in which alginate is used in combination with polyethylene glycol (PEG). Alginate gels act as core materials in this application, while PEG, which exhibits certain useful properties such as protein resistance, low toxicity and low immunogenicity, together with the ability to preserve the biological properties of proteins, acts as a coating membrane. A chitosan/PEG-alginate microencapsulation process applied to biological macromolecules such as albumin or hirudin was reported to be a good candidate for oral delivery of bioactive peptides [77].

In general, drugs with nonfavourable solid-state properties, such as low solubility, benefit from encapsulation in an amorphous gel matrix. The synthesis of alginate bearing cyclodextrin (CD) molecules covalently linked to polymer chains for sustained release of hydrophobic drugs has been reported [78]. Such CD derivatives of alginate are promising as they exhibit cumulative properties of size specificity of CD and transport properties of polymer matrix. Solid preparations based on alginate as oral tablets [79], microcapsules [80], implants [81] and topical delivery systems [82, 83] are currently disposable. The oral route is considered the preferred administration route. Tablets are the most abundant dosage form, due to their convenience, easy of preparation and handling. The simpler tablet formulations are prepared by direct compression of a mechanical mixture of various ingredients, without any need for granulation or coating. Tablets based on alginate have been prepared by direct compression as well as wet or dry granulation and coating with various techniques [79, 84, 85]. In monolithic tablets made from alginate (in which the drug is homogeneously dispersed), drug release is controlled by the formation of a viscous hydrated layer around the tablet, which acts as a diffusion barrier by opposing the penetration of water into the tablet. Water-soluble drugs are mainly released by diffusion across this gel layer, while poorly soluble drugs are mainly released by erosion of the tablet.

Micro- and nano-capsules can be prepared from alginate. Microcapsules (typically > 200 μm) are simply obtained by dropping an aqueous solution of alginate into a gelling solution, either acid (pH < 4) or, more usually, containing calcium chloride ($CaCl_2$) as crosslinking agent. Microspheres of lower dimension (<10 μm) are produced by a water-in-oil emulsification process using an ultrasonicator [86]. To obtain a stable water-in-oil emulsion, a surfactant agent is used. An aqueous $CaCl_2$ solution is then added to the emulsion under stirring to allow ionotropic gelation of the particles.

The main shortcoming of alginate devices is their rapid erosion at neutral pH and low adhesion to mucosal tissues, which is further reduced upon crosslinking. Bioadhesive formulations [85, 87, 88] or formulations with prolonged gastric residence times [89] made from alginate have been reported. In these works, alginate was used in combination with chitosan, polylysine or vegetable oils. Alginate beads for floating drug delivery systems (FDDS) have been prepared [90]. FDDS have lower density than gastric fluids, so their gastric residence time is longer. Floating alginate beads are easily obtained by dropping an alginate solution containing a foaming agent such as $CaCO_3$ or $NaHCO_3$ in $CaCl_2$/acetic acid. The CO_2 gas produced remains entrapped inside the beads, which show low density and high porosity.

Dual drug-loaded alginate beads containing drug in inner and outer layers were prepared by dropping single-layered alginate beads into $CaCl_2$ solution and their drug release characteristics were investigated in simulated gastric fluid followed by intestinal fluid. The beads protected the drug in the gastric fluid with no release of the drug, while a biphasic release (i.e., a linear release for the first 4 hours and another linear phase thereafter) was obtained when the dissolution medium was changed to intestinal fluid [91].

The *in vivo* delivery of antituberculosis drugs was investigated in mice for alginate nanoparticles prepared by cation-induced gelation. A single oral dose achieved therapeutic drug concentrations in the blood plasma for 7–11 days and in organs such as the lungs, liver and spleen for a total of 15 days. The drugs encapsulated into these nanoparticles resulted in significantly higher bioavailability compared to the free drug. Furthermore, in *Mycobacterium tuberculosis*-infected mice, only three oral doses of nanoparticles that were spaced 15 days apart resulted in complete bacterial clearance from specific organs, which is comparable to 45 conventional doses of the free drug [92].

Modification of polysaccharides by introducing acrylic polymer chains is used to obtain a finer control over drug release rate and to improve adhesion to biological substrates [93–99]. Hydrogels based on crosslinked poly(acrylic acid) [100, 101] have been reported to adhere to mucus providing a barrier against irritations and inflammations of membranes of the gastrointestinal system. Acrylic polymers containing amine functionality, such as poly(dimethylaminoethylacrylate), in combination with glycolic residues have been demonstrated to show good bioadhesion and mucoadhesion [102].

Among the possible applications of alginate gel systems, one of the most promising is for cell immobilisation. Gel entrapments allow suspension cells to be cultivated in several types of bioreactors to achieve high cell densities. In cell immobilisation applications, the main drawback of alginate matrix gels is represented by their high density of network, which limits cell growth [70]; moreover, cell anchorage, a strict requirement for survival, is limited on alginate gels, because of its hydrophilic nature. PEG copolymers are used to improve the biocompatibility of polysaccharides. Several PEG-alginate systems for cell entrapment have been reported [103], but not many examples of PEG-alginate copolymers are found in literature [104].

Pectin is moreover marketed as a nutritional supplement for the management of elevated cholesterol [105, 106]. Pectin has been found to alter the characteristics of the fibrin network architecture, suggesting that it may have some antithrombotic effects [107]. Pectin has been investigated as an excipient in many different types of dosage forms such as film coating of colon-specific drug delivery systems when mixed with ethyl cellulose, microparticulate delivery systems for ophthalmic preparations and matrix-type transdermal patches. It has a high potential as a hydrophilic polymeric material for controlled-release drug delivery systems, but its aqueous solubility contributes to premature and fast release of the drug from these matrices [3].

One of the options to reduce the high solubility of pectin in aqueous medium is through chemical modification without affecting favourable biodegradability properties. Pectins can be chemically modified by saponification catalysed by mineral acids, bases, salts of weak acids, enzymes, concentrated ammonium systems and primary aliphatic amines. Calcium salts of pectin have reduced solubility and matrix tablets prepared with calcium pectinate showed very good potential to be used in colon-targeted drug delivery systems. Furthermore, crosslinking of pectin with calcium ions inhibits the release of the incorporated drug from pectin tablets by suppressing both the dissolution and swelling of these systems [108–111].

Depending on the type and structure of the pectin molecule, pectins can gel in various ways. Gelling can be induced by acid or crosslinking with calcium ion or by reaction with alginate. When a pectin solution is titrated with acid, the ionisation of carboxylate groups on pectins is repressed causing pectin molecules to no longer repel each other over their entire chains. The pectins can thus associate over a portion of their chains to form acid-pectin gels. Gel forming has been investigated widely for sustained drug delivery. A mixture of xyloglucan with pectin resulted in an *in situ* gel-forming system with sustained paracetamol drug delivery in rats [110, 112].

Inulin with a high degree of polymerisation was used to prepare biodegradable colon-specific films in combination with Eudragit® RS, which could withstand breakdown by the gastric and intestinal fluids [33]. It was shown in another study where different Eudragit® polymers were formulated into films with inulin that when a combination of Eudragit® RS and Eudragit® RL was mixed with inulin it exhibits better swelling and permeation properties in colonic medium rather than other gastrointestinal media [113]. Methylated inulin hydrogels were developed as colon-specific drug delivery systems and investigated for water uptake and swelling. The hydrogel exhibited a relatively high rate of water uptake and anomalous dynamic swelling behaviour [34].

Inulin derivatised with methacrylic anhydride and succinic anhydride produced a pH-sensitive hydrogel by UV irradiation that exhibited a reduced swelling and low chemical degradation in acidic medium, but a good swelling and degradation in simulated intestinal fluid in the presence of its specific enzyme inulinase [114].

2.3 Conclusion

Hemicelluloses are the second most abundant polysaccharides in nature after cellulose and they are of increasing importance because they are obtained from renewable resources. At the same time, the hemicelluloses can be used for pharmaceutical and medicinal applications due to their valuable properties, possibilities for chemical modification, lack of toxicity, biodegradability and biocompatibility. There are a number of examples concerning the introduction of hemicelluloses in almost all fields of biomedical interest. However, one of the main problems is the structural variability of these polysaccharides, which precludes establishing a correlation between their structure and properties, which forms the basis for ensuring reproducibility of the results. Therefore, it is also worth mentioning that improving the performances of the hemicelluloses will be an opportunity for the medical and pharmaceutical industry, as the time-to-market of the said polymers will be reduced.

References

1. C.E. Beneke, A.M. Viljoen and J.H. Hamman, *Molecules*, 2009, **14**, 7, 2602.
2. M. Charusia, M.K. Chourasia, N.K. Jain, A. Jain, V. Soni and S.K. Jain, *AAPS Pharmaceutical Science and Technology*, 2006, **7**, 3, E1.
3. A. Shirwaikar, A. Shirwaikar, S.L. Prabu and G.A. Kumar, *Indian Journal of Pharmaceutical Sciences*, 2008, **70**, 4, 415.
4. L. Viikari, A. Kantelinen, J. Sundquist and M. Linko, *FEMS Microbiology Reviews*, 1994, **13**, 2/3, 335.
5. I. Spiridon and V.I. Popa in *Polysaccharides; Structural Diversity and Functional Versatility*, 2nd Edition, Ed., S. Dumitriu, Dekker/CRC Press, Boca Raton, FL, USA, 2005, p.475.
6. P. Odonmazig, A. Ebringerova, E. Machova and L. Alfoldi, *Carbohydrate Research*, 1994, **252**, 1, 317.
7. Y.T. Kim, E.H. Cheong, D.L. Wiliams, C.W. Kim and S.T. Lim, *Carbohydrate Research*, 2000, **328**, 3, 331.
8. O. Ishrud and J.F. Kennedy, *Carbohydrate Polymers*, 2005, **59**, 4, 531.
9. M. Rinaudo, *Journal of Intelligent Material Systems and Structures*, 1993, **4**, 2, 210.
10. C.I. Simionescu, V. Rusan and V.I. Popa, Chimia *Algelor Marine (Seaweeds Chemistry)*, Publishing House of Romanian Academy, Bucharest, Romania, 1974, p.95.
11. C.I. Simionescu, V.I. Popa, A. Liga and V. Rusan, *Cellulose Chemistry and Technology*, 1975, **9**, 5, 547.
12. Z-Y. Wang, Q-Z. Zhang, M. Konno and S. Saito, *Journal de Physique II France*, 1993, **3**, 1, 1.
13. W.J. Sime in *Food Gels*, Ed., P. Harris, Elsevier Applied Science, London, UK, 1989, p.47.
14. G.T. Grant, E.S. Morris, D.A. Rees, P.J.C. Smith and D. Tom, *FEBS Letters*, 1973, **32**, 1, 195.

15. G.T. Kulkarni, K. Gowthamarajan, R.R. Dhobe, F. Yohanan and B. Suresh, *Drug Delivery*, 2005, **12**, 4, 201.

16. A.M. Gamal-Eldeen, H. Amer and W.A. Helmy, *Chemico-Biological Interactions*, 2006, **161**, 3, 229.

17. J.P. Doyle, G. Lyons and E.R. Morris, *Food Hydrocolloids*, 2008, **23**, 6, 1501.

18. T. Coviello, F. Alhaique, A. Dorigo, P. Maticardi and M. Grassi, *European Journal of Pharmaceutics and Biopharmaceutics*, 2007, **66**, 2, 200.

19. M.R. Gittings, L. Cipelleti, V. Trappe and D.A. Weitz, *Journal of Physical Chemistry A*, 2001, **105**, 40, 9310.

20. Y. Sudhakar, K. Kuotsu and A.K. Bandyopadhyay, *Journal of Controlled Release*, 2006, **114**, 1, 15.

21. J. Varshosaz, N. Tavakoli and S.A. Eram, *Drug Delivery*, 2006, **13**, 2, 113.

22. T. Durig and R. Fassihi, *Journal of Controlled Release*, 2002, **80**, 1–3, 45.

23. M. Glicksman in *Advances in Food Research*, Volume 11, Eds., E.M. Mrak and G.F. Stewart, Academic Press, New York, NY, USA, 1962, p.109.

24. J. Sujja-areevath, D.L. Munday, P.J. Cox and K.A. Khan, *International Journal of Pharmaceutics*, 1996, **139**, 1/2, 53.

25. C.W. Vendruscolo, I.F. Andreazza, J.L.M.S. Ganter, C. Ferrero and T.M.B. Bresolin, *International Journal of Pharmaceutics*, 2005, **296**, 1/2, 1.

26. K.K. Nishi, M. Antony, P.V. Mohanan, T.V. Anilkumar, P.M. Loiseau and A. Jayakrishman, *Pharmaceutical Research*, 2007, **24**, 5, 971.

27. A. Ramakrishnan, N. Pandit, M. Badgujar, C. Bhaskar and M. Rao, *Bioresource Technology*, 2007, **98**, 2, 368.

28. E. Lu, Z. Jiang, Q. Zhang and X. Jiang, *Journal of Controlled Release*, 2003, **92**, 3, 375.

29. J.N. BeMiller, C. Chen and R.L. Whisttler in *Industrial Gums and Their Derivatives*, Eds., L.R. Whistler and J.N. BeMiller, Academic Press, New York, NY, USA, 1993, p.228.

30. P.Y. Madhav, D.B. Johnston, A.T. Hotchkiss, Jr., and K.B. Hicks, *Food Hydrocolloids*, 2007, **21**, 7, 1022.

31. J.N. Be Miller in *Advances in Carbohydrate Chemistry*, Eds., M.L. Wolfrom and R.S. Tipson, Academic Press, New York, NY, USA, 1967, **22**, p.25.

32. P. Albersheim, H. Neukom and H. Deuel, *Archives of Biochemistry and Biophysics*, 1960, **90**, 1, 46.

33. L. Vervoort and R. Kinget, *International Journal of Pharmaceutics*, 1996, **129**, 1/2, 185.

34. L. Vervoort, G. Vanden Mooter, P. Augustijns and R. Kinget, *International Journal of Pharmaceutics*, 1998, **172**, 1/2, 127.

35. H.W. Lopez, M.A. Levrat, C. Guy, A. Messager, C. Demigne and C. Remesy, *Journal of Nutritional Biochemistry*, 1999, **10**, 9, 500.

36. Z.X. Lu, P.R. Gibson, J.G. Muir, M. Fielding and K. O'Dea, *Journal of Nutrition*, 2000, **130**, 8, 1984.

37. H.J. Freeman, *Canadian Medical Association Journal*, 1979, **121**, 3, 291.

38. R. McPherson in *CRC Handbook of Dietary Fiber in Human Nutrition*, 2nd Edition, Ed., G.A. Spiller, CRC Press, Boca Raton, FL, 1993, p.7.

39. L.W. Doner and K.B. Hicks, *Cereal Chemistry*, 1997, **74**, 2, 176.

40. R.B. Hespel, *Journal of Agricultural and Food Chemistry*, 1998, **46**, 7, 2615.

41. M. Ghoneum and A. Jewett, *Cancer Detection and Prevention*, 2000, **24**, 4, 314.

42. C.K. Simi and E.A. Abraham, *Colloid and Polymer Science*, 2010, **288**, 3, 297.

43. A. Castillo, A. Rascon-Chu, G. Vargas, E. Carvajal-Millan, E. Valenzuela-Soto, R.R. Sotelo-Mundo and A.L. Martinez, *Molecules*, 2009, **14**, 10, 4159.

44. C.M. Berlanga-Reyes, E. Carvajal-Millan, J. Lizardi-Mendoza, A. Rascon-Chu, J.A. Marquez-Escalante and A.L. Martinez-Lopez, *Molecules*, 2009, **14**, 4, 1475.

45. P. Zhang, J.L. Wampler, A.K. Bhunia, K.M. Burkholder, J.A. Patterson and R.L. Whistler, *Cereal Chemistry*, 2004, **81**, 4, 51.

46. M. Alonso-Sande, D. Teijeiro, C. Remunan-Lopez and M.J. Alonso, *European Journal of Pharmaceutics and Biopharmaceutics*, 2008, **72**, 2, 453.

47. F. Alvarez-Mancenido, M. Landin, I. Lacik and R. Martinez-Pacheco, *International Journal of Pharmaceutics*, 2008, **349**, 1/2, 11.

48. X. Wen, T. Wang, L. Li and C. Zhao, *International Journal of Biological Macromolecules*, 2008, **42**, 3, 256.

49. M. Liu, J. Fan, K. Wang and Z. He, *Drug Delivery*, 2007, **14**, 6, 397.

50. E.V. Groman, P.M. Enriquez, C. Jung and L. Josephson, *Bioconjugate Chemistry*, 1994, **5**, 6, 547.

51. S. Gregory and N.D. Kelly, *Alternative Medicine Review*, 1999, **4**, 2, 96.

52. J. Hauer and F.A. Anderer, *Cancer Immunology and Immunotherapy*, 1993, **36**, 4, 237.

53. L.S. Kind, B. Macedo-Sobrinho and D. Ako, *Immunology*, 1970, **19**, 5, 799.

54. R. Wisner, E.G. Amparo, D.R. Vera, J.M. Brock, T.W. Barlow, S.M. Griffey, C. Drake and R.W. Katzberg, *Journal of Computer Assisted Tomography*, 1995, **19**, 2, 211.

55. F. Zulli, F. Suter, H. Blitz, H.P. Nissen and M. Birman, *Cosmetics & Toiletries*, 1996, **111**, 12, 91.

56. E. Chiellini, *Journal of Bioactive and Compatible Polymers*, 2006, **21**, 3, 257.

57. D.H. Paper, H. Vogl, G. Franz and R. Hoffman, *Macromolecular Symposium*, 1995, **99**, 219.

58. K.M. Picker, *Drug Development and Industrial Pharmacy*, 1999, **25**, 3, 329.

59. V.K. Gupta, M. Hariharan, T.A. Wheatley and J.C. Price, *European Journal of Pharmaceutics and Biopharmaceutics*, 2001, **51**, 3, 241.

60. Z. Mohamadnia, M.J. Zohuriaan, K. Kabiri, A. Jamshidi and H. Mobedi, *Journal of Biomaterial Science - Polymer Edition*, 2008, **19**, 1, 47.

61. L.S. Lai and P.H. Lin, *Journal of the Science of Food and Agriculture*, 2004, **84**, 11, 1307.

62. G. Jackson, R.T. Roberts and T. Wainwright, *Journal of Institute of Brewing*, 1980, **86**, 1, 34.

63. J. O'Reilly, *Brew Guard*, 1996, **125**, 7, 22.

64. G. Orive, S.K. Tam, J.L. Pedraz and J. Halle, *Biomaterials*, 2006, **27**, 20, 3691.

65. D. Dufrane, M. Steenberghe, R.M. Goebbels, A. Saliez, Y. Guiot and P. Gianello, *Biomaterials*, 2006, **27**, 17, 3201.

66. C.V. Liew, L.W. Chan, A.L. Ching and P.W. Heng, *International Journal of Pharmaceutics*, 2006, **309**, 1/2, 25.

67. C.G. Thanos, B.E. Bintz, W.J. Bell, H. Qian, P.A. Schneider, D.H. MacArthur and D.F. Emerich, *Biomaterials*, 2006, **27**, 19, 3570.

68. M.G. Sankalia, R.C. Mashru, J.M. Sankalia and V.B. Sutariya, *AAPS Pharmaceutical Science and Technology*, 2005, **62**, 2, 209.

69. T. Nagakura, H. Hirata, M. Tsujii, T. Sugimoto, K. Miyamoto, T. Koriuchi, M. Nagao, T. Nakashima and A. Uchida, *Plastic and Reconstructive Surgery*, 2005, **116**, 3, 831.

70. J.A. Rowley, G. Madlambayan and D.J. Mooney, *Biomaterials*, 1999, **20**, 1, 45.

71. Y. Suzuki, M. Tanihara, K. Suzuki, A. Saitou, W. Sufan and Y. Nishimura, *Journal of Biomedical Material Research*, 2000, **50**, 3, 405.

72. A. Atala, W. Kim, K.T. Paige, C.A. Vacanti and A.B. Retik, *Journal of Urology*, 1994, **152**, 2, 641.

73. Y.M. Elcin, V. Dixit and T. Gitnick, *Artificial Organs*, 2001, **25**, 7, 558.

74. P. Soon-Shiong, M. Otterlie, S. Skjak-Braek, O. Smidsrod, R. Heintz, R.P. Lanza and T. Espevik, *Transplant Proceedings*, 1991, **23**, 1, 758.

75. K. Suzuki, Y. Suzuki, M. Tanihara, K. Ohnishi, T. Hashimoto, K. Endo and Y. Nishimura, *Journal of Biomedical Materials Research*, 1999, **49**, 4, 528.

76. P.C. Edwards and G.M. Mason, *Head & Face Medicine*, 2006, **2**, 1, 12.

77. T. Chandy, D.L. Mooradian and G.H.R. Rao, *Journal of Applied Polymer Science*, 1998, **70**, 11, 2143.

78. W. Pluemsab, N. Sakairi and T. Furuike, *Polymer*, 2005, **46**, 23, 9778.

79. M.A. Bayomi, S.A. Al-Suwayeh and A-R.M. El-Helw, *Drug Development and Industrial Pharmacy*, 2001, **27**, 6, 499.

80. N. Jerry, Y. Anitha, C.P. Sharma and P.D. Sony, *Drug Delivery*, 2001, **8**, 1, 19.

81. E. Fragonas, M. Valente, M. Pozzi-Mucelli, R. Toffanin, R. Rizzo, F. Silvestri and F. Vultur, *Biomaterials*, 2000, **21**, 8, 795.

82. S. Cohen, E. Lobel, A. Trevgoda and Y. Peled, *Journal of Controlled Release*, 1997, **44**, 2, 201.

83. S. Miyazaki, S. Kubo and D. Attwood, *Journal of Controlled Release*, 2000, **67**, 2/3, 275.

84. A.C. Hodsdon, J.R. Mitchell, M.C. Davies and C.D. Melia, *Journal of Controlled Release*, 1995, **33**, 1, 143.

85. C.S. Young, J-H. Jung, J-D. Rhee, C-K. Kim and H-G. Choi, *Drug Development and Industrial Pharmacy*, 2001, **27**, 5, 447.

86. R. Srivastava, J.Q. Brown, H. Zhu and M.J. McShane, *Biotechnology Bioengineering*, 2005, **91**, 1, 124.

87. S. Miyazaki, A. Nakayama, M. Oda, M. Takada and D. Attwood, *International Journal of Pharmaceutics*, 1995, **118**, 2, 257.

88. B.L. Strand, O. Gaserod, B. Kulseng, T. Espevik and G. Skak-Braek, *Journal of Microencapsulation*, 2002, **19**, 5, 615.

89. Y. Murata, N. Sasaki, E. Miyamoto and S. Kawashima, *European Journal of Pharmaceutics and Biopharmaceutics*, 2000, **50**, 2, 221.

90. B.Y. Choi, H.J. Park, S.J. Hwank and J.B. Park, *International Journal of Pharmaceutics*, 2002, **239**, 1/2, 81.

91. B-J. Lee, J-H. Cui, T-W. kim, M-Y. Heo and C-K. Kim, *Archives of Pharmacal Research*, 1998, **21**, 6, 645.

92. Z. Ahmad, R. Pandley, S. Sharma and G.K. Khuller, *Indian Journal of Chest Disease and Allied Sciences*, 2006, **48**, 3, 171.

93. M. Yazdani-Pedram, A. Lagos and P. Jaime Retuert, *Polymer Bulletin*, 2002, **48**, 1, 93.

94. Y. Hu, X. Jiang, Y. Ding, H. Ge, Y. Yuan and C. Yang, *Biomaterials*, 2002, **23**, 15, 3193.

95. M.K. Chun, C.S. Cho and H.K. Choi, *Journal of Controlled Release*, 2002, **81**, 3, 327.

96. S. Ahn, H.K. Choi and C.S. Cho, *Biomaterials*, 2001, **22**, 9, 923.

97. A.H. Shojaei, J. Paulson and S. Honary, *Journal Controlled Release*, 2000, **67**, 2/3, 223.

98. A. Peniche, W. Arguelles-Monal, N. Davidenko, R. Sastre, A. Gallardo and J. San Roman, *Biomaterials*, 1999, **20**, 20, 1869.

99. R.J. Mumper, A.S. Hoffman, P.A. Puolakkainen, L.S. Bouchard and W.R. Gombotz, *Journal of Controlled Release*, 1994, **30**, 3, 241.

100. N.A. Peppas, P. Bures, W. Leobandung and H. Ichikawa, *European Journal of Pharmaceutics and Biopharmaceutics*, 2000, **50**, 1, 27.

101. N.A. Peppas and J.J. Sahlin, *Biomaterials*, 1996, **17**, 16, 1353.

102. A.J. Limer, A.K. Rullay, V.S. Miguel, C. Peinado, S. Keely, E. Fitzpatrick, F.D. Carrington, D. Brayden and D.M. Haddleton, *Reactive and Functional Polymers*, 2006, **66**, 1, 51.

103. D.B. Seifert and J.A. Phillips, *Biotechnology Progress*, 1997, **13**, 5, 569.

104. G.G. d'Ayala, M. Malinconico and P. Lasurienzo, *Molecules*, 2008, **13**, 9, 2069.

105. F. Yamaguchi, S. Uchida, S. Watabe, H. Kojima, N. Shimizu and C. Hatanaka, *Bioscience, Biotechnology, and Biochemistry*, 1995, **59**, 11, 2130.

106. P.A. Judd and A.S. Truswell, *British Journal of Nutrition*, 1985, **53**, 3, 409.

107. J.W. Anderson and T.J. Hanna, *Journal of Nutrition*, 1999, **129**, 7, 1457.

108. S. Bhatia, R. Deshmukh, P. Choudhari and N.M. Bhatia, *Scientia Pharmaceutica*, 2008, **76**, 4, 775.

109. V.R. Sinha, *International Journal of Pharmaceutics*, 2001, **224**, 1/2, 19.

110. L. Liu, M.L. Fishman, J. Kost and K.B. Hicks, *Biomaterials*, 2003, **24**, 19, 3333.

111. M.K. Chourasaia and S.K. Jain, *Drug Delivery*, 2004, **11**, 2, 129.

112. K. Itoh, M. Yahaba, A. Takahashi, R. Tsuruya, S. Miyazaki, M. Dairaku, M. Mikami and D. Attwood, *International Journal of Pharmaceutics*, 2008, **356**, 1/2, 95.

113. A. Akhgari, F. Farahmand, H. Garekani, F. Sadeghi and T.F. Vandamme, *European Journal of Pharmaceutics*, 2006, **28**, 4, 307.

114. F. Castelli, M.G. Sarpietro, D. Micieli, S. Ottim, G. Pitaresi, G. Tripodo, B. Carlisi and G. Giammona, *European Journal of Pharmaceutics*, 2008, **35**, 1/2, 76.

3 Fungal Exopolysaccharides

Robert J. Seviour, Frank Schmid and
Bradley S. Campbell

3.1 Introduction

Under appropriate culture conditions, many fungi synthesise large amounts of noncellulosic chemically diverse exopolysaccharides (EPS). In most cases, such polymers are water soluble, while in some fungi they become organised, at least in part, into a capsular layer around the producing cells. They include heteropolysaccharides, which are often complex and composed of several different sugar monomers (**Table 3.1**), some of which may be negatively charged, and thus are acidic polymers. However, our understanding of the chemical structures of most of these, often being restricted to monomer sugar analyses only, and their possible functions in the organisms synthesising them is less than rudimentary. Some have strong antioxidant properties and are bioactive [1, 2], and deserve more attention. Several are synthesised by the same fungus simultaneously with homopolymers of glucose (glucans - explained later), where the relative amounts produced are determined by several culture parameters including pH [3] and the carbon source used [4, 5].

More is known currently about the exocellular noncellulosic fungal glucans, homopolymers of glucopyranosyl units, which are linked together in several ways [4, 6, 7]. These glucans are the subject of this chapter. Many have unique physical, chemical and biological properties of interest, and are of considerable benefit to us.

The majority of cell walls of the *Mycota* contain glucans as important structural components [8], and some 'fungi-like' organisms contain stored cytoplasmic glucans, but it is not often clear what the biosynthetic relationship between these and the exocellular glucans is. Their chemistries are not necessarily the same [9], but this may reflect postsynthesis structural modification during exocellular glucan export. However, whether fungi elaborate some of their wall glucans into the culture media following a breakdown in the regulation of their synthesis leading to an overproduction is uncertain, and in some cases clearly unsustainable from available experimental data (as for example with pullulan, which is not found in the cell walls of *Aureobasidium pullulans*) [4].

Table 3.1 Examples of fungal exocellular heteropolysaccharides

Fungal source	Linkage types	Monomer composition	Reference
Alternaria solani	Not reported	D-Glucose, D-galactose, D-glucosamine	[236]
Stereum strigosozonatum *Fomes australis* *Trametes lilacinogilva* *Polyporus tumulosus*	Not reported	D-Mannose, D-galactose, xylose, fucose	[237]
Aspergillus nidulans	α-linkages	D-Galactosamine, D-galactose, acetate	[238]
Aureobasidium pullulans	(1,3)-β-, (1,6)-β-	D-Glucose, malate	[239]
Aspergillus flavus	(1,2)-β-	D-Mannose, D-galactose	[240]
Penicillium erythromeles	(1,6)-β-	D-Glucose, malonate	[241]
Penicillium allahabadense	(1,6)-β-	D-Glucose, malonate, D-mannose, D-galactose	[242]
Paecilomyces lilacinus	Not reported	D-Glucose, D-galactose	[243]
Phomopsis foeniculi	(1,6)-α-, (1,6)-β-, (1-5)-β-	D-Mannose, D-galactose	[244]
Aureobasidium pullulans	(1,4)(1,6)-α-, (1,6)-α-	D-Glucose (87–91%), D-mannose	[170]
Tremella mesenterica	(1,3)-α-, (1,2)-β-	D-Mannose backbone, xylose and D-glucuronic acid side chains	[1]
Guignardia citricarpa	Complex	D-Galactose	[5]
Phellinus linteus	Not reported	Varies, D-mannose D-galactose, D-glucose, ribose, D-arabinose, xylose	[192]
Tremella fuciformis	Not reported	D-Mannose, xylose, fucose	[245]
Hericium erinaceus	(1,3)-β-, (1,2)-β-	D-Mannose	[246]
Penicillium sp. F23-2	Not reported	D-Mannose, D-glucose, D-galactose, uronic acid	[2]
Tolypocladium sp.	(1,6)-α-	D-Glucose, D-mannose, D-galactose	[247]

More likely glucans serve other important functions for the producing fungus, although what these might be is not generally clear. In some plant pathogenic necrotrophic wilt causing fungi, like *Phanerochaete chrysosporium*, these glucans may block the vascular bundles leading to plant death [10]. However, not all glucan-producing fungi are similarly pathogenic. Other evidence suggests that fungi synthesise exocellular glucans in times of substrate abundance, to be available to them as an alternative carbon source when more readily metabolisable substrates like glucose are no longer available. Certainly there is considerable evidence that exocellular glucan-degrading enzymes or glucanases are commonly produced by fungi [11, 12]. They are mostly inducible enzymes (although which substrates act as their true inducers is contentious), whose synthesis is frequently regulated by catabolite (glucose) repression [11]. Again, not all exocellular glucan-producing fungi appear to produce these degradative enzymes, and vice versa, and in some their function is clearly different, for example, they may function as primary biochemical offensive weapons in mycoparasitism of other fungi [13–15] or may be involved in fungal autolytic processes [16]. What seems commonplace is that in most fungi for which detailed data are available, the laboratory conditions encouraging EPS production are those representing some environmental stress, often where active vegetative growth is curtailed by imposing unbalanced culture conditions, as discussed later.

3.2 Determining the Structures of Fungal Exocellular Glucans

Most methods available to elucidate fungal glucan structure can be applied to all polysaccharides. These methods have problems associated with them, which without proper consideration can lead to data misinterpretation and incorrect structural interpretations. Thus, many of the early published fungal EPS structures, especially those based on superficial characterisations only, should be viewed cautiously, especially where different structures have been reported for glucans from the same or closely related fungi [17] (**Table 3.2**). Several important pieces of information are needed, and most are sought if complete unambiguous glucan structure elucidation is the aim (**Table 3.2**). This has happened too infrequently with fungal β-glucans in the past. Advances in nuclear magnetic resonance (NMR) technology (see later) mean that many of these techniques are no longer strictly necessary for unambiguous glucan structural elucidation, but nevertheless will be discussed briefly here. The reader is directed to the recent reviews of Stone [7] and Yang and Zhang [18] for further detail.

3.2.1 Purification of Glucans

In many studies of fungal exocellular glucans, the assumption made is that the material, collected usually by ethanol precipitation, from each individual fungus and then analysed, is a single polymer with a uniform constant chemical structure. However, this should not be assumed, since many single cultures of fungi produce several chemically very different EPS, whose relative amounts may change with culture conditions. For example, *A. pullulans* synthesises simultaneously the α-glucan pullulan (see later) and a β-glucan, and possibly others [4, 6, 19], but all the ethanol-insoluble

Table 3.2 Chemical features of examples of exocellular β-glucans isolated from fungi

Source	Structural features	Reference
Sclerotium glucanicum[a]	(1,3)-β-linked main chain with single D-glucopyranosyl groups linked (1,6)-β to every third residue of the main chain	[74] [75]
Claviceps purpurea	(1,3)-β-linked main chain with (1,6)-β-linked side branching (20%)	[73]
Plectonia occidentalis	(1,3)-β-linked (70%) main chain with some (1,6)-β-linked branches (30%)	[248]
Helotium species	(1,3)-β-linked (70%) main chain with some (1,6)-β-linked branches (30%)	[248]
Claviceps fusiformis	(1,3)-β-linked main chain, with single glucopyranosyl units at branch points along it	[83]
Schizophyllum commune	(1,3)-β-linked main chain possessing glucose residues attached by (1,6)-β-linkages every third residue	[249]
Pythium debaryanum	Highly branched glucan possessing mainly (1,3)-β-linkages and small numbers of (1,6)-β-linkages	[250]
Monilia fructicola	(1,3)-β-linked glucan possessing glucose residues attached by (1,4)-β-linkages every third unit on the main chain	[96]
Coriolus versicolor	Highly branched (1,3)(1,6)-β-glucan	[94]
Monilinia fructigena	(1,4)-β-linked main chain with an appreciable proportion of (1,2)-β-linkages; possible α- and β-linkages present	[97]
Monilinia fructigena	(1,3)-β-linked main chain with single (1,6)-β-glucosyl side chains on every second D-glucose unit	[98]
Botrytis cinerea	(1,3)-β-linked main chain (66%) with some (1,6)-β-linked branches (23%) and fewer (1,4)-β-linked branches (11%)	[95]
Auricularia auricula-judae[a]	(1,3)-β-linked (70%) main chain with some (1,6)-β-linked branches	[130]

Table 3.2 Continued		
Source	Structural features	Reference
Botrytis cinerea	(1,3)-β-linked backbone with single (1,6)-β-linked glucosyl residues attached, on average, to every third unit of the main chain	[85]
Sclerotium rolfsii[a]	(1,3)-β-linked main chain with single D-glucopyranosyl groups linked (1,6)-β to every third residue of the main chain	[46]
Pestalotia species 815[a]	(1,3)-β-linked D-glucosyl residues, and three out of every five D-glucosyl substituted at O-6, mostly with D-glucosyl groups, or very short (1,6)-β-linked oligosaccharides	[47]
Ulocladium atrum[a]	(1,3)-β-linked main chain, with (1,6)-β-linked side chains on every second D-glucose unit	[251] [252]
Phanerochaete chrysosporium	A backbone of (1,3)-β-linked residues with a branch in the O-6 position on approximately every second D-glucose residue and some internal (1,6)-β-linked D-glucose residues	[253]
Botrytis cinerea	(1,3)-β-linked main chain with (1,6)-β-linked side branching on every fifth backbone residue	[92]
Laetisaria arvalis[a]	Main chain of (1,3)-β-linked glucose residues with single (1,6)-β-linked D-glucose linked to every third glucose residue in the chain	[75]
Glomerella cingulata[a]	(1,3)-β-linked chain with occasional (1,6)-β-linkages; variable length (1–4 (1,3)-β-linked D-glycosyl residues) branches at the O-6 position	[254]
Drechslera spicifera[a]	(1,3)-β-linked backbone with single (1,6)-β-linked D-glucose residues attached, on average, to two out of every five backbone residues	[255]
Pleurotus species	(1,3)-β-glucan with no branching	[256]
Cryptoporus volvatus[a]	(1,3)-β-linked backbone with single (1,6)-β-linked D-glucose residues attached, on average, to every fourth backbone residue	[257]
Ganoderma lucidum	(1,3)-β-linked backbone with suggested single (1,6)-β-linked side chains	[190]
Pleurotus species[a]	(1,3)-β-glucan branched at C_6 every two or three residues along the main chain	[93]
Phytophthora parasitica[a]	(1,3)-β-linked main chain substituted with mono-, di- and oligosaccharides attached by (1,6)-β-linkages	[65]

Table 3.2 Continued

Source	Structural features	Reference
Phytophthora parasitica[a]	(1,3)-β-linked main chain substituted with mono-, di- and oligosaccharides attached by (1,6)-β-linkages	[65]
Microdochium nivale	Linear (1,4)-β-glucan	[99]
Epicoccum nigrum[a]	(1,3)-β-linked backbone with single (1,6)-β-linked D-glucose residues attached, on average, to two out of every three backbone residues	[36] [35]
Daedalea quercina	β-linked glucan	[258]
Phoma herbarum CCFEE 5080	Highly branched (1,3)(1,6)-β-glucan	[89]
Thelephora terrestris	(1,3)-β- and (1,6)-β-glucans	[259]
Poria cocos[a]	(1,3)-β-glucan	[260]
Botryosphaeria rhodina[a]	(1,3)-β-glucan with (1,6)-β-linked mono- or disaccharide residues attached, on average, to every fifth backbone residue	[91] [86] [82]
Botryosphaeria rhodina strains[a]	Linear (1,6)-β-glucan	[82]
Akanthomyces pistillariiformis BCC2694[b] *Cordyceps dipterigena* BCC2073[b] *Phytocordyceps* species BCC2744[b]	(1,3)-β-linked main chain substituted with mono-, di- and oligosaccharides attached by (1,6)-β-linkages	[261]
Aureobasidium pullulans[a]	(1,3)-β-glucan backbone with single (1,6)-β-glucopyranosyl side-branched units every two residues (major structure) and a (1,3)-β-glucan backbone with single (1,6)-β-glucopyranosyl side-branching units every three residues (minor structure)	[48]
Acremonium species.[b]	(1,3)-β-linked backbone with suggested (1,6)-β-linked side chains, whose frequency varies with species	[51]
Grifola frondosa[a]	(1,3)-β-linked main chain possessing glucose residues attached by (1,6)-β-linkages every third residue	[48]

[a] ^{13}C-NMR was used in structural elucidation
[b] Solid-state NMR was used in structural elucidation

EPS material precipitated is often assumed to be exclusively pullulan [4, 6]. Yet Campbell and co-workers [20], Zheng and co-workers [21] and Orr and co-workers [22], and others reviewed by Gibbs and Seviour [4], have shown that its composition can change markedly during its batch culture, where little pullulan may eventually remain. Consequently, an initial purification and separation should be carried out in all cases. As glucans are uncharged, gel permeation chromatography is generally able to fractionate individual glucans by their size [18], although problems with their solubility and often polydisperse nature can make this task difficult, and it is not often undertaken.

3.2.2 Determination of Monosaccharide Composition

Polysaccharides consist of monosaccharide residues linked with glycosidic bonds [23]. Monomer composition gives important basic structural information, and total acid hydrolysis is required to identify these, where glycosidic linkages between the glycosidic oxygen and the anomeric carbon are cleaved [24]. The conditions chosen must cleave all glycosidic linkages but not destroy the resulting monosaccharides. However, this requirement is rarely met. Acids most commonly used are HCl, H_2SO_4, trifluoroacetic acid (TFA) and formic acid [25]. Acid concentrations, reaction temperatures and times required for hydrolysis vary depending on polysaccharide structure, and hydrolysis conditions required for low-molecular-weight oligosaccharides are not always appropriate to fully hydrolyse their polysaccharide equivalents probably due to the tertiary and quaternary conformations some glucans adopt [25].

Branched *Claviceps*-type glucans (**Table 3.2**) like scleroglucan and schizophyllan (sonifilan) (**Figure 3.1**) require stronger hydrolysis conditions than unbranched examples, which may reflect different rate constants for hydrolysis of (1,3)-β- and (1,6)-β-linkages [26], and their tertiary conformation. Data suggest that hydrolysis occurs preferentially at nonreducing residues, and is slower at residues containing side branches [25]. It is evident that no single method is suitable for complete hydrolysis of all glucans. Published methods may have to be assessed and adapted to ensure total hydrolysis of the glucan of interest. Subsequent separation, detection and identification of individual monosaccharides are readily achieved by gas chromatography-mass spectrometry (GC-MS) after forming volatile alditol acetate derivatives [27–29].

3.2.3 Determination of Anomeric Form of Glucopyranose by Infrared Spectroscopy

Infrared spectroscopy identifies the anomeric configurations of unknown glucans as either α or β [18]. Distinctive absorption bands for the α-anomer

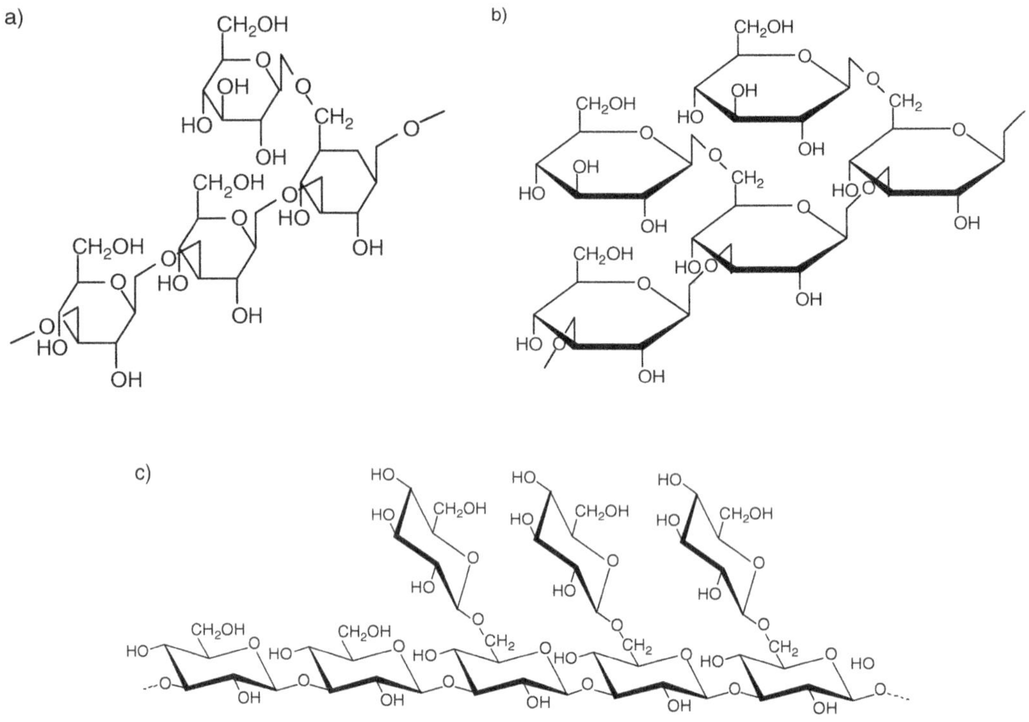

Figure 3.1 (a) Average repeating unit of many fungal β-glucans including schizophyllan and scleroglucan showing the (1,3)-β-linked backbone with (1,6)-β-linked single branches on every third backbone residue; (b) average repeating unit of epiglucan from *Epicoccum* species showing the (1,3)-β-linked backbone with two (1,6)-β-linked single branches on every three backbone residues and (c) average repeating unit of pestalotan from *Pestalotia* species (1,3)-β-linked backbone with (1,6)-β-linked side branches attached to three of every five backbone monomers. Side branches are predominantly monosaccharides with the occasional oligosaccharides

of glucopyranosyl occur at 844 ± 7 cm^{-1}, whereas the absorption bands for the β-anomer occur at 891 ± 7 cm^{-1} [30, 31]. Glucans are analysed either as a KBr disk in which the finely ground glucan is dispersed, or by first dissolving the glucan in water or Nujol, and then drying it to produce a thin film [30, 31]. With either method, no residual water should be present as this interferes by OH stretches, broadening peaks [30, 31]. However, NMR (see **Section 3.5.2**) is now a preferred method as the positions of the anomeric carbons are identified less equivocally [18, 32, 33].

3.2.4 Determination of Linkage Positions by Methylation Analysis

Methylation analysis reveals the positions of substituted carbon atoms in each monosaccharide unit in the polymer [34]. Ideally all free hydroxyl groups will become fully methylated. After complete acid hydrolysis, partially methylated sugars are reduced, acetylated and identified by GC-MS [28, 34]. However, undermethylation is a common problem [35, 36], and leads to artefacts subsequently identified as false linkage types [27, 37]. Current methylation methods are derivations and improvements of procedures outlined by Purdie and Irvine [38], Haworth [39] and Hakomori [40]. All depend on the formation of a suitable anion for deprotonation of the hydroxyl groups, and a methylating agent like methyl iodide or methyl sulfate. Most are problematic and repeated methylation steps are generally needed.

Hakomori [40] used the methyl sulfinyl carbanion [41], but it does have problems. It degrades rapidly in the presence of water and CO_2 if present during its preparation or methylation, which may lead to undermethylation. It is also unstable, and so has a finite storage time [42, 43]. *tert*-Butoxide [42] and potassium hydride [43] replacing sodium hydride give more stable carbanions, and markedly improve the yields of methylated products [27].

Ciucanu and Kerek [27] questioned the function of the methyl sulfinyl anion and proposed that HO^- and H^- ions are the basic agents in methylation. Their procedure uses solid sodium hydroxide, dimethyl sulfoxide (DMSO) and methyl iodide, and has superseded the Hakomori method as the procedure of choice for most polysaccharides [44]. Reaction times are decreased markedly, the method can be carried out at room temperature and nonsugar products are eliminated [27].

Positions of linkages in the original glucan are indicated by the carbons substituted by acetoxyl groups. For example, a 1,3,5-tri-O-acetyl-2,4,6-tri-O-methyl glucitol indicates that carbons 1 and 3 are involved in glucosidic linkages. Even though the Ciucanu and Kerek [27] method is preferred for fungal glucans, their solubility directly determines the degree of methylation achievable [35, 36], and several solvents have been used to achieve complete solubilisation. Generally, the lower the polarity of the solvent, the lower the yield of methylated glucan [27]. However, improved NMR methodology has made such analyses less imperative.

3.2.5 Nuclear Magnetic Resonance Spectroscopy

NMR spectroscopy ensures that structural elucidation of glucans is less time consuming and more informative, and is now used routinely. ^{13}C-NMR can determine the repeating structure of a glucan by distinguishing between all nonequivalent carbons within each unit. Assignments of carbons are referenced to those of glucose itself. All those not involved in glucosidic linkages generally occur in the same location as those in glucose, while those that are involved are found 4–10 ppm downfield from their positions relative to glucose [32, 33]. Carbon atoms adjacent to or near linkages also show slight chemical shifts [18, 25]. **Table 3.2** gives examples where ^{13}C-NMR has been used to determine the structures of fungal β-glucans, and a typical NMR spectrum of a branched β-glucan, epiglucan from *Epicoccum* species [36], is given in **Figure 3.2a**. The β-anomeric form of C_1 occurs at 102–104 ppm, C_2

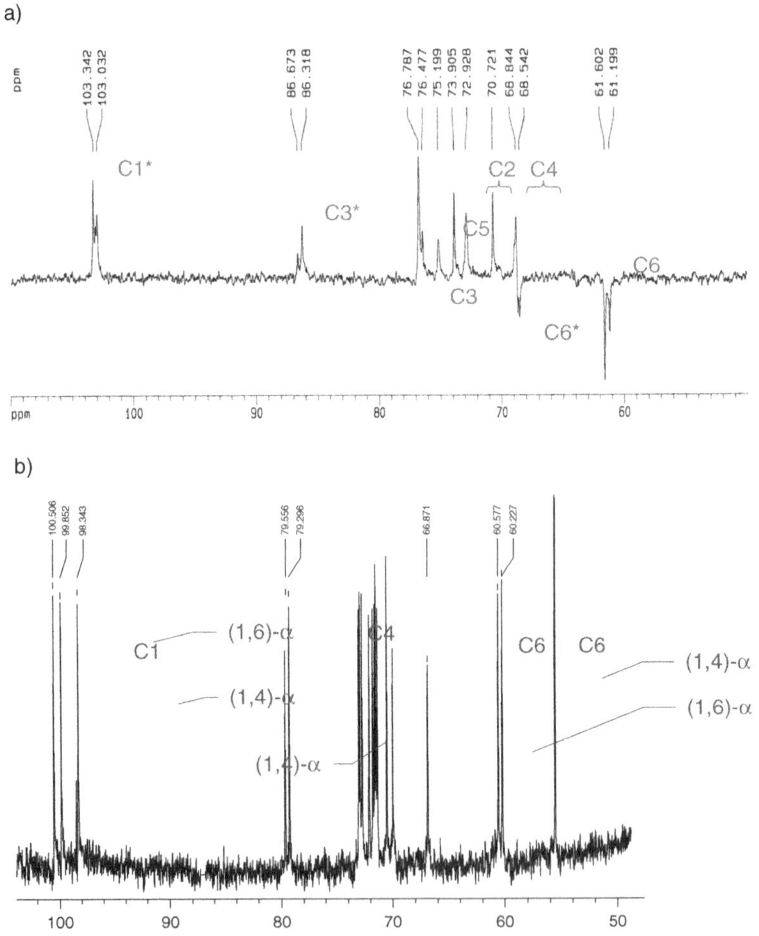

Figure 3.2 (a) DEPT 135 (distortionless enhancement by polarisation transfer) ^{13}C-NMR spectrum of epiglucan from *Epicoccum* strain F19 (* denotes carbons involved in glucosidic linkages) and (b) ^{13}C-NMR (DEPT) spectrum of pullulan from *Aureobasidium pullulans*

at 72–75 ppm, C_3 unsubstituted at 75–77 ppm, C_3 substituted at 86–88 ppm, C_4 at 68–70 ppm, C_5 at 75–78 ppm, C_6 unsubstituted at 60–62 ppm and C_6 substituted at 66–71 ppm [32, 33].

The exact assignment of carbons, and the precise structural elucidation, is complicated by β-glucan solubility and the temperature at which data are collected [18, 32, 45]. Both need to be optimised on an individual basis. The major problem is that many of the carbons occur in the same structural environment, leading to overlapping signals [32]. It is exacerbated by the fact that only small differences are found in the chemical shifts of the carbon atoms with glucose [32], especially when attempting to resolve C_2, C_3, C_4 and C_5 atoms, as they all occur in the narrow region of 70–77 ppm [32, 45]. Shifts for carbons in glucosidic linkages are well defined. They occur 4–10 ppm downfield from their nonsubstituted form, and are usually well separated from other signals [32, 33]. However, the substituted C_6 atoms in (1,3)(1,6)-β-glucans can be difficult to assign as they appear in the region of 66–71 ppm, and may overlap with signals from the aforementioned carbons in this region.

Glucan insolubility and inappropriate temperature essentially cause the same problems, in that if the carbons within the glucan cannot fully relax, then line broadening results and resolution is lost [32]. Thus scleroglucan, a (1,3)(1,6)-β-glucan, has been fully assigned in DMSO as solvent, but signals were lost when dissolved in deuterated water (D_2O), since it formed a gel at room temperature. Even at elevated temperatures (90 °C), no signals were obtained [46]. A structure for scleroglucan (**Figure 3.1**) was fully assigned by constructing it from the spectra of other linear (1,3)-β- and (1,6)-β-glucans [46], which is a common practice [32]. The published spectrum for chemically modified pestalotan [47] shows clearly the effects of line broadening. Although peaks could be assigned to individual carbon groups, it was not possible to distinguish individual carbons belonging to the monosaccharides in the repeating structure.

To overcome these problems, other approaches have been used increasingly to aid in spectral interpretation. They include a combination of one-, two- or three-dimensional ^{1}H- and ^{13}C-NMR techniques (see review by Yang and Zhang [18]) like attached proton test (APT), distortionless enhancement by polarisation transfer (DEPT), insensitive nuclei enhanced by polarisation transfer (INEPT), correlation spectroscopy (COSY), total correlation spectroscopy (TOCSY), nuclear Overhauser effect spectroscopy (NOESY), heteronuclear single-quantum correlation (HSQC), heteronuclear multiple-bond correlation (HMBC), HSQC-TOCSY, heteronuclear two-bond correlation (H2BC), NOESY-TOCSY and TOCSY-NOESY, which by themselves can provide unequivocal chemical structural resolution of the repeating unit of a glucan [32, 48, 49].

Alternatively, solid-state ^{13}C-NMR can be applied to those glucans that are not easily solubilised, as with scleroglucan [50] and *Acremonium* glucans [51]. Although peak broadening makes interpretation of spectra problematic, the data obtained with these were consistent with both having (1,3)-β-linked backbones with (1,6)-β-linked side branches.

^{13}C-NMR has also been used successfully to confirm the structure of pullulan [52–57] (**Figure 3.2b**). The anomeric C region shows three signals corresponding to the (1,4)-α-linkage (99.9, 100.5 ppm) and (1,6)-α-linkage (98.3 ppm) in pullulan. Splitting of the C_4 (79.3, 79.6 ppm) and C_6 (60.2, 60.6 ppm) resonances of the (1,4)-α-glycosidic linkages reflects their sensitivities to the type of linkage at C_1. Thus, the C_6 signals are those of the two linkage types of the (1,4)-α-linkages, while that at 66.9 ppm (**Figure 3.2b**) corresponds to C_6 of the (1,6)-α-linked D-glucose.

3.2.6 The Periodate Oxidation Reaction

Oxidation of a glucan with periodate cleaves its glucopyranosyl residues only at their vicinal hydroxyl groups, and so available cleavage sites depend on which carbons are involved in linkages [58–60]. In linear (1,3)-β-glucans, such groups are present only at the reducing and nonreducing ends. Oxidation at the former consumes 1 mol of periodate, while 2 mol are consumed at the latter with the formation of 1 mol of formic acid [25]. Quantification of the periodate consumed and reaction products formed can be used to calculate the average chain length of a linear glucan [61]. If it contains (1,6)-β side branches, this effectively increases the proportion of nonreducing end residues, and so each side branch will consume 2 mol of periodate and produce 1 mol of formic acid.

Several reviews have been published evaluating periodate oxidation [59, 60, 62, 63]. Oxidants used are periodic acid, sodium metaperiodate ($NaIO_4$) (the most commonly used oxidant) and potassium metaperiodate [58]. The pH must be within a range where it protects the blocking groups such that hydrolysis of already formed acetyl or ester linkages is not possible, which would expose further reactive groups [58]. A dilute solution of $NaIO_4$ in water produces a pH of around 4 [63] although a pH between 3 and 5 is suitable [59].

Most periodate oxidations can be performed at room temperature [58, 59], and a lower temperature reduces any possible overoxidation [58]. Batra and co-workers [64] used a low temperature (4 °C) to oxidise a (1,3)(1,6)-β-glucan produced by *Sclerotium rolfsii*, although room temperature has also been successful [65, 66].

3.2.7 Smith Degradation

Smith degradation is an extension of periodate oxidation, where the polyaldehyde product is reduced to a polyalcohol. This is then subjected to mild acid hydrolysis, which selectively cleaves acetal but not glucosidic linkages [60, 67–69]. Smith degradation of (1,3)(1,6)-β-glucans involves reducing the oxidised glucan with sodium borohydride followed by mild acid hydrolysis to yield glycerol, glycolaldehyde and a now linear glucan [25]. Determining the type of acid, its optimum concentration and reaction time is complex. HCl, H_2SO_4 and TFA between 0.005 and 0.5 M appear to be most popular [60, 64, 67, 70–72]. If hydrolysis conditions are mild enough, the only detectable products should be glycerol and glycolaldehydes from the original nonreducing residue [67]. Once glucose, arising from internal glucan residues, is detected, then either the acid concentration is too high or the reaction has been incubated for too long, or both [67].

3.2.8 Periodate Oxidation and Smith Degradation

When periodate oxidation is combined with Smith degradation, further structural information, like glucan branching frequency, can be obtained [60], and has been used to help elucidate, for example, (1,3)(1,6)-β-glucan structure of *Claviceps* glucan [73], scleroglucan [74] and the glucan from *Laetisaria arvali* [75]. However, this usually requires more than one round of oxidation/degradation and a comparison of reaction products after each step [25]. Nonreducing ends are cleaved between C_2/C_3 and C_3/C_4, consuming a total of 2 mol of periodate and producing 1 mol of formic acid. The resulting polyaldehyde undergoes borohydride reduction forming a polyalcohol. Transformed nonreducing ends are now susceptible to mild acid hydrolysis and can be removed selectively [69] to give a linear (1,3)-β-glucan, glycerol (1 mol) and glycolaldehyde (1 mol).

Such oxidation/degradation reactions can clarify structural ambiguities of branched (1,3)-β-glucans as in *Omphalia lapidescens* [72]. Schmid [35] also used it to help resolve structural ambiguities in the branching frequency of the repeating unit of epiglucan from *Epicoccum* species where the data, consistent with those from ^{13}C-NMR [36], supported single side branches on average at two of every three backbone glucosyl residues (**Figure 3.1**).

3.2.9 Enzymatic Analysis

Digestion of native or periodate-oxidised exocellular glucans with appropriate glucanases [11, 35, 36, 51, 75, 76] can yield much useful structural information

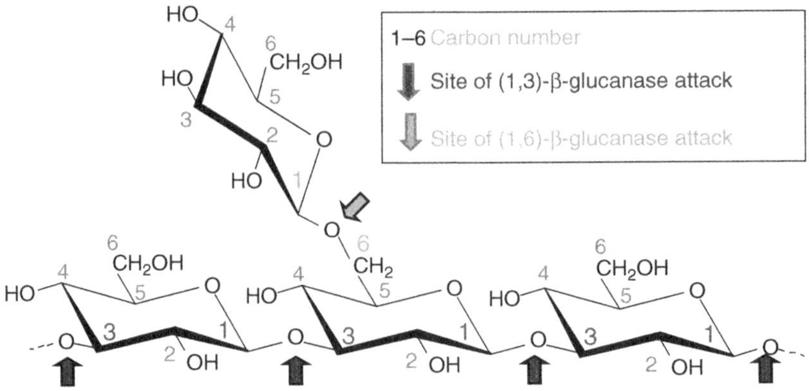

Figure 3.3 Mode of attack of β-glucanase on branched (1,3)(1,6)-β-glucans: exo-(1,3)-β-glucanase attack generates glucose and gentiobiose in a ratio reflecting branching frequency

after chromatographic analyses of the degradation products. For branched (1,3)(1,6)-β-glucans, their linkage types and branching structure and frequency are revealed with highly purified glycoside hydrolases like exo-(1,3)-β-glucanases, as applied to epiglucan from *Epicoccum nigrum* [36, 77] and scleroglucan from *Sclerotium glucanicum* [78], *Acremonium* species [51] and *S. rolfsii* [79]. Such enzymes sequentially attack the glucan from the nonreducing end [11] to generate glucose and gentiobiose (**Figure 3.3**), and after complete enzymatic hydrolysis, their ratios reflect chain branching frequencies.

Pullulanase (pullulan (1,6)-α-glucanohydrolase, CAZy family 41) hydrolyses the (1,6)-α-linkages in pullulan to generate maltotriose (**Figure 3.1**), and it has been used widely not only to quantify the pullulan produced by *A. pullulans* under different culture conditions [4, 6] but also to monitor the detailed chemical composition of the pullulan produced in submerged cultures under changing conditions [20, 22, 80].

3.3 Chemistry of Exocellular Fungal Glucans

Many of the criticisms pertaining to our current understanding of the chemistry of fungal heteropolysaccharides are equally applicable to the exocellular noncellulosic fungal glucans, since many have not been characterised sufficiently to allow their detailed chemical structures to be resolved adequately. However, an increasing application of two-dimensional NMR [18, 48] now permits their unambiguous structural elucidation. All glucans can be grouped into two main categories on the basis of how their individual glucopyranosyl residues are chemically linked, which to a large extent also determines their physical properties [81], as discussed later. In many, it is a β-linkage, and these fluoresce strongly and specifically with aniline blue [25], while a smaller known number have α-glycosidic linkages serving to join the glucopyranosyl residues together. Each will be discussed separately.

3.3.1 Fungal β-Glucans

A small number of fungi including *Botryosphaeria rhodina* [82] and *Guignardia citricarpa* [5] have been reported to synthesise pustulan-like exocellular glucans under certain cultural conditions. These were first characterised in the lichen *Umbilicaria pustulata*, and were shown to consist entirely of unbranched (1,6)-β-linked glucopyranosyl residues. They are similar to the glucans widely found in fungal cell walls and may derive from them [8].

However, most fungal exocellular β-glucans consist of a backbone chain of glucopyranosyl residues linked by (1,3)-β-linkages. Often single (1,6)-β-linked glucopyranosyl residues arise from it. The extent of this branching varies considerably, and directly affects the physicochemical properties of the glucan [81] and its potential industrial applications. Many glucans that have been characterised extensively share a similar structure [7], with a single glucopyranosyl residue attached to the (1,3)-β-linked glucan backbone at every third glucose residue (**Figure 3.1**). This is the structure proposed for the *Claviceps* glucan [83], the glucan from *L. arvalis* [75], scleroglucan from *S. glucanicum* [64, 74], schizophyllan from *Schizophyllum commune* [84], cinerean from *Botrytis cinerea* [85], botryosphaeran from *B. rhodina* [82, 86] and grifolan from *Grifola frondosa* [48]. All these examples have been characterised extensively using combinations of the analytical methods discussed above (**Table 3.2**). Other less well characterised glucans probably share this structural feature, although while their structures also contain a glucan backbone of (1,3)-β-linked glucopyranosyl residues the extent of its branching is unknown, as with *Acremonium* spp. [51, 87] (**Table 3.2**). Claims have been made that the frequency of branching and side chain length can vary in scleroglucan depending on culture conditions and *Sclerotium* strain used [88]. However, because of how branching frequencies are assessed for these glucans, where the values determined experimentally are often rounded up to the nearest whole number, they probably reflect average values. There is certainly no evidence to suggest that overall branching frequencies are as regular as the data might imply.

Some (1,3)(1,6)-β-glucans appear to show markedly different branching frequencies from those described above, although whether these arise from experimental artefacts is not always clear. Thus, in *Phoma herbarum* [89], three out of every five backbone glucopyranosyl residues appeared to carry a single attached glucopyranose residue, while with the epiglucan from *Epicoccum* species [36, 77] and the (1,3)(1,6)-β-glucan in the basidiocarps of *Flammulina velutipes* [90], two out of every three backbone residues are substituted by a single glucopyranose residue (**Figure 3.1**). On the other hand, Barbosa and co-workers [91] and Vasconcelos and co-workers [82] both claim that the botryosphaeran from their isolate of *B. rhodina* was branched on average at one every five backbone glucopyranosyl residues, indicating structural features different from those reported in other strains of the same fungus [82, 86] (**Table 3.2**). In fact, such variations have been described frequently (**Table 3.2**). For example, cinerean from *B. cinerea* has been proposed to branch on one in every three [85] or one in every five [92] glucopyranosyl backbone residues. On the other hand, Gutiérrez and co-workers [93] showed that the glucans from six *Pleurotus* species were all structurally much more similar, with side branching occurring every second or third glucopyranosyl residue. Oligosaccharide side chain length may also vary. Thus, for pestalotan from *Pestalotia* species [47], on the basis of methylation, Smith degradations and analyses of products of enzymatic hydrolysis, it was claimed that three out of every five glucosyl residues were linked to single glucopyranose residues, and a few short (1,6)-β-linked oligosaccharide chains were also present (see **Figure 3.1**). Similarly, coriolan from *Coriolus versicolor* [94] and *Phytophthora parasitica* glucan [65] were considered to exist with mono-, di- and oligosaccharide branches attached by (1,6)-β-linkages to the glucan backbone.

The published data also suggest that the linkage types in some fungal exocellular β-glucans may vary from the branched (1,3)- and (1,6)-β-linkage types discussed so far. Thus, the various structures published for cinerean from *B. cinerea* differ not only in their branching frequencies but also in their linkage types and length. For example, Leal and co-workers [95] described a structure consisting of a (1,3)-β-linked backbone with both single (1,4)-β- (11%) and (1,6)-β-linked (23%) side chains, quite different (see above) from that proposed later by Dubourdieu and co-workers [85] and Pielken and co-workers [92]. Three structurally different exocellular glucans have been reported for *Monilinia fructigena*. Based solely on methylation analysis, Feathers and Malek [96] found a (1,3)-linked glucan containing 33% (1,4)-linked side branches, and no conformational evidence (i.e., whether α- or β-anomers) was presented. Periodate oxidation data of Archer and co-workers [97] suggested that it consisted of a (1,4)-linked backbone containing small amounts (6%) of (1,2)-linked side branches, an unusual structure not reported elsewhere for any other fungal glucan. No analytical data were reported from methylation or NMR analyses to support the periodate oxidation data, but when Santamaria and co-workers [98] used periodate oxidation together with Smith degradation and enzymatic digestion, and infrared analysis, they could show that the *M. fructigena* glucan had the more usual

(1,3)-β-linked backbone with single (1,6)-β-linked gluocopyranosyl residues attached to every second backbone residue. However, a small number (1%) of (1,4)-linkages were also detected. Exocellular fungal (1,4)-β-glucans have been reported rarely, although a linear (1,4)-β-glucan was isolated from *Microdochium nivale* and rigorously characterised [99], suggesting that structurally more diverse exocellular glucans may well be isolated and characterised in the future.

3.3.2 Fungal α-Glucans

Fewer fungi are known that produce exocellular α-glucans. Pullulan synthesised by *A. pullulans* is the best known example [4, 6, 19, 80, 100]. A structurally identical α-glucan has been reported from the basidiocarps of *Tremella mesenterica* [101], and EPS from both *Cyttaria darwinii* [102] and *Cyttaria harioti* [103] contained a pullulan. Pullulan has also been reported from *Cryphonectria parasitica* [104]. The EPS from *Rhodotorula bacarum* was also described as pullulan, an interpretation based solely on its susceptibility to pullulanase attack [105].

There is general agreement, based on data from the analytical protocols described earlier including ^{13}C-NMR (**Figure 3.2b**), that pullulan from *A. pullulans* is a linear polymer of (1,4)-α-linked maltotriose units joined by (1,6)-α-linkages [52, 55], with apparently randomly substituted (1,6)-α-linked maltotetraose subunits, as shown in **Figure 3.4**. However, other structures, for which less experimental support exists, have been proposed (reviewed by Gibbs and Seviour [4] and Singh and co-workers [6]). The content of maltotetraose units seems to vary with the strain and the culture conditions used [22], but was considered to be no more than 7% of the total glucan [6]. Yet, the two different pullulans from *C. parasitica* are reported to contain much higher levels of maltotetraose substitutions [106]. Furthermore, pullulan is clearly not the only EPS material characterised from cultures of *A. pullulans* [4, 21, 22]. A branched (1,3)-β-glucan, whose structure is uncertain, and aubasidan, a (1,3)(1,6)-β-glucan with (1,4)-α-linked glucopyranosyl residues, have also been reported [6]. The molecular

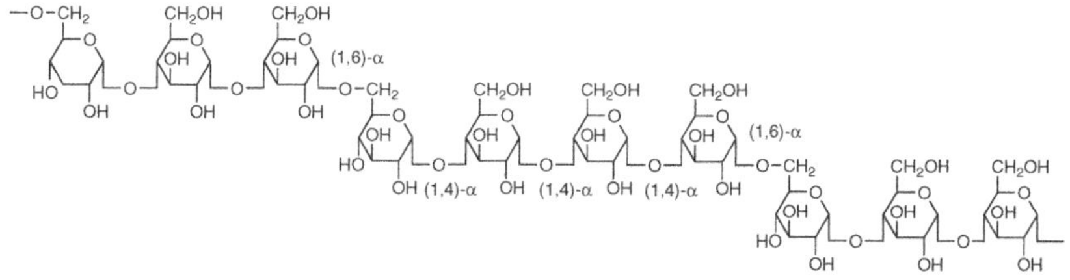

Figure 3.4 The commonly accepted linear structure of pullulan, showing an occasional maltotetraose substitution in a repeating maltotriose backbone.

weight of aubasidan may also change with culture conditions like pH [107]. In many studies, no attempt was made to see whether these or others were present in the total ethanol-insoluble material harvested from cultures of this fungus. Instead all the precipitatable material was assumed to be a single polymer and was referred to collectively as 'pullulan' [4].

3.4 Relationship between Chemical and Physical Properties of Fungal Exocellular Glucans

3.4.1 Fungal β-Glucans

As Gidley and Nishinari [81] emphasise, there is still much to learn about the relationships between the chemical and physicochemical properties of these glucans. However, several of their important physicochemical behavioural properties can now be explained in structural terms. Thus, their inherent chain stiffness explains why their solutions have very high intrinsic shear rate-dependent viscosities. Furthermore, these glucans adopt a helical chain shape determined by the geometrical configuration of their glycosidic linkages and the potential to form triple helices [81]. Experimental data show that this is also the case with linear chains of the bacterial (1,3)-β-glucan curdlan and the branched (1,3)-β-glucans including *Acremonium* β-glucans [51], scleroglucan, lentinan, botryosphaeran [108] and schizophyllan. However, the (1,6)-β-linked side chains protruding to the outside of the helix prevent lateral aggregation of the chain [81], which occurs with curdlan. The solubility of individual (1,3)-β-glucans depends on the ability of the solvent to overcome the strong hydrogen bonding stabilising the triple helix conformation, their degree of polymerisation (DP) and the extent of any side branching [81]. Thus, while curdlan is insoluble in water, many (1,3)-β-glucans with single (1,6)-linked side branches distributed along the backbone dissolve readily [81], and solubility is thought to increase as the frequency of side branching increases. However, this is not always the case, as seen with epiglucan [36]. Examples of such soluble glucans are scleroglucan and schizophyllan, both with high DP of approximately 100 [109]. Some water-insoluble (1,3)-β-glucans can only be dissolved in alkaline (NaOH) or polar aprotic solvents (DMSO and *N-N*-dimethylformamide) [25]. Essentially, both disrupt the hydrogen bonding between the constituent chains in the triple helix, releasing the (1,3)-β-glucan chains as single random coils, although how they achieve this differs [25, 81].

Hydrogen bonding can also be weakened by heating an aqueous glucan suspension, and upon cooling some β-glucans may form stable gels [25, 81]. Curdlan-type β-glucans produce a weak gel after cooling from 50–60 °C, and a pressure-resistant irreversible gel when cooled from 95 °C [110]. Organisation of the chains in the weak gel is thought to arise from junction zones formed after interaction between surface chains of microfibrils that have become partially solubilised [111]. In an irreversible gel, the triple helices may separate totally, and on cooling the helices reform and may include chains that arise from other triple helices [112].

Branched (1,3)(1,6)-β-glucans such as scleroglucan and schizophyllan may form very weak gels at room temperature, a property that does not change even after being heated to above 55 °C [81]. It appears likely that although the branched glucans may form triple helices in solution, these cannot aggregate subsequently to form stable gels because of interferences from their (1,6)-linked side residues [51, 81]. Although the addition of other chemicals to these glucans can generate much stronger gels [81], they are more commonly exploited for their rheological properties and high intrinsic viscosities, as discussed below, which appear to be independent of pH, temperature and presence of electrolytes [109, 113–116]. Problems associated with the non-Newtonian rheological properties of the culture media occurring during their large-scale production will be discussed later.

3.4.2 Fungal α-Glucans

Solutions of pullulan from *A. pullulans* are also viscous, displaying a non-Newtonian pseudoplastic rheology at higher concentrations [115, 117, 118]. However, the high proportion (33%) of (1,6)-α-D-glucosidic linkages makes it unlikely that pullulan will assume a regular helical or ribbon conformation, since the three degrees of rotational freedom inherent in this linkage confer a much higher level of flexibility on this molecule in solution [119]. The regular alternation of (1,6)-α- and (1,4)-α-linkages is responsible for the two most distinctive physical properties of pullulan: its water solubility and structural flexibility. These features make it suitable for a wide range of medical and industrial applications, including the formation of strong oxygen-impermeable films and fibres [4, 6, 19, 100, 120], as detailed later. An infrequent substitution of an occasional extra (1,4)-α-D-glucosidic linkage located in an occasional maltotetraose residue is unlikely to alter the chemical or physical properties of pullulan. However, the pullulans from *C. parasitica* have much higher maltotetraose substitution levels [106], and thus should be much more rigid and more viscous. Available data suggest this to be the case [121].

3.5 Biological Activity of Fungal (1,3)-β-Glucans

Basidiocarps from more than 600 species of *Basidiomycota* are claimed to have antitumour properties [122], and historically many Asian societies have harvested them for eating and preparation of traditional medicines. However, only quite recently have these antitumour properties been attributable to β-glucans and their derivatives [123]. Schizophyllan (*SPG, sizofilan or sizofiran*) from *S. commune* and lentinan from basidiocarps of *Lentinus edodes* are both approved in Japan for the clinical treatment of several cancers [122]. In addition, there is now well-documented evidence for the effectiveness of β-glucan in treating a range of infectious microbially mediated diseases in humans, and in lowering blood pressure and cholesterol levels in humans [124].

The antitumour and other biological activities of β-glucans could be attributed to their powerful immunomodulatory effects, affecting both our innate and adaptive immune systems. They do not attack and kill cancer or invasive microbial cells directly. Instead, because they are not synthesised by humans, our immune systems recognise them as non-self molecules and hence become stimulated [124–126]. How this happens is becoming clearer, but much is still poorly understood about the signal transduction mechanisms involved [126]. Host cells possess receptors known as 'pattern recognition receptors' to detect innately such non-self molecules [125]. Several receptors have been identified, and include dectin-1 (a lectin), with a carbohydrate recognition domain, which binds specifically to β-glucans. Other receptors thought to be involved include complement receptor 3 (C_3), scavenger receptors, lactosylceramide and toll-like receptors. Several receptors may be activated by a single β-glucan [124, 126].

There is still no general agreement as to which chemical or physicochemical properties of β-glucans are most important in determining their immunomodulatory activities [122, 124, 127–129]. This may reflect an inadequate chemical characterisation of the glucan preparations used in many of these experiments, or where basidiocarps have been used as sources, contamination with other polymers during their extraction [122]. Orally administered glucans may also be chemically and/or enzymatically modified in the gut before absorption [124].

Despite these concerns, it seems clear that individual β-glucans differ markedly in their effectiveness as immunomodulators, and so it is possible to suggest, with due caution, that high biological activity may be associated with the following:

(1) A (1,3)-β-linked glycosidic backbone, since α-glucans like pullulan are biologically inactive [124].

(2) The presence of (1,6)-β-linked glycosidic and other side chains, and high branching frequency [129, 130], although the explanation for this is unclear. Furthermore, although both scleroglucan and schizophyllan have the same branching frequencies (1:3), scleroglucan has a higher receptor binding capacity than schizophyllan [131].

(3) A triple helical conformation [132], although some (1,3)(1,6)-β-glucans with triple helical arrangements are ineffective and their activity became apparent once their conformation changed from a triple helix to a single helical conformation [133, 134]. With schizophyllan, a single helical conformation gave higher biological activity than that shown by the native triple helical form [132].

(4) A high molecular weight, which might reflect the formation of triple helices, since low-molecular-weight glucans form no helices as the chains are too short [129, 135, 136].

(5) A high water solubility [129], which reflects their branching frequencies, the likelihood of triple helix formation and to some extent their DP.

3.6 Biosynthesis of Fungal Exocellular Glucans

3.6.1 Assembly of β-Glucans

The mechanisms of biosynthesis of exocellular (1,3)-β-glucans in fungi, including process enzymology and assembly and transport into the exocellular environment, are unclear. They are based largely on our understanding of intracellular or cell wall (1,3)-β-glucan synthesis in fungi not producing exocellular glucans, where the assumption is that their origin in those that produce exocellular glucans may be wall glucans. Hence both are probably synthesised in a similar way [137]. However, no direct link has ever been reported regarding their possible site of synthesis or assembly. Any discussion on the synthesis of (1,3)-β-glucan chains needs to consider the donor molecules and acceptors/initiators involved, the mode of chain termination (which determines its DP and physicochemical properties) and the direction of chain elongation. There is insufficient space here to discuss all of these in detail, and the reader is encouraged to read the reviews on fungal and yeast cell wall β-glucan synthesis by Adams [8] and Nogami and Ohya [138].

The two mechanisms by which polymers like β-glucans are thought to elongate are referred to as 'headward' or 'tailward' growth [25]. In headward growth, the growing polymer is activated at the reducing end, and is added to an activated saccharide [25]. No evidence has been published that fungal exocellular or cell wall β-glucans follow this mechanism of elongation. Conversely, elongation by the tailward growth mechanism occurs at the nonreducing end. There is some evidence that cell wall (1,3)-β-glucans in *Saccharomyces cerevisiae* elongate from the nonreducing end by the transfer of glucose from UDP-Glc [139, 140]. Pulse-labelling experiments with [^{14}C]glucose and repeated Smith degradation suggested that scleroglucan also elongates from the nonreducing end [64], and with the same approach Schmid and co-workers [77] showed convincingly that epiglucan from *E. nigrum* is synthesised in the same direction.

As already mentioned, many exocellular fungal (1,3)-β-glucans contain some side branching (see **Table 3.2**). There are no convincing data to show how and when these side branches are inserted into the elongating glucan chain. Several mechanisms (see **Figure 3.5**) have been proposed and are all based on an elongation of the newly synthesised glucan from the nonreducing end [64, 141]. In mechanism 1, the

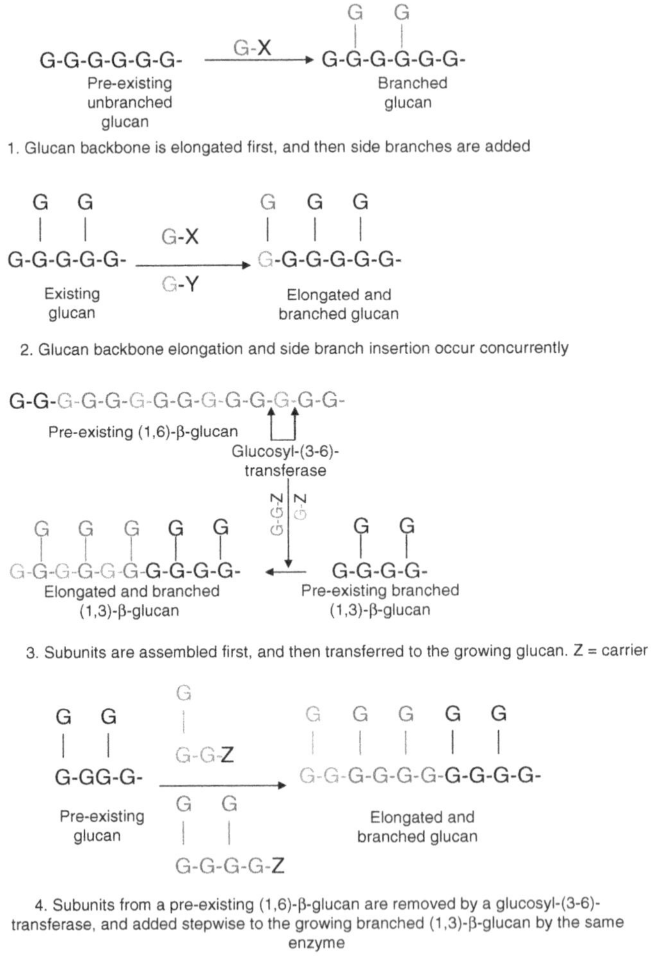

Figure 3.5 Possible mechanisms by which single side chain residues may be incorporated into a growing (1,3)-β-glucan. Examples given here assume that elongation occurs from the nonreducing end of the newly synthesised β-glucan. G refers to glucose residues. X and Y refer to a nucleoside diphosphate saccharide activator. Z refers to an enzyme activator such as a glucosyltransferase

(1,3)-β-linked backbone is synthesised first, and then activated side branch residues are added later to this backbone. It was proposed that (1,3)-β-glucans in *S. commune* have single and chains of (1,6)-β-linked residues attached to them after they are transported into the cell wall, although the enzymatic mechanisms by which this occurs are yet to be elucidated [116].

The next three mechanisms, although possible, are not easily distinguished by any current experimental methods. Mechanism 2 involves the concurrent addition of glucose residues to the (1,3)-β-linked backbone and the (1,6)-β-linked side branches as the chain elongates. In mechanism 3, subunits containing both (1,3)-β- and (1,6)-β-linkages arc assembled first, and then transferred to the growing glucan as a pre-existing oligosaccharide. Mechanism 4 uses a pre-existing glucan chain, where residues are removed by enzymatic action and transferred stepwise by the same enzyme to the elongating glucan. Of these three mechanisms, mechanism 2 has received some support from Batra and co-workers [64] for scleroglucan assembly, where branch point residues and branch residues were postulated as being inserted simultaneously. The same mechanism probably holds true for epiglucan synthesis by *E. nigrum* [77], and suggests that all branched (1,3)-β-glucans may be assembled in a similar way. However, it is not possible to distinguish experimentally between mechanisms 2 and 3 (**Figure 3.5**), which requires an understanding of the detailed enzymology of β-glucan synthesis, an area of research much neglected.

No enzymes have been isolated from any fungus that can synthesise a branched glucan like scleroglucan, and where single residue branches are attached to an existing glucan backbone. A branching enzyme was isolated from *Candida albicans*, which removed laminaribiose residues from an existing (1,3)-β-glucan or laminarioligosaccharide, and transferred these to the nonreducing ends of a newly synthesised (1,3)-β-glucan via (1,6)-β-linked intrachain branch points [142]. A similar enzyme, a (1,3)-β-glucanosyltransferase, was reported in *Aspergillus fumigatus*, which again introduced (1,6)-β-linked intrachain branches into an existing (1,3)-β-glucan [143], and this might provide some clues as to the mechanisms for side chain insertion. Another possibility proposed by Balint and co-workers [141] involves two glucosyltransferases, where one attaches (1,3)-β-linked residues and the other inserts the (1,6)-β-linked branch residues, but again there is no experimental support for this hypothesis.

3.6.2 Assembly of Pullulan

Little is known about how pullulan is synthesised in *A. pullulans* either. All the evidence suggests that pullulan is not a component of the cell walls of this fungus

[4, 6, 100]. Instead it may be synthesised at the cytoplasmic membrane and then transported outside the cell by an unidentified lipid carrier [144], although Campbell and co-workers [145] suggested that pullulan may be synthesised and initially stored intracellularly. No 'pullulan synthase' enzyme(s) has ever been detected or characterised from *A. pullulans*, but presumably the enzyme must exist. How pullulan is assembled is also unclear. It is not branched like glycogen or starch, suggesting that its synthesis must be different. Whether the maltotetraose units are inserted during the elongation of the largely maltotriose backbone chain, as with the side branches in the (1,3)(1,6)-β-glucan, or are inserted randomly or in a predetermined fashion (and controlled by ?) is also unknown. Equally unclear is why the extent of maltotetraose substitutions in pullulan can vary between strains, and why culture conditions can apparently affect this [22].

3.7 Applications of Fungal Glucans

3.7.1 Fungal β-Glucans

It has already been mentioned that fungal β-glucans are bioactive polymers with potent antimicrobial and antitumour properties, and some of the structural features thought to affect their bioactivity have been discussed [124, 126]. Their applications in health care explain the current high level of interest in their biological distribution and production [146]. However, their rheological features (detailed above) also provide them with attractive functional properties, some of which have been exploited industrially. For example, scleroglucan is produced commercially under several trade names by companies using *Sclerotium* spp. where its non-Newtonian pseudoplastic behaviour and very high inherent viscosity retained over a wide range of temperatures, ionic strengths and pH values are exploited for enhanced oil recovery [109]. It has also been used for the same reason as a thickener and stabiliser in the food and cosmetic industries, and as a ceramic binder.

3.7.2 Fungal α-Glucans

Pullulan is also produced on a commercial basis in Japan and China, and several hundred patents have been issued for it and its derivatives for application in the food, pharmaceutical and other industries worldwide. These have been reviewed extensively by Leathers [80, 100] and Singh and co-workers [6]. It is nontoxic, nonirritant, tasteless and edible, and has excellent adhesive properties. Like scleroglucan, its viscosity (which is relatively much lower) is unaffected by changes in pH, ionic strength and temperature, and so it is also used as an effective thickening and binding agent. Its ability to form oxygen-impermeable films makes it suitable as an edible wrapping for foods susceptible to oxidation, as a coating for tablets and pills and paper, and as a blood plasma substitute, among many other applications [6].

Except in the case where they have unique properties, the full industrial potential of all fungal glucans will probably not be realised until the economics of their production become competitive with those of other natural polysaccharides, and this might explain why so much effort has been directed at yield optimisation studies. Some of this work is discussed below.

3.8 Factors Affecting Exocellular Fungal Glucan Production

Several reviews discuss the influence of varying strain and culture conditions on fungal exocellular glucan yields, but generally without explaining why or how the changes occur [4, 6, 17, 109]. The interested reader is directed towards those reviews for more detail. There are attractive advantages in using submerged cultures. Large-scale production under carefully controlled and monitored conditions should ensure a more standardised product of known composition and less risk of chemical contamination [146]. The increasing use of β-glucans in medicine also demands that they are produced at high purity in high yields at a competitive cost. Furthermore, rational strategies to improve glucan yield can be undertaken, and the possibility to manipulate their chemical and physical properties exists by changing culture conditions to generate polymers with more desirable and industrially attractive properties [22, 86]. However, diversion of carbon into metabolites other than those sought [109] or formation of products affecting their recovery and purification (e.g., melanins with *A. pullulans* [6, 147, 148]) may need to be considered in selecting the fungal strain or culture conditions used for any large-scale process.

This requires using fermenters and axenic cultures, where operational parameters can be monitored and the physical and chemical environment controlled precisely. However, with filamentous fungi, such culture systems are not without their problems [114, 115, 149]. Their non-Newtonian rheology from the highly viscous nature of the culture medium with filamentous fungi, exacerbated by the presence of high levels of exocellular glucans, may prevent effective mixing. Hence heterogeneity can be created within large reactors, affecting heat and especially oxygen mass transfer rates for aerobic organisms [115, 150]. These factors in turn can modify culture behaviour and metabolite synthesis adversely, although as discussed below, exocellular glucan synthesis in some fungi seems to be enhanced under oxygen-limiting conditions [114, 115, 146, 150]. Precise pH control may be difficult for the same reasons. Because fermenter configuration affects bulk mixing capacity with all its consequences, and may determine fungal morphology, which itself can influence glucan production, the choice of vessel design is a critical consideration in any large-scale fungal metabolite production process [109, 114, 115, 149].

3.8.1 General Considerations

Before establishing large-scale fermentation processes, optimisation of many culture conditions for exocellular glucan production by fungi is attempted usually by changing medium components or operating parameters, often in an arbitrary manner [17, 109, 114]. However, response surface methodology (RSM) and other statistical approaches are used more commonly now in this exercise [6, 146]. RSM aims to optimise culture conditions more efficiently since it identifies and allows for possible interactions between operating factors. Shake-flask batch cultures are employed initially to screen simultaneously a wide range of such parameters, although results from these experiments must be interpreted with caution. For example, there are fewer provisions than exist in fermenters to measure or more importantly control influential environmental parameters like dissolved oxygen (DO) or pH of the medium, which with pH can change with medium N source and its concentration. Both are known to affect glucan yields in many fungi [4, 6, 17, 109, 116, 151]. Similarly, increasing agitation speeds will increase oxygen mass transfer rates, but possibly affect culture morphology from shear effects. Again each may influence glucan synthesis individually, with the latter especially relevant to *A. pullulans* and pullulan production [4, 6]. Consequently, resolving which of the several possible interacting cultural parameters play a direct role in glucan production is challenging and requires careful and sound experimental design.

Furthermore, the data published from most of these shake-flask (and in fact fermenter-based) studies are entirely descriptive. They fail to interpret the results mechanistically by seeking answers as to why glucan yields might differ from any imposed changes in culture conditions. Thus, much of the literature is very confusing and full of unexplained contradictions. The same criticisms apply equally, for example, to why individual strains of *A. pullulans* differ so markedly in their EPS or pullulan productivities. Explanations for these events are urgently required if process optimisation is to proceed in a rational scientific manner. In addition, and as discussed above, too often the detailed chemical nature of the glucan/s is not reported, with the untested assumption being that a single polysaccharide only is recovered from the culture medium in the ethanol precipitation step.

Even with this unsatisfactory state of affairs, it is evident from the literature that no single culture medium recipe or set of operating conditions will support optimum yields of exocellular β-glucan in all fungi, and so they must be determined on an individual organism basis. Consequently, the following discussion will try to seek general principles from work where the chemical nature of the EPS is known in most cases with some certainty.

An often expressed view is that glucan production is favoured by conditions where vegetative growth rates of the organism are limited by scarcity of some essential nutrients in media high in carbon (C). Yet whether changes in fungal growth rates, which are affected by a wide range of parameters including nitrogen (N) source, culture pH and DO, can explain the striking differences in glucan yields recorded from fungi is largely unexplored, since very few studies have used chemostat-grown cultures for such studies [22, 152, 153]. Without much published evidence, it seems probable that most glucan-producing natural isolates of fungi are heterokaryons (e.g., *Epicoccum* [35, 36]), whose performance under different culture conditions may reflect differential expression of their genetic heterogeneity. For reasons that are also not clear, much higher exocellular β-glucan yields have been obtained in some shake-flask cultures than in well-mixed laboratory-scale fermenters [35, 154], even though the greater shear with the latter should more efficiently remove any surface-associated β-glucan material, a feature seen with many fungi, into solution [92, 109, 154–156].

3.8.2 Carbon Source

3.8.2.1 Fungal β-Glucans

Yields of fungal exocellular β-glucans and sometimes their branching frequencies [157] are affected by both the carbon source and its initial concentration [17, 109, 146], and many studies have reported the influence of varying both parameters. Those supporting highest biomass yields do not always give optimal exocellular glucan production, and vice versa. Using cheaper C sources like raw and hydrolysed starch has been popular, and may support yields comparable to those obtained with more expensive C sources, as with scleroglucan [158]. The preferred C source may also be strain specific [159], and even when different species of the same fungus prefer the same C source, they may not produce the same final yields of glucan. More interesting is why fungi prefer some C substrates over others for β-glucan production, which is still totally unexplained in biochemical/regulatory terms. In some, the C source used can affect markedly the chemical composition of the EPS synthesised [5].

Studies on the effect of varying initial C concentrations have often shown an increase in β-glucan yields with increasing initial C concentration [154, 160], but then a decrease as concentrations are increased further [17]. Although yield reductions can be explained readily if C sources are present in limiting concentrations (see later), no convincing explanations have been forthcoming as to why nonlimiting concentrations might inhibit production [17, 35]. However, some enzymes involved in β-glucan synthesis require UDP-Glc for their activation, and are active only over a defined sugar concentration range. So once this range is exceeded, inhibitory effects will be apparent [25].

Work with fermenter-grown cultures often shows that β-glucan production decreases or ceases before the C source is exhausted from the medium, as with *Acremonium persicinum* and *Epicoccum* [35, 159], when other factors may have become limiting. These factors might include intracellular N reserves needed for continued biomass production [161], since high β-glucan yields in *S. glucanicum* and *S. commune* can be directly linked to biomass yields [156, 162].

3.8.2.2 Fungal α-Glucans

A. pullulans can also utilise a wide range of C sources for EPS and biomass production, but while sucrose is generally considered the best substrate, no explanations are available as to why. Furthermore, its positive effect seems in some cases to be strain specific [163, 164]. In many shake-flask experiments, run with no pH or DO control, the C concentrations chosen by individual investigators seem empirical, making data comparisons difficult [4, 6, 80]. However, as with some β-glucan-producing fungi, neither C-limiting conditions nor very high initial sucrose levels, while claiming to increase the molecular weight of the pullulan formed [165], provide favourable conditions [166–168]. Fed-batch systems have been used successfully to overcome the adverse effects of high sucrose levels [167]. Some C sources may also favour the production of exoheteropolysaccharides other than pullulan [169, 170]. As with β-glucans, using cheap agroindustrial wastes for pullulan production has received considerable attention [6, 80], especially since yields can be competitively attractive [171–174]. However, its molecular weight, purity and other physicochemical properties important in its industrial application are often compromised with these substrates.

3.8.3 Nitrogen Source

3.8.3.1 Fungal β-Glucans

It is widely accepted that N limitation and/or high C:N ratio conditions are required to achieve high β-glucan yields in fungi [35, 154, 160, 175], and several reviews have been published which emphasise this point [17, 109, 116, 151]. This effect has been shown not only in shake-flask experiments, but also in more carefully controlled fermenter cultures, where any pH variation reflecting the N status of the medium can be negated [17, 109, 151]. Yet not all fungi show this feature [83], and no persuasive mechanisms are available to explain how N limitation might encourage β-glucan formation, and how changes in growth rate affect β-glucan yields. Preferential diversion of C into biomass production under nonlimiting N conditions may be one explanation, but without any convincing direct supportive metabolic or molecular evidence [17]. Similar negative effects of increasing ammonium (NH_4^+) concentrations on antibiotic production in fungi [176] and curdlan production by the bacterium *Agrobacterium tumefaciens* [177] are well documented. Biomass levels can continue to increase after N in the medium becomes exhausted, which may indicate that cells are rapidly assimilating but then storing N (but in what form?) in the early stages of the fermentation, and using it for subsequent growth [17]. However, once cultures enter stationary phase following C exhaustion, β-glucan production rates often slow down.

Screening the influence of different N sources at varying initial concentrations on yields has again generally used shake-flask cultures [17], where such data need additional cautious interpretation because of the influence the N source has on the pH of the medium if uncontrolled. Generally, NH_4^+ salts cause a drop in the pH of the medium and nitrate (NO_3^-) salts increase the pH of the medium [17, 109].

It seems from both shake-flask and pH-controlled fermenter experiments that fungal β-glucan production is favoured when NO_3^- is the preferred N source [35, 178], and is often inhibited (but see Buck and co-workers [83]) in the presence of NH_4^+ salts at the same pH [17, 109, 159, 179].

Several studies, including those of Kim and co-workers [180, 181, 183] and Hwang and co-workers [182], have screened β-glucan yields with several different N sources but all failed to mention how their levels were standardised in the various medium (e.g., as g/l NO_3^- or g/l total N), thus making comparisons within their and similar studies challenging. The chosen N source may also affect fungal morphology, as in *Epicoccum*, where small smooth compact pellets were obtained with NH_4^+ while with NO_3^- pellets were loose and with hairy edges [35]. The relevance of this difference to epiglucan production is unclear except that a smooth pelleted growth form seems to favour glucan production in many fungi (see below). However, further N supplementation following its initial exhaustion from the medium increased markedly both biomass and epiglucan yields, but pellet morphology did not change [35].

Organic N sources have also been screened for the same purpose, and in several fungi some support yields greater than those achieved with NO_3^- [159, 182, 184], and chemically undefined corn steep powder has been shown to be a suitable source for glucans from *Paecilomyces sinclairii* and *Cordyceps militaris* NG3 [180, 181]. However, bearing in mind the earlier comments about the importance of N limitation in affecting β-glucan yields, surprisingly the initial N concentrations used were not always reported. High scleroglucan yields have also been supported by yeast extract and casein hydrolysate [109] and by phosphorus limitation [162].

3.8.3.2 Fungal α-Glucans

The role of different N sources and concentrations in affecting pullulan production has likewise received considerable attention [4, 6, 20]. While most studies have employed inorganic N sources (NH_4^+ and NO_3^-), amino acids have also been used successfully [185, 186]. As with C sources, several high-N waste products have been screened, and some support yields equivalent to those achieved with NH_4^+ [6]. The general belief is that NH_4^+ N-limiting conditions are essential for pullulan production in *A. pullulans* [4, 6, 21, 22], as with fungal β-glucan production, although some published data appear to question this [187]. The same requirement seems to apply to NO_3^- limitation [20, 22]. EPS synthesis commences only following NO_3^- or NH_4^+ exhaustion from the medium [188], but whether this is an essential prerequisite is becoming less clear [20, 22], and fermenter configuration and the N concentration may play an equally important role in influencing yields.

The pH of the medium and cell morphology change with N source, and both are known to influence markedly pullulan synthesis in *A. pullulans* [4]. So identifying which of these variables is in fact responsible for yield changes demands that pH-controlled fermenters are used for such studies. This has not always occurred. Much of the published data suggest that NO_3^- is the preferred N source for pullulan production, where regardless of its concentration, cell morphology is totally unicellular, compared to the predominantly mycelial forms usually seen with NH_4^+ [4]. The possible relevance of this morphological switch to pullulan production is discussed next.

As with the fungal β-glucans, there is no clear explanation as to how NH_4^+ and NO_3^- levels might control exocellular glucan production in *A. pullulans*. The striking switch of C flux into biomass production at the expense of pullulan synthesis at higher initial NH_4^+ levels [188] has led to speculation that NH_4^+ (and

NO_3^-?) may regulate the activity of key enzyme/s (like phosphofructokinase?) dictating the direction of C flow. Yet little experimental evidence exists to support this attractive idea, and Bulmer and co-workers [189] suggested instead that NH_4^+ acts at the level of protein synthesis, presumably affecting production of a pullulan synthase. Again direct evidence in support of this is lacking. With no information on how pullulan biosynthesis is regulated, any role for N will remain contentious. Understanding the biochemical mechanisms involved in the assimilation of inorganic N sources supplied at different levels in the medium into amino acids and subsequent effects on fungal amino acid pools in *A. pullulans* would be well worth pursuing. Equally intriguing is why exhaustion of either NH_4^+ or NO_3^- does not lead to an immediate cessation of cell growth and thus pullulan synthesis, and why both continue as long as C is available. It must be that *A. pullulans* stores N intracellularly, but as with the fungal β-glucans how and in what form is unknown [17, 114].

The levels and type of N source supplied can have an impact in other ways on this fermentation. Thus, a role for pullulan-degrading enzyme/s in deciding the final pullulan yields under different N conditions seems feasible. These as yet uncharacterised enzymes appear to be synthesised only at high NO_3^- (not high NH_4^+ where little pullulan seems to be synthesised [22]) levels, and only in continuously stirred tank reactor (CSTR)-grown cultures but not airlift-grown cultures [20, 21]. Changes in the maltotriose composition of the pullulan produced under these different conditions may determine directly its susceptibility to any enzymatic degradation [22].

3.8.4 Culture pH

3.8.4.1 Fungal β-Glucans

Surprisingly, only limited literature is available on the possible effects of pH on fungal β-glucan production, probably because of the predominant use of shake-flask culture studies for screening. Culture pH is affected not only by the N source used, as mentioned already, but also by acid/alkaline formation, nutrient uptake, oxidation/reduction and decreases in the buffering capacity of the medium [17, 109]. Reliable published data are available for few studies, and with *Sclerotium* spp. and scleroglucan, a two-stage fermentation consisting of an initial period at pH 2 for biomass production and then a change in the pH to 5.5 for glucan production was more productive than when maintaining a constant pH [109]. A similar strategy was used equally successfully by Lee and co-workers [190] and Wu and co-workers [191] for *Ganoderma lucidum* and *Auricularia auricula* β-glucans, respectively.

Stasinopoulos and Seviour [159] also found that β-glucan production in *A. persicinum* was influenced by pH in an airlift fermenter. Highest yields were obtained at pH 7.5 and decreased above and below that value. Yet in studies with *A. auricula* [191], *B. rhodina* [178], *G. frondosa* [184], *Phellinus linteus* [192] and *Agaricus blazei* [3], much lower pH values favoured β-glucan production, and with *A. blazei*, the glucan molecular weight increased. Park and co-workers [193], Kim and co-workers [180, 181], Fang and Zhong [194], Hwang and co-workers [182] and Xu and co-workers [195] all showed that varying initial pH values in shake-flask cultures did not affect the amount of β-glucan produced by different fungi, but these experiments need to be conducted in fermenters with strict pH control to be helpful.

3.8.4.2 Fungal α-Glucans

Obtaining evidence for any direct role of culture pH in pullulan production is equally problematic since pH also influences the culture morphology of *A. pullulans* [4]. As with much of the work on β-glucan, no attempt has been made in many studies to control medium pH, but the early observations of Catley [166] that pullulan production is highest at low pH (which may reflect culture morphology) have generally been confirmed. This outcome seems to vary with the nature of the N source [4, 185]. Both pullulan molecular weight and its contribution to the total EPS material produced may also be affected by the culture pH [165].

3.8.5 Dissolved Oxygen

3.8.5.1 Fungal β-Glucans

The level of DO available to fungal cells during any fermentation process producing exocellular β-glucans will be affected by a combination of fermenter configuration and its mixing system, changes to broth rheology arising from β-glucan accumulation and the morphology of the producing fungus [4, 17, 109]. The possible adverse effects of high DO levels on submerged fungal cultures are well documented [196]. Again from the experimental design used, much of the published data are difficult to interpret and compare, and although it often suggests that low DO levels are required to obtain high glucan yields, there is no clear agreement on what constitutes a low DO. Furthermore, pH control was not always implemented in many studies and so in those studies, its possible influence on glucan yields cannot be discounted. Nevertheless, a need for oxygen-limiting conditions for enhanced β-glucan yields has been reported for both *P. chrysosporium* [197] and *S. commune* [198]. Further evidence that β-glucan formation is favoured at low DO has also been shown with *A. persicinum* [159], *C. militaris* [199], *G. lucidum* [175, 200] and *P. sinclairii* [201].

Rau and co-workers [156] found that both scleroglucan and schizophyllan yields in *Sclerotium* species and *S. commune*, respectively, were highest under oxygen-limiting conditions, and proposed, with no direct support, that high DO levels encouraged biomass formation. Once oxygen became limiting, β-glucan production was then favoured [156]. On the other hand, Wang and McNeil [202] showed that scleroglucan yields actually decreased at low DO, and under such conditions the C source was channelled preferentially into biomass formation. Again no explanation for this switch was provided, although it may reflect an indirect influence on glucan production following changes in fungal growth rate. Here DO levels were controlled by changing agitation speed, which inevitably alters several other important physical parameters, and these by themselves may affect glucan yields. On the other hand, epiglucan yields increased in *Epicoccum* with constant high DO levels [35], making interpretation of the impressively high yields of this β-glucan obtained in shake-flask culture even more difficult. When DO was controlled at both 10 and 75% saturation independent of stirrer speed, the epiglucan yields were much less than when left uncontrolled, but no explanation was forthcoming. Aeration rates have also been proposed to change the chemical composition of fungal exocellular β-glucans [184] and their molecular weight [175].

3.8.5.2 Fungal α-Glucans

Most studies with *A. pullulans* have failed to comment on any possible role of DO levels in affecting pullulan yields, even though these often fall to zero in CSTR batch fermentations, an event coinciding with the onset of pullulan synthesis [20, 22, 203]. The available data support claims of yield enhancement under both high and low DO levels [4]. Thus, while Ono and co-workers [204], Audet and co-workers [205] and McNeil and Kristiansen [206] have reported increased EPS (whose composition was not reported) yields at high agitation rates, Wecker and Onken [207] suggested that optimum production occurred with low DO levels. Gibbs and Seviour [208] precisely controlled DO levels independent of any changes in agitation rate and obtained reduced EPS and pullulan yields at high (>75%) DO levels. Furthermore, these yields increased markedly when low DO levels (7.5 and 15% saturation) were maintained during the early stages of the fermentation. Any role for fungal morphology in these experiments was not discussed.

3.8.6 Fermenter Configuration

Relatively little information is available on the possible influence of fermenter configuration on α-glucan or β-glucan yields in fungi [4, 109, 115]. One main function of a fermenter is to achieve efficient homogeneous mixing of the medium, promoting mass transfer of oxygen and other substrates to the cells [115, 156]. Two configurations are most commonly used for fungal fermentations. One is the CSTR and the other the airlift fermenter. Each is based on a different principle of mixing, which represents a high- and low-shear regime, respectively [17, 109, 114, 149, 150]. However, with most available published data, it is difficult to separate the complex individual effects of shear/mixing/mass transfer or DO levels and biomass morphology on glucan yields, as the experiments have not been designed generally to differentiate between each of them.

3.8.6.1 High-shear Configurations

A stirrer system that imparts a high shear stress upon the medium uses Rushton turbines, which pump the medium out radially from the turbine [109, 114, 115, 146, 149]. In this system, oxygen and heat mass transfer rates increase and mixing times decrease as stirrer speeds increase [206], and fungal morphology is often quite different from that seen in low-shear systems [114, 146, 149, 150].

3.8.6.2 Low-shear Configurations

Low-shear fermenter configurations can be categorised conveniently into two groups. One is the airlift fermenter, which uses differences in hydrostatic pressure or density to achieve fluid mixing [114, 115, 149, 150, 209]. Air is injected through a sparger into the bottom of a riser tube, decreasing the effective density of the medium there. As the bubbles rise to the top, they are released in the head space; the medium becomes denser and it then descends to the bottom of the vessel via a downcomer or an external loop. In a CSTR, low-shear configurations rely on modifications of the stirrer systems and the impellers. The low-shear impellers available include axial flow and helical ribbon stirrers [115, 156]. Both operate in the same way by pumping the fluid from the top to the bottom of the fermenter at reduced liquid stress and against the flow of air [156].

3.8.6.3 Fungal β-Glucans

With fungi producing scleroglucan type of polysaccharides, a high shear stress has been used to explain decreases in glucan yields [109], although generally no attempt has been made to determine directly any possible influence from corresponding changes in DO levels or shifts in fungal morphology. A few examples will illustrate this. Exocellular glucan yields of *C. militaris* were highest in a CSTR fitted with

Rushton turbines run at a stirrer speed of 150 rpm, and with a tight compact pelleted morphology [210]. Higher and lower stirrer speeds decreased yields, a trend thought to reflect marked changes in morphology (see below) from compact to less compact pellets [210]. No DO profiles were provided. Kim and co-workers [201] attempted to differentiate between any effects of aeration rate and stirrer speed on β-glucan yields in *P. sinclairii* by keeping the former constant and varying the latter. They then compared their data with those from experiments run with constant stirrer speeds and changing aeration rates. In contrast to other similar studies, they found that high glucan yields required not only high agitation speeds (250 rpm) but also high air flow rates (3.5 vvm). Again it is difficult to interpret their data as they failed to mention either the constant stirrer speed used for the aeration experiments or the constant aeration rate in experiments when stirrer speeds were changed, and no attempts were made to control DO independent of stirrer speed.

The effects of low-shear systems on fungal β-glucan production appear to be equally complex. Thus, when Rau and co-workers [156] compared yields in fermenters fitted with either a helical ribbon stirrer or axial flow impellers, they found higher schizophyllan yields with axial flow impellers. This was explained as some shear stress being necessary to remove the attached glucan from the surfaces of the *S. commune* cells, allowing increased mass transfer of nutrients and oxygen to the cell. A similar requirement for shear stress to remove attached glucan and so increase glucan yields was also proposed for *S. glucanicum* [109]. Yet Rau and co-workers [156] and Gura and Rau [211] found improved schizophyllan yields after Rushton turbines were replaced by low-shear fan and marine propeller mixing systems, and large diameter fan impellers operating at low speed were the most productive systems they tried. Similar trends were noted by Crognale and co-workers [212] with botryosphaeran production in *B. rhodina*.

Glucan yields by *A. persicinum* in an airlift fermenter at constant pH exceeded those obtained in a CSTR under the same pH conditions, possibly reflecting differences in oxygen mass transfer rates [159]. Higher exocellular glucan yields by *B. rhodina* were also supported with an airlift configuration, but other culture conditions were not optimised [213], while scleroglucan yields from *S. glucanicum* grown in an airlift fermenter with an external loop and run at constant pH were comparable to those obtained in a CSTR with a high-shear agitation system. These decreased at higher aeration rates in the airlift reactor [214]. Yet with *Epicoccum* species, epiglucan yields were lower than those obtained in a CSTR at low stirrer speeds [35], although pellet sizes in both were of a similar size. Slight yield improvement was noticed when *Epicoccum* was grown at lower agitation rates in the novel assisted airlift configuration described by Gibbs and Seviour [215]. Here pellet size did not change markedly when agitation rates were increased, but epiglucan yields decreased [35].

3.8.6.4 Fungal α-Glucans

Few studies have used reactors like low-shear airlift systems instead of CSTR to produce pullulan [117, 215–217]. Higher yields have been reported in airlift- than CSTR-grown cultures with NO_3^- as N source, an outcome thought to reflect the lower oxygen mass transfer rates there [216], although Audet and co-workers [205] showed increased productivity in a reciprocal plate bioreactor at higher DO levels. While impeller design had no obvious influence on EPS yields in a CSTR, Gibbs and Seviour [215] could enhance the yield markedly in *A. pullulans* grown in an assisted airlift reactor with a single Rushton turbine placed at the base of the stirrer shaft inside the draft tube to encourage gas dispersal and small bubble size. An axial flow impeller placed near the top ensured efficient mixing.

3.8.7 Fungal Morphology

3.8.7.1 Fungal β-Glucans

Fungal morphology in submerged culture (pelleted *versus* dispersed filamentous growth) is affected by a variety of cultural parameters, which include inoculum strength, medium composition and fermenter configuration (see above) [146, 149]. It is generally believed that a pelleted fungal morphology may produce higher β-glucan yields than diffuse mycelial cultures, as suggested for example for *A. persicinum* [159], *Sclerotium* spp. [218], *Phellinus* species [192], *B. rhodina* [212] and *Epicoccum purpurascens* [219]. Formation of β-glucans in *C. militaris* and *P. sinclairii* was also favoured by tight compact pellets [199, 201]. Pellet size was not affected by aeration rate in *S. commune* grown in a high-shear stirred tank reactor [198]. However, when cultures were grown in bubble column reactors under controlled conditions, pellet size clearly affected schizophyllan production – pellets of large diameter resulted in increased yields, but only under low aeration conditions. Morphological changes in response to increases in aeration/agitation rates in CSTR have been recorded for several fungi. Thus, in *Paecilomyces tenuipes* [220], compact pellets were transformed to diffuse mycelium, a switch associated with a fall in the yield. Both Stasinopoulos and Seviour [159] and Schmid [35] have also shown that glucan yield decreased in response to increases in CSTR agitation speed, which again coincided with a switch from pellets to diffuse mycelial growth in both *A. persicinum* and *Epicoccum* spp.

3.8.7.2 Fungal α-Glucans

A. pullulans has a very complex life cycle [221], and the commonly used descriptive term yeast- mycelial 'dimorphic' is not appropriate for it. Instead it is highly polymorphic since several morphologically distinct unicellular forms are produced depending on the strain and the culture conditions, which include agitation speed, pH and N source. These events are discussed in detail by Gibbs and Seviour [4]. One question that has attracted considerable interest [4] is which morphological form(s) is responsible for pullulan production? Catley [222] first reported that the onset of pullulan production appeared to coincide with a shift in morphology from mycelium to unicellular blastospores, one of many unicellular forms produced during the life cycle of *A. pullulans.*

Many other studies have reported that EPS yields are higher under conditions where *A. pullulans* grows as unicells, but there is no information as to which unicellular form is connected with high yields. No evidence exists for the presence of pullulan in the cell walls of any of the morphological forms of this fungus [4], although Campbell and co-workers [145] showed intracellular electron transparent material in unicells which conjugated with gold-labelled pullulanase. This suggests that pullulan might be synthesised inside these cells, and further work [145] led them to suggest that chlamydospores and swollen cells are the most likely candidates involved in the production of pullulan, a view expressed earlier by Simon and co-workers [223] from less direct histochemical evidence. However, a medium with high NH_4^+ levels, where mycelium dominates, can also support high EPS levels [4] and light microscopy clearly shows that mycelial forms of this fungus are often surrounded by a capsular layer [203]. Whether this is pullulan has not been resolved, although Orr and co-workers [22] showed that at high NH_4^+ levels in a CSTR, *A. pullulans* produced an EPS that contained no pullulan. Its appearance in the medium clearly coincided with a total switch from unicellular to predominantly mycelial forms, but its detailed chemical structure was not elucidated.

3.8.8 The Influence of Other Media Components on β-Glucan Yields

Several studies have shown that addition of vegetable oils and fatty acids to culture media as antifoams can improve β-glucan yields in fungi, often quite substantially. Thus Stasinopoulos and Seviour [224] showed impressive increases in β-glucan yields in *A. persicinum* after supplementation of the media with both palmitic and oleic acids; similar results were observed with *C. militaris* [225], but not with *Epicoccum* and epiglucan production [35]. Linoleic acid appeared to inhibit production in each case as did the antifoam polypropylene glycol 2025 in *A. persicinum* [226], but again not in *Epicoccum* species [35]. Olive oil addition had a similar effect in *G. lucidum* [227]. How and why these fatty acids stimulate or otherwise EPS synthesis is unclear. Whether they change the chemical composition of the fungal cytoplasmic membrane or act as paraffin oil leading to an increase in oxygen mass transfer rates [212] remains to be seen. The latter seems unlikely as oxygen-limiting conditions often stimulate exocellular β-glucan yields. Survase and co-workers [88] reported that addition of lysine and uridine monophosphate substantially improved scleroglucan yields in *S. rolfsii*, but they did not explain the mechanism involved, and whether other β-glucan-producing fungi respond in the same way is not known.

3.9 The Role of Fungal Glucanases in Affecting Glucan Yields

3.9.1 Fungal β-Glucans

As mentioned earlier, many fungi elaborate into the culture medium often multiple isoforms of (1,3)-β- and (1,6)-β-glucanases. Each is capable of degrading exocellular branched (1,3)(1,6)-β-glucans [11, 12], allowing the producing cells to reutilise them potentially as the sole carbon and energy sources. Some of these enzymes are wall bound [154]. Their shared feature of being inducible enzymes whose synthesis in glucan-synthesising fungi is regulated by catabolite (glucose) repression (but see Crognale and co-workers [154]) means that they have a potential to impact markedly final β-glucan yields once glucose becomes exhausted from the medium during batch fermentation. Glucose fed-batch systems may eliminate any such enzymatic influence, and some published data suggest that these enzymes have no effect on final glucan yields [154]. However, convincing evidence now exists that yields may fall as a consequence of such enzyme activity in several β-glucan-producing fungi, including *B. cinerea* [228] and *S. glucanicum* [155], where up to 90% of the glucan was degraded, and *A. persicinum* [229, 230]. In *S. glucanicum* and *A. persicinum*, β-glucosidases are also probably involved in the degradation of oligoglucoside products of the (1,3)-β- and (1,6)-β-glucanase activities. Similar studies are needed on other β-glucan-producing fungi, and the regulation of β-glucan synthesis needs to be elucidated.

3.9.2 Fungal α-Glucans

Many studies have reported that EPS yields and/or the molecular weight of pullulan falls in later stages of *A. pullulans* fermentations [20, 22, 217, 231–235], an observation consistent with the production of pullulan-degrading enzymes by this fungus. Saha and co-workers [233] and West and Strohfus [235] have described a pullulan-degrading enzyme activity (described as a glucoamylase B) in the culture filtrates of their *A. pullulans* strain, but its role in pullulan disappearance is not clear. Campbell and co-workers [20] and Orr and co-workers [22] both showed that pullulan was degraded in CSTR- but not airlift-grown cultures in medium containing high N levels, and only after glucose became exhausted from the medium. This is again consistent with the synthesis of pullulan-degrading enzymes regulated by catabolite (glucose) repression. Yet no activity could be detected against commercial pullulan as substrate, but a crude exocellular amylolytic activity against it increased at this time. Glucose was also the major pullulan digestion product, consistent with some glucoamylase activity [20]. However, *A. pullulans* failed to grow on pullulan and on its own EPS synthesised under these conditions as the sole carbon source. Surprisingly, amylolytic activity was detected in the culture filtrates of *A. pullulans* grown on glucose, suggesting that it is synthesised constitutively by *A. pullulans*. This implies that some novel enzyme may be involved in the disappearance of pullulan.

3.10 Conclusion

This chapter attempts to summarise what we know currently about fungal EPS, and highlights some areas where future work is needed. Their current applications and unrealised industrial potential earmark these as metabolites of considerable future economic and social importance. Only when we understand better how their synthesis is regulated will we be able to optimise their production rationally and not empirically, as we are now able to do with many other microbial metabolites. Acquiring this information is a challenge, but if the aim is to produce tailor-made EPS with enhanced bioactive and more attractive physicochemical properties, then such challenges need to be met. So now is an opportune time to try to elucidate their biosynthetic pathways and their regulation, both of which seem feasible with the rapid advances in practical tools for the genetic manipulation of fungi.

References

1. S. de Baets, S. du Laing, C. Francois and E.J. Vandamme, *Journal of Industrial Microbiology and Biotechnology*, 2002, **29**, 4, 181.

2. H.H. Sun, W.J. Mao, Y. Chen, S.D. Guo, H.Y. Li, X.H. Qi, Y.L. Chen and J. Xu, *Carbohydrate Polymers*, 2009, **78**, 1, 117.

3. C.H. Shu and M.Y. Lung, *Process Biochemistry*, 2004, **39**, 931.

4. P.A. Gibbs and R.J. Seviour in *Polysaccharides in Medicinal Applications*, Ed., S. Dumitriu, Marcel Dekker, Inc., New York, NY, USA, 1996, p.59.

5. G.L. Sassaki, J.C. Ferreira, C. Glienke-Blanco, G. Torri, F. De Toni, P.A.J. Gorin and M. Iacomini, *Carbohydrate Polymers*, 2002, **48**, 4, 385.

6. R.S. Singh, G.K. Saini and J.F. Kennedy, *Carbohydrate Polymers*, 2008, **73**, 4, 515.

7. B.A. Stone in *Chemistry, Biochemistry, and Biology of (1-3)-β-Glucans and Related Polysaccharides*, Eds., A. Bacic, G.B. Fincher and B.A. Stone, Elsevier, Inc., Burlington, MA, USA, 2009, p.5.

8. D.J. Adams, *Microbiology*, 2004, **150**, 2029.

9. M.d.L.C. da Silva, E.K. Fukuda, A.F.D. Vasconcelos, R.F.H. Dekker, A.C. Matias, N.K. Monteiro, M.S. Cardoso, A.M. Barbosa, J.L.M. Silveira, G.L. Sassaki and E.R. Carbonero, *Carbohydrate Research*, 2008, **343**, 4, 793.

10. K. Ruel and J.P. Joseleau, *Applied and Environmental Microbiology*, 1991, **57**, 374.

11. K. Martin, B.M. McDougall, S. McIlroy, Jayus, J. Chen and R.J. Seviour, *FEMS Microbiology Letters*, 2007, **31**, 2, 168.

12. S.M. Pitson, R.J. Seviour and B.M. McDougall, *Enzyme and Microbial Technology*, 1993, **15**, 178.

13. R.C. Amey, P.R. Mills, A. Bailey and G.D. Foster, *Fungal Genetics and Biology*, 2003, **39**, 264.

14. S. Djonović, M.J. Pozo and C.M. Kenerley, *Applied and Environmental Microbiology*, 2006, **72**, 12, 7661.

15. M. Montero, L. Sanz, M. Rey, A. Llobell and E. Monte, *Journal of Applied Microbiology*, 2007, **103**, 4, 1291.

16. J.L. Copa-Patiño, I.F. Monistrol, F. Laborda and M.I. Pérez-Leblic, *Transactions of the British Mycological Society*, 1987, **88**, 3, 317.

17. R.J. Seviour, S.J. Stasinopoulos, D.P.F. Auer and P.A. Gibbs, *Critical Reviews in Biotechnology*, 1992, **12**, 3, 279.

18. L. Yang and L-M. Zhang, *Carbohydrate Polymers*, 2009, **76**, 3, 349.

19. K.I. Shingel, *Carbohydrate Research*, 2004, **339**, 447.

20. B.S. Campbell, B.M. McDougall and R.J. Seviour, *Enzyme and Microbial Technology*, 2003, **33**, 1, 104.

21. W. Zheng, B.S. Campbell, B.M. McDougall and R.J. Seviour, *Bioresource Technology*, 2008, **99**, 16, 7480.

22. D. Orr, W. Zheng, B.S. Campbell, B.M. McDougall and R.J. Seviour, *Journal of Applied Microbiology*, 2009, **107**, 2, 691.

23. S.C. Jong in *Biopolymers, Polysaccharides II: Polysaccharides from Eukaryotes*, Eds., E.J. Vandamme, S. De Baets and A. Steinbüchel, Wiley-VCH Verlag GmbH, Germany, 2002, p.159.

24. J.N. BeMiller, *Advances in Carbohydrate Chemistry*, 1967, **22**, 25.

25. B.A. Stone and A.E. Clarke, *Chemistry and Biology of (1-3)-β-Glucans*, LaTrobe University Press, Australia, 1992.

26. M.L. Wolfrom, A. Thompson and C.E. Timberlake, *Cereal Chemistry*, 1963, **40**, 82.

27. I. Ciucanu and F. Kerek, *Carbohydrate Research*, 1984, **131**, 209.

28. G.G.S. Dutton, *Advances in Carbohydrate Chemistry and Biochemistry*, 1974, **30**, 9.

29. P.J. Harris, R.J. Henry, A.B. Blakeney and B.A. Stone, *Carbohydrate Research*, 1984, **127**, 59.

30. S.A. Barker, E.J. Bourne and D.H. Whiffen, *Methods of Biochemical Analysis*, 1956, **3**, 213.

31. A.J. Mitchell and G. Scurfield, *Australian Journal of Biological Sciences*, 1970, **23**, 345.

32. P.K. Agrawal, *Phytochemistry*, 1992, **31**, 10, 3307.

33. J.H. Bradbury and G.A. Jenkins, *Carbohydrate Research*, 1984, **126**, 125.

34. B. Lindberg, *Methods in Enzymology*, 1972, **28**, 178.

35. F. Schmid, *Studies on Extracellular β-Glucans from the Fungi Epicoccum nigrum and Acremonium spp.*, La Trobe University, Bendigo, Australia, 2004. [Ph.D. Thesis]

36. F. Schmid, B.A. Stone, B.M. McDougall, A. Bacic, K.L. Martin, R.T.C. Brownlee, E. Chai and R.J. Seviour, *Carbohydrate Research*, 2001, **331**, 163.

37. A. Isogagi, A. Ishizu, J. Nakano, S. Eda and K. Kato, *Carbohydrate Research*, 1985, **138**, 99.

38. T. Purdie and J.C. Irvine, *Journal of the Chemical Society*, 1903, **83**, 1021.

39. N. Haworth, *Journal of the Chemical Society*, 1915, **107**, 8.

40. S. Hakomori, *Journal of Biochemistry*, 1964, **55**, 205.

41. E.J. Corey and A.J. Chaykovsky, *Journal of the American Chemical Society*, 1962, **84**, 3782.

42. J. Finne, T. Krusius and H. Rauvala, *Carbohydrate Research*, 1980, **80**, 336.

43. L.R. Phillips and B. Fraser, *Carbohydrate Research*, 1981, **90**, 149.

44. P.W. Needs and R.R. Selvendran, *Carbohydrate Research*, 1993, **245**, 1.

45. P.A.J. Gorin in *Fungal Polysaccharides*, Eds., P.A. Sandford and M. Matsuda, American Chemical Society, Washington, DC, USA, 1980, p.159.

46. M. Rinaudo and M. Vincendon, *Carbohydrate Polymers*, 1982, **2**, 135.

47. A. Misaki, K. Kawaguchi, H. Miyaji, H. Nagae, S. Hokkoku, M. Kakuta and T. Sasaki, *Carbohydrate Research*, 1984, **129**, 209.

48. R. Tada, Y. Adachi, K. Ishibashi and N. Ohno, *Carbohydrate Research*, 2009, **344**, 3, 400.

49. H. van Halbeek, *Current Opinion in Biotechnology*, 1994, **4**, 697.

50. M. Bardet, A. Rousseau and M. Vincedon, *Magnetic Resonance in Chemistry*, 1993, **31**, 887.

51. F. Schmid, F. Separovic, B.M. McDougall, B.A. Stone, R.T.C. Brownlee and R.J. Seviour, *Carbohydrate Research*, 2007, **342**, 16, 2481.

52. C. Arnosti and D.J. Repeta, *Starch/Stärke*, 1995, **47**, 2, 73.

53. A. Lazaridou, C.G. Biliaderis, T. Roukas and M. Izydorczyk, *Applied Biochemistry and Biotechnology*, 2002, **97**, 1, 1.

54. A. Lazaridou, T. Roukas, C.G. Biliaderis and H. Vaikousi, *Enzyme and Microbial Technology*, 2002, **31**, 1/2, 122.

55. D.D. McIntyre and H.J.V. Calgary, *Starch/Stärke*, 1993, **45**, 11, 406.

56. F. Youssef, C.G. Biliaderis and T. Roukas, *Applied Biochemistry and Biotechnology*, 1998, **74**, 1, 13.

57. F. Youssef, T. Roukas and C.G. Biliaderis, *Process Biochemistry*, 1999, **34**, 4, 355.

58. J.M. Bobbitt, *Advances in Carbohydrate Chemistry*, 1956, **11**, 1.

59. E.A. Davidson, *Carbohydrate Chemistry*, Holt, Rinehart and Winston, Inc., New York, NY, USA, 1967.

60. B. Lindberg, J. Lönngren and S. Svensson in *Advances in Carbohydrate Chemistry and Biochemistry*, Eds., R.S. Tipson and D. Horton, Academic Press, New York, NY, USA, 1975, p.185.

61. G.W. Hay, B.A. Lewis and F. Smith, *Methods in Carbohydrate Chemistry*, 1965, **5**, 377.

62. R.D. Guthrie, *Advances in Carbohydrate Chemistry and Biochemistry*, 1961, **16**, 105.

63. E.L. Jackson in *Organic Reactions*, Ed., R. Adams, John Wiley & Sons, New York, NY, USA, 1944, p.341.

64. K.K. Batra, J.H. Nordin and S. Kirkwood, *Carbohydrate Research*, 1969, **9**, 221.

65. C. Gandon and M. Bruneteau, *Carbohydrate Research*, 1998, **313**, 259.

66. R.A. Reis, C.A. Tischer, P.A.J. Gorin and M. Iacomini, *FEMS Microbiology Letters*, 2002, **210**, 1, 1.

67. G.G.S. Dutton and K.B. Gibney, *Carbohydrate Research*, 1972, **25**, 99.

68. I.J. Goldstein, J.K. Hamilton and F. Smith, *Abstracts of Papers of the American Chemical Society Meeting*, 1959, **135**, 6252.

69. B. Lindberg, *Chemical Society Reviews*, 1981, **10**, 409.

70. M. Bruneteau, I. Fabre, J. Perret, G. Michel, P. Ricci, J.P. Joseleau, J. Kraus, M. Schneider, W. Blaschek and G. Franz, *Carbohydrate Research*, 1988, **175**, 137.

71. J. Defaye, S. Kohlmünzer, K. Sodzawiczny and E. Wong, *Carbohydrate Research*, 1988, **173**, 316.

72. K. Saito, M. Nishijima and T. Miyazaki, *Chemical and Pharmaceutical Bulletin*, 1990, **38**, 1745.

73. A.S. Perlin and W.A. Taber, *Canadian Journal of Chemistry*, 1963, **41**, 2278.

74. J. Johnson, S. Kirkwood, A. Misaki, T.E. Nelson, J.V. Scaletti and F. Smith, *Chemistry and Industry*, May 1963, 820.

75. S. Aouadi, A. Heyraud, F. Seigle-Murandi, R. Steiman, J. Kraus and G. Franz, *Carbohydrate Polymers*, 1991, **16**, 155.

76. Y. Adachi, N.N. Miura, N. Ohno, H. Tamura, S. Tanaka and T. Yadomae, *Carbohydrate Polymers*, 1999, **39**, 3, 225.

77. F. Schmid, B.A. Stone, R.T.C. Brownlee, B.M. McDougall and R.J. Seviour, *Carbohydrate Research*, 2006, **341**, 3, 365.

78. G. Brigand in *Industrial Gums*, Eds., R.L. Whistler and J.N. BeMiller, Academic Press, New York, NY, USA, 1993, p.461.

79. J.I. Fariña, S.C. Viñarta, M. Cattaneo and L.I.C. Figueroa, *Journal of Applied Microbiology*, 2009, **106**, 1, 221.

80. T.D. Leathers, *Applied Microbiology and Biotechnology*, 2003, **62**, 5/6, 468.

81. M.J. Gidley and K. Nishinari in *Chemistry, Biochemistry and Biology of (1-3)-β-Glucans and Related Polysaccharides*, Eds., A. Bacic, G.B. Fincher and B.A. Stone, Elsevier, Inc., Burlington, MA, USA, 2009, p.47.

82. A.F.D. Vasconcelos, N.K. Monteiro, R.F.H. Dekker, A.M. Barbosa, E.R. Carbonero, J.L.M. Silveira, G.L. Sassaki, R. da Silva and M.D.C. da Silva, *Carbohydrate Research*, 2008, **343**, 14, 2481.

83. K.W. Buck, A.W. Chen, A.G. Dickerson and E.B. Chain, *Journal of General Microbiology*, 1968, **51**, 337.

84. K. Tabata, W. Ito, T. Kojima, S. Kawabata and A. Misaki, *Carbohydrate Research*, 1981, **89**, 121.

85. D. Dubourdieu, P. Ribereau-Gayon and B. Fournet, *Carbohydrate Research*, 1981, **93**, 294.

86. M.d.L.C. da Silva, N.L. Izeli, P.F. Martinez, I.R. Silva, C.J.L. Constantino, M.S. Cardoso, A.M. Barbosa, R.F.H. Dekker and G.V.J. da Silva, *Carbohydrate Polymers*, 2005, **61**, 1, 10.

87. S.J. Stasinopoulos and R.J. Seviour, *Mycological Research*, 1989, **92**, 1, 55.

88. S.A. Survase, P.S.S. Saudagar and R.S. Singhal, *Bioresource Technology*, 2007, **98**, 2, 410.

89. L. Selbmann, S. Onofri, M. Fenice, F. Federici and M. Petruccioli, *Research in Microbiology*, 2002, **153**, 585.

90. F.R. Smiderle, E.R. Carbonero, C.G. Mellinger, G.L. Sassaki, P.A.J. Gorin and M. Iacomini, *Phytochemistry*, 2006, **67**, 19, 2189.

91. A.M. Barbosa, R.M. Steluti, R.F.H. Dekker, M.S. Cardoso and M.L. Corradi da Silva, *Carbohydrate Research*, 2003, **338**, 1691.

92. P. Pielken, P. Stahmann and H. Sahm, *Applied Microbiology and Biotechnology*, 1990, **33**, 1.

93. A. Gutiérrez, A. Prieto and A.T. Martínez, *Carbohydrate Research*, 1996, **281**, 143.

94. T. Miyazaki, T. Yadomae, M. Sugiura, H. Ito, K. Fujii, S. Naruse and M. Kunihisa, *Chemical and Pharmaceutical Bulletin*, 1974, **22**, 1739.

95. J.A. Leal, P. Rupérez and B. Gomez-Miranda, *Transactions of the British Mycological Society*, 1979, **72**, 172.

96. M. Feathers and A. Malek, *Biochimica et Biophysica Acta*, 1972, **264**, 103.

97. S.A. Archer, J.R. Clamp and D. Migliore, *Journal of General Microbiology*, 1977, **102**, 157.

98. F. Santamaria, F. Reyes and R. Lahoz, *Journal of General Microbiology*, 1978, **109**, 287.

99. U. Schweiger-Hufnagel, T. Ono, K. Izumi, P. Hufnagel, N. Morita, H. Kaga, M. Morita, T. Hoshino, I. Yumoto, N. Matsumoto, M. Yoshida, M.T. Sawada and H. Okuyama, *Biotechnology Letters*, 2000, **22**, 183.

100. T.D. Leathers in *Biopolymers*, Eds., S. De Baets, E.J. Vandamme and A. Steinbuechel, Wiley-VCH, Weinheim, Germany, 2002, p.1.

101. H.J. Jennings and I.C.P. Smith, *Journal of the American Chemical Society*, 1973, **95**, 2, 606.

102. E.M. Oliva, A.F. Cirrelli and R.M. De Lederkremer, *Carbohydrate Research*, 1986, **158**, 262.

103. N.D. Waksman, R.M. Lederkremer and A.S. Cerezo, *Carbohydrate Research*, 1977, **59**, 2, 505.

104. M.M. Corsaro, C. De Castro, A. Evidente, R. Lanzetta, A. Molinaro, M. Parrilli and L. Sparapano, *Carbohydrate Polymers*, 1998, **37**, 2, 167.

105. Z. Chi and S. Zhao, *Enzyme and Microbial Technology*, 2003, **33**, 2/3, 206.

106. F. Delben, A. Forabosco, M. Guerrini, G. Liut and G. Torri, *Carbohydrate Polymers*, 2006, **63**, 4, 545.

107. N.S. Madi, B. McNeil and L.M. Harvey, *Journal of Chemical Technology and Biotechnology*, 1996, **65**, 4, 343.

108. E.C. Giese, R.F.H. Dekker, A.M. Barbosa and R. da Silva, *Carbohydrate Polymers*, 2008, **74**, 4, 953.

109. Y. Wang and B. McNeil, *Critical Reviews in Biotechnology*, 1996, **16**, 3, 185.

110. W.S. Fulton and E.D.T. Atkins in *Fibre Diffraction Methods*, Eds., A.D. French and K.H. Gardner, America Chemical Society, Washington, DC, USA, 1980, p.385.

111. R.H. Maechessault, Y. Deslandes, K. Ogawa and P.R. Sundararajan, *Canadian Journal of Chemistry*, 1977, **55**, 300.

112. T. Harada, A. Koreeda, S. Sato and N. Kasai, *Journal of Electron Microscopy*, 1979, **28**, 147.

113. I. Giavasis, L.M. Harvey and B. McNeil in *Biopolymers*, Eds., E.J. Vandamme, S. De Baets and A. Steinbüchel, Wiley-VCH, Weinheim, Germany, 2002, p.37.

114. P.A. Gibbs, R.J. Seviour and F. Schmid, *Critical Reviews in Biotechnology*, 2000, **20**, 1, 17.

115. B. McNeil and L.M. Harvey, *Critical Reviews in Biotechnology*, 1993, **13**, 4, 275.

116. U. Rau in *Biopolymers, Polysaccharides II: Polysaccharides from Eukaryotes*, Eds., E. Vandamme, S. De Baets and A. Steinbüchel, Wiley-VCH, Weinheim, Germany, 2002, p.61.

117. T. Roukas and F. Mantzouridou, *Journal of Chemical Technology and Biotechnology*, 2001, **76**, 4, 371.

118. K. Toda, Y. Gotoh, T. Asakura, I. Yabe and H. Furuse, *Journal of Bioscience and Bioengineering*, 2000, **89**, 3, 258.

119. G.S. Buliga and D.A. Brant, *International Journal of Biological Macromolecules*, 1987, **9**, 2, 77.

120. A.A. LeDuy, L. Chaplin, J.E. Zajic and J.H.T. Luong in *Encyclopaedia of Polymer Science and Engineering*, Eds., H.F. Mark, N.M. Bikales, G. Overburger and G. Menges, Wiley and Sons, New York, NY, USA, 1988, p.650.

121. A. Forabosco, G. Bruno, L. Sparapano, G. Liut, D. Marino and F. Delbe, *Carbohydrate Polymers*, 2006, **63**, 4, 535.

122. S. Wasser, *Applied Microbiology and Biotechnology*, 2002, **60**, 3, 258.

123. B.Z. Zaidman, M. Yassin, J. Mahajna and S.P. Wasser, *Applied Microbiology and Biotechnology*, 2005, **67**, 453.

124. J. Chen and R.J. Seviour, *Mycological Research*, 2007, **111**, 6, 635.

125. G.D. Brown and S. Gordon, *Cellular Microbiology*, 2005, **7**, 4, 471.

126. G.D. Brown and D.L. Williams in *Chemistry, Biochemistry, and Biology of (1-3)-β-Glucans and Related Polysaccharides*, Eds., A. Bacic, G.B. Fincher and B.A. Stone, Elsevier, Inc., Burlington, MA, USA, 2009, p.579.

127. J.A. Bohn and J.N. BeMiller, *Carbohydrate Polymers*, 1995, **28**, 3.

128. W.M. Kulicke, A.I. Lettau and H. Thielking, *Carbohydrate Research*, 1997, **297**, 2, 135.

129. D.B. Zekovié, S. Kwiatkowskié, M.M. Vrvié, D. Jakovljevié and C.A. Moran, *Critical Reviews in Biotechnology*, 2005, **25**, 4, 205.

130. A. Misaki, M. Kakuta, T. Sasaki, M. Tanaka and H. Miyaji, *Carbohydrate Research*, 1981, **92**, 115.

131. A. Mueller, J. Raptis, P.J. Rice, J.H. Kalbfleisch, R.D. Stout, H.E. Ensley, W. Browder and D.L. Williams, *Glycobiology*, 2000, **10**, 4, 339.

132. B.H. Falch, T. Espevik, L. Ryan and B.T. Stokke, *Carbohydrate Research*, 2000, **329**, 3, 587.

133. W. Blaschek, J. Keasbauer, J. Kraus and G. Franz, *Carbohydrate Research*, 1992, **231**, 293.

134. G. Kogan in *Studies in Natural Products Chemistry: Bioactive Natural Products, Part D*, Ed., A. Rahman, Elsevier, Amsterdam, 2000, p.107.

135. T. Kojima, K. Tabata, W. Itoh and T. Yanaki, *Agricultural and Biological Chemistry*, 1986, **50**, 231.

136. H. Saito, T. Ohki, N. Takasuka and T. Sasaki, *Carbohydrate Research*, 1977, **58**, 293.

137. C. Clavaud, V. Aimanianda and J-P. Latgea in *Chemistry, Biochemistry, and Biology of (1-3)-β-Glucans and Related Polysaccharides*, Eds., A. Bacic, G.B. Fincher and B.A. Stone, Elsevier, Inc., Burlington, MA, USA, 2009, p.387.

138. S. Nogami and Y. Ohya in *Chemistry, Biochemistry, and Biology of (1-3)-β-Glucans and Related Polysaccharides*, Eds., A. Bacic, G.B. Fincher and B.A. Stone, Elsevier, Inc., Burlington, MA, USA, 2009, p.259.

139. J. Ruiz-Herrera, *Antonie Van Leeuwenhoek*, 1991, **60**, 73.

140. G.J.P. Dijkgraaf, H. Li and H. Bussey in *Biopolymers, Vol. 6, Polysaccharides II: Polysaccharides from Eukaryotes*, Eds., E.J. Vandamme, S. De Baets and A. Steinbüchel, Wiley-VCH, Weinheim, Germany, 2002, p.179.

141. S. Balint, V. Farkas and S. Bauer, *FEBS Letters*, 1976, **64**, 44.

142. R.P. Hartland, G.W. Emerson and P.A. Sullivan, *Proceedings of the Royal Society B: Biological Sciences*, 1991, **246**, 155.

143. I. Mouyna, R.P. Hartland, T. Fontaine, M. Diaquin, C. Simenel, M. Delepierre, B. Henrissat and J.P. Latgé, *Microbiology*, 1998, **144**, 3171.

144. B.J. Catley and W. McDowell, *Carbohydrate Research*, 1982, **103**, 65.

145. B.S. Campbell, A-B.M. Siddique, B.M. McDougall and R.J. Seviour, *FEMS Microbiology Letters*, 2004, **232**, 2, 225.

146. M.L. Fazenda, R.J. Seviour, B. McNeil and L.M. Harvey, *Advances in Applied Microbiology*, 2008, **63**, 33.

147. D.K. Kachhawa, P. Bhattacharjee and R.S. Singhal, *Carbohydrate Polymers*, 2003, **52**, 25.

148. R.S. Singh, G.K. Saina and J.F. Kennedy, *Carbohydrate Polymers*, 2009, **78**, 1, 89.

149. M. Papagianni, *Biotechnology Advances*, 2004, **22**, 3, 189.

150. F. Garćia-Ochoa and E. Gomez, *Biotechnology Advances*, 2009, **27**, 2, 153.

151. U. Rau in *Methods in Biotechnolgy – Carbohydrate Biotechnology Protocols*, Ed., C. Bucke, Humana Press Inc., Totowa, 1999, p.43.

152. M. Reeslev, T. Strom, B. Jensen and J. Olsen, *Mycological Research*, 1997, **101**, part 6, 650.

153. P. Wood and R.J. Seviour, *World Journal of Microbiology and Biotechnology*, 1994, **10**, 14.

154. S. Crognale, M. Bruno, M. Fidaleo, M. Moresi and M. Petruccioli, *Journal of Applied Microbiology*, 2007, **102**, 3, 860.

155. P. Rapp, *Journal of General Microbiology*, 1989, **135**, 2847.

156. U. Rau, E. Gura, E. Olszewski and F. Wagner, *Journal of Industrial Microbiology*, 1992, **9**, 19.

157. M.S. Deshpande, V.B. Rale and J.M. Lynch, *Enzyme and Microbial Technology*, 1992, **14**, 7, 514.

158. L. Selbmann, S. Crognale and M. Petruccioli, *Letters in Applied Microbiology*, 2002, **34**, 51.

159. S.J. Stasinopoulos and R.J. Seviour, *Applied Microbiology and Biotechnology*, 1992, **36**, 465.

160. W.Y. Lee, Y. Park, J.K. Ahn, K.H. Ka and S.Y. Park, *Enzyme and Microbial Technology*, 2007, **40**, 2, 249.

161. M.J. Wilcock, B.M. McDougall and R.J. Reviour, *Current Topics in Botanical Research*, 1993, **1**, 237.

162. S. Taurhesia and B. McNeil, *Journal of Chemical Technology and Biotechnology*, 1994, **59**, 2, 157.

163. L.H. Gibson and R.W. Coughlin, *Biotechnology Progress*, 2002, **18**, 3, 675.

164. R.F. Sena, M.C. Costelli, L.H. Gibson and R.W. Coughlin, *Brazilian Journal of Chemical Engineering*, 2006, **23**, 4, 507.

165. K.Y. Lee and Y.J. Yoo, *Biotechnology Letters*, 1993, **15**, 10, 1021.

166. B.J. Catley, *Applied Microbiology*, 1971, **22**, 4, 650.

167. Y.C. Shin, Y.H. Kim, H.S. Lee, Y.N. Kim and S.M. Byun, *Biotechnology Letters*, 1987, **9**, 9, 621.

168. T.P. West and B. Reed-Hamer, *Microbios*, 1991, **67**, 117.

169. P. Cescutti, R. Pupulin, F. Delben, M. Abbate, M. Dentini, L. Sparapano, R. Rizzo and V. Crescenzi, *Carbohydrate Research*, 2002, **337**, 1203.

170. J.W. Lee, W.G. Yeomans, A.L. Allen, F. Deng, R.A. Gross and D.L. Kaplan, *Applied and Environmental Microbiology*, 1999, **65**, 12, 5265.

171. C. Israilides and A. Philippoussis, *Biotechnology and Genetic Engineering Reviews*, 2003, **20**, 247.

172. C. Israilides, B. Scanlon, A. Smith, S.E. Harding and K. Jumel, *Carbohydrate Polymers*, 1994, **25**, 3, 203.

173. C. Israilides, A. Smith, B. Scanlon and C. Barnett, *Biotechnology and Genetic Engineering Reviews*, 1999, **16**, 309.

174. C.J. Israilides, A. Smith, J.E. Harthill, C. Barnett, G. Bambalov and B. Scanlon, *Applied Microbiology and Biotechnology*, 1998, **49**, 5, 613.

175. C. Hsieh, M-H. Tseng and C-J. Liu, *Enzyme and Microbial Technology*, 2006, **38**, 1–2, 109.

176. F.R. Schmidt in *Mycota*, Ed., H.D. Osiewacz, Springer-Verlag, Berlin, 2002, p.69.

177. M. McIntosh, B.A. Stone and V.A. Stanisich, *Applied Microbiology and Biotechnology*, 2005, **68**, 2, 163.

178. L. Selbmann, F. Stingele and M. Petruccioli, *Antonie van Leeuwenhoek*, 2003, **84**, 2.

179. J.I. Fariña, F. Siñeriz, O.E. Molina and N.I. Perotti, *Biotechnology Letters*, 1998, **20**, 825.

180. S.W. Kim, H.J. Hwang, C.P. Xu, Y.S. Na, S.K. Song and J.W. Yun, *Letters in Applied Microbiology*, 2002, **34**, 389.

181. S.W. Kim, C.P. Xu, H.J. Hwang, J.W. Choi, C.W. Kim and J.W. Yun, *Biotechnology Progress*, 2003, **19**, 428.

182. H.J. Hwang, S.W. Kim, C.P. Xu, J.W. Choi and J.W. Yun, *Journal of Applied Microbiology*, 2003, **94**, 708.

183. S.W. Kim, H.J. Hwang, C.P. Xu, J.M. Sung, J.W. Choi and J.W. Yun, *Journal of Applied Microbiology*, 2003, **94**, 120.

184. B.C. Lee, J.T. Bae, H.B. Pyo, T.B. Choe, S.W. Kim, H.J. Hwang and J.W. Yun, *Enzyme and Microbial Technology*, 2004, **35**, 5, 369.

185. D.P.F. Auer and R.J. Seviour, *Applied Microbiology and Biotechnology*, 1990, **32**, 637.

186. B. Reed-Hamer and T.P. West, *Microbios*, 1994, **80**, 323, 83.

187. M. Reeslev, J.C. Nielsen, J. Olsen, B. Jensen and T. Jacobsen, *Mycological Research*, 1991, **95**, 2, 220.

188. R.J. Seviour and B. Kristiansen, *European Journal of Applied Microbiology and Biotechnology*, 1983, **17**, 178.

189. M.A. Bulmer, B.J. Catley and P.J. Kelly, *Applied Microbiology and Biotechnology*, 1987, **25**, 362.

190. K.M. Lee, S.Y. Lee and H.Y. Lee, *Journal of Bioscience and Bioengineering*, 1999, **88**, 6, 646.

191. J. Wu, Z.Y. Ding and K.C. Zhang, *Enzyme and Microbial Technology*, 2006, **39**, 4, 743.

192. H-J. Hwang, S-W. Kima, J-W. Choi and J-W. Yun, *Enzyme and Microbial Technology*, 2003, **33**, 309.

193. J.P. Park, S.W. Kim, H.J. Hwang and J.W. Yun, *Letters in Applied Microbiology*, 2001, **33**, 76.

194. Q.H. Fang and J.J. Zhong, *Process Biochemistry*, 2002, **37**, 769.

195. C.P. Xu, S.W. Kim, H.J. Hwang, J.W. Choi and J.W. Yun, *Process Biochemistry*, 2003, **38**, 1025.

196. Z.H. Bai, L.M. Harvey and B. McNeil, *Critical Reviews in Biotechnology*, 2003, **23**, 267.

197. C.G. Dosoretz, A.H-C. Chen and H.E. Grethlein, *Applied Microbiology and Biotechnology*, 1990, **34**, 131.

198. C.H. Shu, P.F. Chou and I.C. Hsu, *Journal of Chemical Technology and Biotechnology*, 2005, **80**, 12, 1383.

199. J.P. Park, Y.M. Kim, S.W. Kim, H.J. Hwang, Y.J. Cho, Y.S. Lee, C.H. Song and J.W. Yun, *Process Biochemistry*, 2002, **37**, 1257.

200. Y.J. Tang and J.J. Zhong, *Enzyme and Microbial Technology*, 2003, **32**, 3–4, 478.

201. S.W. Kim, H.J. Hwang, C.P. Xu, J.W. Choi and J.W. Yun, *Letters in Applied Microbiology*, 2003, **36**, 321.

202. Y. Wang and B. McNeil, *Biotechnology Letters*, 1995, **17**, 3, 257.

203. B.S. Campbell, *Exopolysaccharide Production by the Fungus Aureobasidium pullulans*, La Trobe University, Bendigo, Australia, 2008. [Ph.D. Thesis]

204. K. Ono, N. Yasuda and S. Ueda, *Agricultural and Biological Chemistry*, 1977, **41**, 11, 2113.

205. J. Audet, M. Lounes and J. Thibault, *Bioprocess Engineering*, 1996, **15**, 209.

206. B. McNeil and B. Kristiansen, *Biotechnology Letters*, 1987, **9**, 2, 101.

207. A. Wecker and U. Onken, *Biotechnology Letters*, 1991, **13**, 3, 155.

208. P.A. Gibbs and R.J. Seviour, *Applied Microbiology and Biotechnology*, 1996, **46**, 5–6, 503.

209. Y. Chisti and U.J. Jauregui-Haza, *Biochemical Engineering Journal*, 2002, **10**, 2, 143.

210. J.P. Park, Y.M. Kim, S.W. Kim, H.J. Hwang, Y.J. Cho, Y.S. Lee, C.H. Song and J.W. Yun, *Letters in Applied Microbiology*, 2002, **34**, 433.

211. E. Gura and U. Rau, *Journal of Biotechnology*, 1993, **27**, 193.

212. S. Crognale, M. Bruno, M. Moresi and M. Petruccioli, *Enzyme and Microbial Technology*, 2007, **41**, 1–2, 111.

213. L. Selbmann, S. Crognale and M. Petruccioli, *Journal of Applied Microbiology*, 2004, **96**, 1074.

214. Y. Wang and B. McNeil, *Journal of Chemical Technology and Biotechnology*, 1995, **63**, 215.

215. P.A. Gibbs and R.J. Seviour, *Applied Microbiology and Biotechnology*, 1998, **49**, 2, 168.

216. P.A. Gibbs and R.J. Seviour, *Biotechnology Letters*, 1992, **14**, 6, 491.

217. T. Roukas and G. Serris, *Applied Biochemistry and Biotechnology*, 1999, **80**, 1, 77.

218. P.A. Sandford, *Advances in Carbohydrate Chemistry and Biochemistry*, 1979, **36**, 265.

219. M. Michel, R.J. Seviour and L.M. Pethica, *Biotechnology Letters*, 1987, **9**, 741.

220. C-P. Xu and J.W. Yun, *Enzyme and Microbial Technology*, 2004, **35**, 33.

221. S. Ramos and I.G. Garcia-Acha, *Transactions of the British Mycological Society*, 1975, **64**, 1, 129.

222. B.J. Catley, *Journal of General Microbiology*, 1973, **78**, 1, 33.

223. L. Simon, C. Caye-Vaugien and M. Bouchonneau, *Journal of General Microbiology*, 1993, **139**, 979.

224. S.J. Stasinopoulos and R.J. Seviour, *Biotechnology and Bioengineering*, 1990, **36**, 778.

225. J.P. Park, S.W. Kim, H.J. Hwang, Y.J. Cho and J.W. Yun, *Enzyme and Microbial Technology*, 2002, **31**, 3, 250.

226. S.J. Stasinopoulos, R.J. Seviour and D.P.F. Auer, *Letters in Applied Microbiology*, 1989, **8**, 91.

227. F.C. Yang, Y.F. Ke and S.S. Kuo, *Enzyme and Microbial Technology*, 2000, **27**, 3–5, 295.

228. K. Stahmann, K. Schimz and H. Sahm, *Journal of General Microbiology*, 1993, **139**, 2833.

229. S.M. Pitson, R.J. Seviour and B.M. McDougall, *Enzyme and Microbial Technology*, 1997, **21**, 182.

230. S.M. Pitson, R.J. Seviour and B.M. McDougall, *Mycological Research*, 1997, **101**, 2, 153.

231. Y. Göksungur, S. Dağbağli, A. Uçan and U. Güvenç, *Journal of Chemical Technology and Biotechnology*, 2005, **80**, 7, 819.

232. T.J. Pollock, L. Thorne and R.W. Armentrout, *Applied and Environmental Microbiology*, 1992, **58**, 3, 877.

233. B.C. Saha, R.W. Silman and R.J. Bothast, *Current Microbiology*, 1993, **26**, 267.

234. T.P. West and B. Reed-Hamer, *FEMS Microbiology Letters*, 1993, **113**, 345.

235. T.P. West and B. Strohfus, *Journal of Basic Microbiology*, 1996, **36**, 5, 377.

236. J.L. Goatley, *Canadian Journal of Microbiology*, 1968, **14**, 1063.

237. V.J. Bender, *Australian Journal of Biological Sciences*, 1975, **28**, 227.

238. J.A. Leal and P. Rupérez, *Transactions of the British Mycological Society*, 1978, **70**, 115.

239. G. Leal-Serrano, P. Ruperez and J.A. Leal, *Transactions of the British Mycological Society*, 1980, **75**, 1, 57.

240. A. Kardosová and J. Rosík, *Collection of Czechoslovak Chemical Communications*, 1981, **46**, 4323.

241. P. Rupérez, B. Gómez-Miranda and J.A. Leal, *Transactions of the British Mycological Society*, 1983, **80**, 313.

242. P. Rupérez, B. Gomez-Miranda and J.A. Leal, *Canadian Journal of Microbiology*, 1984, **30**, 1157–1162.

243. J.M. Sarkar, *Biotechnology Letters*, 1986, **8**, 769.

244. M.M. Corsaro, C. De Castro, A. Evidente, R. Lanzetta, A. Molinaro, L. Mugnai, M. Parrilli and G. Surico, *Carbohydrate Research*, 1998, **308**, 3–4, 349.

245. E.J. Cho, J.Y. Oh, H.Y. Changand and J.W. Yun, *Journal of Biotechnology*, 2006, **127**, 1, 129.

246. H.H. Lee, J.S. Lee, J.Y. Cho, Y.E. Kim and E.K. Hong, *Journal of Microbiology and Biotechnology*, 2009, **19**, 5, 455.

247. J-K. Yan, L. Li, Z-M. Wang and J-Y. Wu, *Carbohydrate Polymers*, 2010, **79**, 1, 125.

248. E.N. Davis, R.A. Rhodes and H.R. Shulke, *Applied Microbiology*, 1965, **13**, 267.

249. S. Kikumoto, T. Miyajima, K. Kimura, S. Okubo and N. Komatsu, *Journal of the Agricultural Chemical Society of Japan*, 1971, **45**, 162.

250. T. Miyazaki and M. Yamada, *Chemical and Pharmaceutical Bulletin*, 1971, **19**, 813.

251. A.T. Martínez, M.J. Martínez and G. Almendros, *Soil Biology and Biochemistry* 1986, **18**, 469.

252. A. Gutiérrez, M.J. Martínez, G. Almendros, F.J. González-Vila and A.T. Martínez, *Science of the Total Environment* 1995, **167**, 315.

253. B. Bes, B. Pettersson, H. Lennholm, T. Iversen and K.E. Eriksson, *Biotechnology and Applied Biochemistry*, 1987, **9**, 310.

254. K. Gomaa, J. Kraus, F. Rosskopf, H. Roeper and G. Franz, *Carbohydrate Research*, 1992, **217**, 153.

255. S. Aouadi, A. Heyraud, F. Seigle-Murandi, R. Steiman and B. Fournet, *Carbohydrate Polymers*, 1992, **17**, 177.

256. P.J. Burns, P. Yeo, T. Keshavarz, S. Roller and C.S. Evans, *Enzyme and Microbial Technology*, 1994, **16**, 566.

257. S. Kitamura, T. Hori, K. Kurita, K. Takeo, C. Hara, W. Itoh, K. Tabata, A. Elgsaeter and B.T. Stokke, *Carbohydrate Research*, 1994, **263**, 111.

258. M. Manzoni and M. Rollini, *Biotechnology Letters*, 2001, **23**, 1491.

259. C.A. Osaku, G.L. Sassaki, G.T. Zancan and M. Iacomini, *FEMS Microbiology Letters*, 2002, **216**, 145.

260. Y. Jin, L. Zhang, Y. Chen, P.C.K. Cheung and L. Chen, *Carbohydrate Research*, 2003, **338**, 1507.

261. P. Methacanon, S. Madla, K. Kirtikara and M. Prasitsil, *Carbohydrate Polymers*, 2005, **60**, 2, 199.

Acknowledgements

This review is dedicated to the memory of Professor Bruce A. Stone, whose contributions to our understanding of β-glucans were remarkable.

4 Pullulan for Biomedical Uses

Isabelle Bataille, Anne Meddahi-Pellé,
Catharine Le Visage, Didier Letourneur and
Frédéric Chaubet

4.1 Introduction

Pullulan is a commercially available polysaccharide purified from the fermentation medium of the fungus-like yeast *Aureobasidium pullulans* (originally *Pullularia pullulans*) [1]. It was first described by Bauer in 1938 [2] and its structure was investigated by Bender and co-workers [3] and Wallenfels and co-workers [4], who named it pullulan. Hayashibara Company (Okayama, Japan) started its commercial production in 1976 [5] and pullulan films were available in 1982. Pullulan has no carcinogenic, mutagenic and toxicological activities [6]. Today a film-based oral care product containing pullulan is commercialised in many countries under the brand name Listerine [7] and capsules (NPcaps® from Capsugel for instance) are used instead of gelatin for addressing a variety of cultural and dietary requirements, including those of vegetarians, diabetics and patients with restricted diets.

Pullulan has numerous uses, as a filler for low-calorie food and beverages [8], in manufacturing for the production of adhesives, cosmetics, binders, thickeners and coating agents [9–12], in electronics and optics [13–15] where it is used because of its film- and fibre-forming properties [5, 16], in chromatography as a molecular weight standard [17–19] and in pharmaceuticals [20–25]. Similar to dextran, pullulan can be used as a plasma expander [26–28]. Pullulan films as thin as 5–60 μm are obtained by drying pullulan solutions [29]. They are clear and highly oxygen-impermeable with excellent mechanical properties [30–32]. Pullulan films are considered as edible packaging and have applications in food industry [7, 33]. As a biodegradable and biocompatible biopolymer, pullulan has achieved wide regulatory acceptance with its proven safety record. In the United States, pullulan has 'Generally Regarded As Safe' status [34]. Pullulan is a unique biopolymer with many properties and hundreds of patented applications; however, its commercial underdevelopment might be due, in large part, to its relatively high price, as stated by Singh and co-workers [7]. A number of reviews on pullulan have been published in the last 15 years [7, 35–39] and we will mainly focus on the biomedical applications of pullulan after a brief introduction of its structure and rheological properties.

4.2 Structure and Rheological Properties

4.2.1 Structure and Enzymatic Degradation

The pullulan structure consists of a chain of D-glucopyranosyl units that alternate regularly between one α,1→6 and two α,1→4 linkages (**Figure 4.1**). Catley and co-workers [40] reported the presence of chain fragments resistant to the action of enzymes and such resistance was attributed to the presence of maltotetraose residues distributed randomly along the pullulan chain. These structural abnormalities may be present in the pullulan chain to a maximum extent of 7% and they do not affect the overall physicochemical properties of the polysaccharide [35, 40]. The purified pullulan is white and nonhygroscopic and decomposes at 250–280 °C [7]. It has a number-average molecular weight (M_n) of about 100–200 kDa and a weight-average molecular weight (M_w) of about 360–480 kDa, with values of polymolecularity (M_w/M_n) ranging from 2.1 to 4.1 [7, 41, 42]. It readily dissolves in hot and cold water forming viscous solutions that do not gel and shows an optical rotatory activity of +192° in a 100 g/l water solution [5, 35]. The good water solubility of pullulan can be related to the lack of crystalline zones within the polymer. Pullulan is for example much more water-soluble than amylose (which has only α,1→4 linkages) because of α,1→4 and α,1→6 bond alternation. Pullulan backbone conformation in solution is the result of α,1→6 linkages: C_1-O-$CH_2(C_6)$ linkage (α,1→6) is more flexible than C_1-O-C_4 linkage in which C_1 and C_4 belong to glucopyranose cycles. Consequently, pullulan backbone is much more flexible than amylase

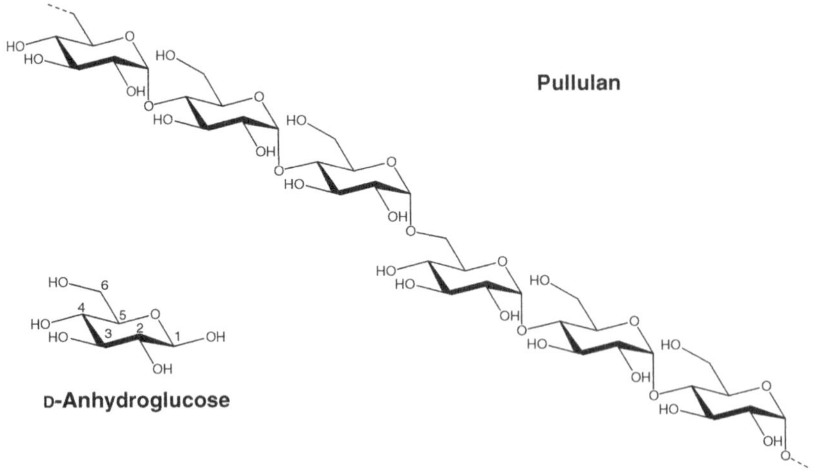

Figure 4.1 Schematic structure of pullulan. The monomer of pullulan is anhydroglucose. The numbers depict the position of the carbon atoms as defined by the nomenclature of carbohydrates

backbone and adopts a random coil conformation in aqueous solutions [43]. It is insoluble in organic solvents with the exception of dimethyl sulfoxide (DMSO) and *N,N*-dimethylformamide (DMF) [5, 44].

Pullulan is mainly obtained from *A. pullulans* and *P. pullulans*, but other pullulan-like polysaccharides have been isolated from the fungi *Tremella mesenterica* [45], *Cyttaria harioti* and *Cyttaria darwinii* [46], *Cryphonectria parasitica* [47], *Teloschistes flavicans* [48] and *Rhodotorula bacarum* [49] and they differ in the ratios of the component glycosidic linkages [46, 48, 50]. The complete assignment of carbon and proton resonances of pullulan from *A. pullulans* has been published by Arnosti and Repeta in 1995 [51]. Enzymes purified from bacteria and fungi have been used to investigate the structure of pullulan. The products of the enzymatic hydrolysis of pullulan are presented in **Table 4.1**. Four main different types of enzymes can degrade pullulan [52]: (1) glucoamylases (EC 3.2.1.3) or 1,4-α-D-glucan glucanohydrolases, which hydrolyse pullulan from nonreducing ends to produce glucose; (2) pullulanases (EC 3.2.1.41) or α-dextrin 6-glucanohydrolases, which hydrolyse the α,1→6 glucosidic linkages to produce maltotriose; (3) isopullulanases (EC 3.2.1.57) or pullulan 4-glucanohydrolases, which hydrolyse the α,1→4 glucosidic linkages to produce isopanose (6-*O*-α-maltosyl-glucose); and (4) neopullulanases (EC 3.2.1.157) or pullulan 4-D-glucanohydrolases, which hydrolyse the α,1→4 glucosidic linkages to produce panose (6-*O*-α-glucosyl-maltose). All enzymes that hydrolyse pullulan have been found in bacteria with the exception of isopullulanases, which are produced by only fungi [53]. A fifth type named pullulan hydrolase type III was more recently isolated from the hyperthermophilic archaeon *Thermococcus aggregans* [54]. Glucoamylases are exoglucanases, and are also known as glucan-1,4-α-glucosidase, amyloglucosidase, exo-1,4-α-glucosidase, γ-amylase, lysosomal α-glucosidase and acid maltase [52]. They cleave off terminal 1→4-linked α-D-anhydroglucose residues from the nonreducing end of the chain releasing β-D-glucose. Pullulanases can be subdivided into pullulanases I and II. Type I pullulanases, also named true pullulanases, are debranching enzymes and were first used by Bender and Wallenfels in 1961 [55] and characterised later [56]. They specifically hydrolyse the α,1→6-D-glucosidic linkages in pullulan and in branched oligosaccharides to maltotriose and linear oligosaccharides, respectively, by an end mechanism of action [57]. Type II pullulanases, or amylopullulanases, have both α-amylase and pullulanase activity, cleaving also α,1→4 and α,1→6 glucosidic linkages in starch and related polysaccharides. Isopullulanases (also called pullulan 4-glucanohydrolase) are type II pullulan hydrolases and they hydrolyse α,1→4-D-glucosidic linkages of pullulan to produce isopanose, but have no effect on starch. Neopullulanases are referred as type I pullulan hydrolases [52].

Bruneel and Schacht [58] investigated the degradation of pullulan and pullulan derivatives by α-amylase, β-amylase, lysosomal enzymes (tritosomes), calf serum

Table 4.1 The four types of enzymes that hydrolyse pullulan and the reaction products from pullulan represented schematically

	...→A→B→C→A→B→C→A→B→C→ ... (pullulan)
Glucoamylases (EC 3.2.1.3)	... A B C A B C A B C ... (D-glucose)
Pullulanases (EC 3.2.1.41)	... A→B→C A→B→C A→B→C ... (maltotriose)
Isopullulanases (EC 3.2.1.57)	... B→C→A B→C→A ... (isopanose)
Neopullulanases (EC 3.2.1.157)	... C→A→B C→A→B ... (panose)

Arrows point out the glycosidic linkages. A→B→C corresponds to a maltotrisose unit: Glc-α-(1,4)-Glc-α-(1,4)-Glc-α-(1,6)
Adapted from M. Doman-Pytka and J. Bardowski, Critical Reviews in Microbiology, 2004, **30**, *2, 107 [52]*

and liver homogenate. Pullulan derivatives were pullulan-[*N*-2(hydroxypropyl)] carbamate, pullulan monosuccinate esters and pullulan successively oxidised with periodate and reduced, which were synthesised as previously described [59–61]. Pullulan fractions with molecular weights ranging from 32 to 100 kDa were used. Different degrees of derivatisation were obtained and submitted to the action of the enzymes. The results indicated that both native and modified pullulan samples were slowly degraded by α-amylase, depending on the degree of derivatisation and the nature of the substituent, as pullulan was almost completely hydrolysed by β-amylase (although Vihinen and Mäntsälä [62] stated that pullulan could not be degraded either by α-amylase or by β-amylase). Eventually, it was found that pullulan was much slowly degraded by tritosomes or calf serum as compared to amylases. More recently, Ball and co-workers [63] studied the effect of substitution at C_6 on the enzymatic degradability of the modified pullulan. Hydroxyls on the C_6 position were substituted by chlorine (6-chloro-6-deoxypullulan) or azide (6-azido-6-deoxypullulan). Conformational flexibility of glucosyl residue allowed some internal substitution of chlorine or azide residues by the hydroxyl on C_3. 6-Chloro-6-deoxypullulan and 3,6-anhydropullulan were highly resistant to hydrolysis by the four different types of pullulanase. 6-Azido-6-deoxypullulan was resistant to three types but susceptible to hydrolysis by the fourth enzyme, isopullulanase. Neopullulanase was strongly inhibited by 6-chloro-6-deoxypullulan and 6-azido-6-deoxypullulan, and the other pullulanases much less so. In an independent study, Ohe and co-workers [64] confirmed the inhibition of pullulanase by 3,6-anhydropullulan derived from C_6-sulfated pullulans.

4.2.2 Rheology of Pullulan Solutions and Films

Pullulan fermentation broths may generally be considered as non-Newtonian fluids, which may be modelled by a power law [65]. Consequently, pullulan viscosity itself is a key parameter during pullulan production. Because of viscosity changes, oxygen and mass transfer modifications that occur during fermentation may influence the properties of the final product [66, 67]. By combining the techniques of confocal microscopy and optical tweezers, an image of the viscosity distribution around a pullulan-producing cell of *A. pullulans* can be obtained [68]. A compromise has thus to be found between polysaccharide quality, namely its final viscosity, and the produced quantity. For example, mixing device geometry may influence both parameters, all things being equal [69]. Once isolated from its fermentation broth and purified, pullulan exhibits a viscous behaviour in aqueous solutions. Diffusing wave spectroscopy of pullulan solutions ($M_w = 10^5$ g/mol, concentrations up to 400 g/l) evidenced a marked viscous behaviour ($G'' > G'$) and critical concentrations C^* and C^{**} were evaluated [70]. C^* is considered as the first critical overlap concentration above which polymer molecules become interpenetrated. C^{**} may be represented as the transition between the semidilute and the concentrated domains. C^* decreases in the case of gamma-ray irradiation of pullulan, since pullulan acquires the solution properties of a polyanion [71].

Pullulan solutions contain short microfibrils, as observed by electron microscopy [72]. Pullulan solutions are able to stabilise turmeric oleoresin emulsions for applications in food industry [73]. In the same field, pullulan enhances the viscosity of frozen sucrose solutions [74]. However, unlike xanthan, pullulan has only a slight effect on the stabilisation of cottonseed protein isolate emulsions [75]. Two-phase systems based on pullulan/sodium dodecyl sulfate mixtures containing sodium chloride were investigated by means of rheology coupled with optical observations. Morphological observations (from droplets to co-continuous structures and strings) could thus be correlated to rheological data under shear [76, 77].

Pullulan molecular characteristics such as molecular weight and molecular size distribution seem to largely influence solution rheology. These parameters have an impact on the mechanical behaviour of pullulan films obtained by casting solutions of pullulan/plasticiser mixtures and subjected to large strains during mechanical testing [78]. Pullulan is well known for its excellent film-forming properties [79, 80]. This ability of membrane forming led to pullulan being used as a matrix in ion-exchange membranes after crosslinking with isocyanate or glutaraldehyde [81]. Films were also prepared from cyanoethylpullulan/cyanoethyl poly(vinyl alcohol) (PVA) mixtures co-crosslinked with poly(vinylidenefluoride-hexafluoropropylene). The presence of pullulan/PVA mixture enhanced the overall properties (i.e., thermal stability and dynamical mechanical properties) of the films [82]. Crosslinking between pullulan and PVA was successfully performed using glyoxal in DMSO and the films obtained were homogeneous and had higher tensile strengths and moduli than the blends [83]. Hydroxypropylmethylcellulose (HPMC)/pullulan blends allowed the formation of films with good thermal and mechanical properties as long as pullulan and HPMC are miscible. This condition is satisfied when the HPMC content is more than 50% in the blend [84].

4.3 Biological Properties of Pullulan and Some Derivatives in Solution

As pullulan shares similar features with dextran, some studies have been carried out with pullulan to develop blood plasma substitutes [26]. Seibutsu and Kenkyujo [28] claimed that pullulan with a narrow molecular weight distribution with $M_w/M_n = 1.2$ and M_w about 60 kDa was suitable for intravenous injection, since Igarashi and co-workers [27] had defined the proper molecular weight as ranging from 30 to 90 kDa. More recently, Shingel and co-workers [35, 71, 85, 86] reported narrow molecular weight anionic fractions from gamma-irradiated pullulan and subsequent efforts to validate their use as blood plasma substitutes. Early experiments performed in a dog model showed a rapid recovery of the main haemodynamic parameters when blood loss was made up for the same volume of polysaccharide fraction [35]. The main difficulties in the development of pullulan as plasma substitute are a dramatic increase in the blood pressure for concentrations above 60 g/l (precluding the use of fractions with molecular weight above 150 kDa) and a rapid hepatic clearance of pullulan fractions with molecular weight lower than 15 kDa [35]. Indeed, compared to dextran, low-molecular-weight fractions have a very short half-life due to a quick hepatic uptake. This uptake markedly decreases when pullulan is co-administered with asialofetuin and arabinogalactan evidencing an endocytosis via asialoglycoprotein receptors [87–93]. This was confirmed by the investigations of Tanaka and co-workers [94], who compared the internalisation of arabinogalactan, pullulan, dextran and mannan in rat liver parenchymal and nonparenchymal cells using ^{125}I- or fluorescein isothiocyanate-labelled polysaccharides [94]. Dextran uptake occurs via fluid-phase endocytosis, arabinogalactan and pullulan are internalised to liver parenchymal cells, whereas mannan is internalised to nonparenchymal cells indicating that receptor-mediated endocytosis plays a role in the biodisposition of these polysaccharides as drug carriers.

4.3.1 Chemical Modifications

Pullulan can be easily derivatised in order to extend its applications by grafting different chemical structures on the backbone. Nine hydroxyl groups are available for substitution reactions on the repeating unit. They are distinguished by their position on the glucosidic moiety (**Figure 4.1**): OH-2, OH-3, OH-4 and OH-6, with the ratios 3, 3, 1 and 2, respectively. Moreover, their relative reactivities also depend on the polarity of the solvent and of the reagents. Pullulan hydroxyls were submitted to numerous chemical reactions leading to a large number of derivatives; the schematic structures of the derivatives are summarised in **Table 4.2**. Since Tian and co-workers [135] performed the graft copolymerisation of butyl acrylate onto pullulan in 1992, pullulan has also been modified by substituting OH groups with monomers such as methacrylate, acrylamide, L-lactide or oxazoline derivatives in order to prepare copolymerised hydrogels endowed with tunable amphiphilic, ionic, thermoresponsive or chaperone-like properties to develop drug delivery carriers [131–134, 136–138]. Some studies are presented later.

4.3.1.1 Carboxymethylation

Carboxymethylation is by far the most widespread reaction performed on neutral polysaccharides in order to allow further chemical modifications or to favour solubilisation in aqueous solutions. Hydroxyl groups are activated as alcoholate in an alkaline aqueous solution to allow the nucleophilic substitution of chloride from monochloroacetic acid. In general, degree of substitution (DS) up to 1.0 can be obtained in one step, and it is necessary to proceed in several steps to reach higher DS. Glinel and co-workers performed the carboxymethylation of pullulan in an alcohol-water mixture leading to carboxymethylpullulans (CMP) with a DS from 0.6 to 1.2. A wide set of CMP oligomers obtained from acid hydrolysis were studied by high-resolution nuclear magnetic resonance (NMR) spectroscopy [99]. The DS and the relative distribution of carboxymethyl groups at OH-2, OH-3, OH-4 and OH-6 of glucose residues were determined by ^{1}H-NMR measurements. The authors observed, as for dextran, that the substitution at C_2 was predominant, and decreased in the order $C_2>C_3>C_6>C_4$. Taking into account the availability of each OH group in the parent pullulan, the order of relative reactivity of hydroxyl groups was found to be OH-2 > OH-4 > OH-6 > OH-3.

Several studies have been conducted to understand and to optimise the behaviour of self-assembled or crosslinked hydrophobised CMP in water solution [106, 111, 130, 139–144]. The thickening properties of pullulan can be improved after carboxymethylation and esterification with alkyl chains. Thus, hydrophobically modified CMP samples were prepared by reaction of small amounts of C_{16} alkylamine on carboxylic groups of the corresponding polyacid. In dilute solution, polymers aggregated intermolecularly and displayed a compact globular structure. A collapse of the aggregates occurred by the addition of NaCl or ethanol [130].

Table 4.2 Schematic chemical structures of the most common pullulan derivatives		
Type of reaction	**Schematic chemical structure of the substituted pullulan (P-OH)**	**References**
Etherification	P-O-CH_3 (permethylation)	[95]
	P-O-$(CH_2)_{2-3}$-CH_3 (alkylation)	[96–98]
	P-O-CH_2-COOH (carboxymethylation)	[99]
	P-O-$(CH_2)_{2-3}$-CH_2-NH_3^+ (cationisation)	[100, 101]
	P-O-CH_2-CH_2-CN (cyanoethylation)	[15, 102]
	P-O-$(CH_2)_{1-4}$-Cl (chloroalkylation)	[103]
	P-O-CH_2-CH_2-(S=O)-CH_3 (sulfinylethylation)	[104]
	P-O-CH_2-CH_2-CH_2-SO_3Na	[105]
	P-O-CH_2-CH_2-N$(CH_2CH_3)_2$	[106, 107]
	P-O-CH_2-CH_2-$N^+(CH_2CH_3)_2$-CH_2-CH_2-N$(CH_2CH_3)_2$	[106, 107]
Esterification	P-O-CO-CH_3 (acetylation)	[108–110]
	P-O-CO-$(CH_2)_{2-14}$-CH_3 (alkoylation)	[111]
	P-O-CO-CH_2-Cl (chloroacetylation)	[103]
	P-O-CO-CH_2-CH_2-COOH (succinoylation)	[60]
	PA-O-CO-CH_2-CH_2-CO-sulfodimethoxine[a]	[108]
	P-O-CO-CH_2-CH_2-CO-cholesterol[b]	[112]
	P-abietate	[113]
	P-stearate	[114]
	PA-folate[a]	[115]
	P-cinnamate	[116]
	P-biotin	[117]
	P-O-SO_2-CH_3	[118]
Type of reaction	**Schematic chemical structure of the substituted pullulan (P-OH)**	**References**
Urethane derivatives	P-O-CO-NH-CH_2-CH(OH)-CH_3 P-O-CO-NH-CH_2-CH_2-NH_3^+	[60]
	P-O-CO-NH-R (R = phenyl or hexyl)	[119]
	P-O-CO-NH-phenyl	[120]
Urethane derivative/ amidification	P-O-CO-NH-$(CH_2)_6$-NH-CO-cholesterol	[121]
Chlorination	P-CH_2-Cl (C_6 substitution)	[63, 118, 122]

Table 4.2 Continued

Sulfation	P-O-SO_3Na	[123–125]
Azido-pullulan	P-CH_2-N_2	[63]
Oxidation	P-COOH (C_6 oxidation)	[126, 127]
	Glycosidic ring opening (periodate oxidation)	[59]
CMP/hydrazone derivatives[c]	P-O-CH_2-CO-NH-doxorubicin	[128]
	P-O-CH_2-CO-NH-antibody	[129]
CMP/amidification[c]	P-O-CH_2-CO-NH-CH_2-$(CH_2)_{14}$-CH_3	[130]
Copolymerisation	P-O-CH_2-O-CO-C$(CH_3)_2$-R with R = poly(methacrylate), poly(methylmethacrylate), poly(hydroxyethylmethacrylate) and poly(sulfopropylmethacrylate)	[131] [132, 133] [134]
	P-O-CH_2-O-CO-C$(CH_3)_2$-R with R = poly(butylmethacrylate)	[135]
	P-O-CH_2-O-CO-C$(CH_3)_2$-R with R = poly(*N*-isopropylacrylamide)	[131, 136]
	P-O-[CO-CH(CH_3)-O]$_N$-H (polylactide)	[137]
	P-O-poly(2-isopropyl-2-oxazoline)	[138]

[a]*Derivatives prepared from pullulan acetate (PA). Substitution occurred on some of the remaining free hydroxyl groups*

[b]*Derivative prepared from succinoylated pullulan. Substitution occurred on carboxylic acid*

[c]*Derivatives prepared from carboxymethylpullulan (CMP). Substitution occurred on carboxylic groups*

In the last years, the tuning of the sequestering properties of such pullulan-based macromolecular structures has been examined for application in the pharmaceutical field of drug delivery. CMP grafted with alkyl chains have been prepared differing in the length of the chains (C_8–C_{16}) and in the DS of carboxymethyl and alkyl residues on the anhydroglucose units. The capacity to solubilise nonpolar molecules was demonstrated even with low DS in alkyl chains [111]. At the same time, a set of self-assembling CMP substituted with C_8 alkyl chain were found to be able to solubilise docetaxel *in vitro*. Cytotoxicity studies revealed that the CMP-C_8-docetaxel complex was equipotent to the commercial docetaxel against cancer cells. Furthermore, in the absence of the drug, CMP-C_8 appeared less cytotoxic against macrophages than the Tween® 80/ethanol-water formulation used as commercial docetaxel vehicle [141].

Novel ionic and amphiphilic CMP-based hydrogels were also synthesised and evaluated for controlled-release entrapment applications. Studies on these macromolecular structures evidenced, in addition to the chemical links, the existence of inter- and/or intra-molecular interactions in aqueous solutions of amphiphilic and modified CMP that were capable of inducing specific conformations at the air-solution interface and in the bulk, where hydrophobic nano- and micro-domains may be formed [139, 140]. Dulong and co-workers prepared pullulan-based hydrogels sensitive to ionic strength and pH from crosslinked CMP by the reaction of a carboxylate group with an alcohol function of the polysaccharide and subsequent substitution with octyl chains. The grafting degree influenced the conformation of the modified polymer in solution and led to the formation of hydrophobic clusters in the hydrogels. The loading of a hydrophobic molecule was controlled by the grafting degree of the hydrogels [142]. The physicochemical properties (viscosity, surface tension, zeta potential, size measurement) of pullulan, diethyl amino ethyl (DEAE)-pullulan and alkylated derivatives were studied in aqueous salt-free solutions at pH 3, 8 and 11 [145]. DEAE-pullulan presented a strong polyelectrolyte character at pH values below 8, which evolved to a marked amphiphilic behaviour in alkaline media, as alkylated DEAE-pullulan derivatives were strongly amphiphilic regardless of the pH. The low intrinsic viscosity values (Fuoss and Fedors models) evidenced the degradation of the cationic amphiphilic pullulan derivatives. Glinel and co-workers obtained compact nanoparticles from hydrophobically modified pullulans in two ways: (1) neutral derivatives were obtained by direct esterification of pullulan with a perfluoroalkyl carboxylic acid (C_8F_{17}-CH_2CH_2COOH) and (2) ionic derivatives were obtained by amidation of CMP with two perfluoroalkylamines (C_7F_{15}-C_8F_{17}-CH_2-NH_2). The molar hydrophobic contents ranged from 1.1 to 4.8% with respect to the anhydroglucose units [144]. The strength of the self-association of the modified polymers could be tuned by the length of the hydrophobic moiety. As stated by Shingel [35], such pullulan conjugates have the ability to transport oxygen, similar to perfluoro-based polymers, and also exhibit biocompatibility in new blood plasma substitutes, similar to pullulan.

The capacity of hydrophobised pullulan and CMP derivatives to form multilayer films was investigated not only in the field of drug delivery, but also to prepare biocompatible surfaces. Guyomard and co-workers investigated the loading and release properties towards a hydrophobic dye of multilayer films based on interpolyelectrolyte interactions between poly(ethyleneimine) and hydrophobically modified CMP derivatives. The loading capacity of the films as well as their release behaviour could be tuned by varying the grafting degree of CMP chains with decyl groups and the composition of the surrounding solution [146]. Besides, CMP and chitosan were grown by layer-by-layer assembly onto brushes of magnetic nanowires endowed with tailored surface properties [147].

CMP can also be considered as a promising polymeric carrier for various drugs since an introduction of negative charges into pullulan backbone results in its prolonged retention within the organism [92]. Horie and co-workers carried out studies on CMP conjugates with sialyl LewisX (CMP-SLe(x)). In an acute inflammation situation in mice, they observed an accumulation of the conjugates in spleen and lymph nodes in contrast to CMP conjugated to other oligosaccharides [148]. The analysis of the distributions of the different CMP conjugates in the tissues evidenced a strong binding of CMP-SLe(x) to the inflamed area and a high selectivity for E-selectin [149]. When the DS in carboxymethyl groups was higher than 0.5, preferential accumulation in the lymph nodes as well as the spleen was observed in a rat model. This effect dramatically decreased with a low DS (0.025) and with structurally modified SLe(x). The authors concluded that CMP-SLe(x) could be used as a drug delivery carrier to target drugs to the peripheral lymphoid tissues [150]. The antitumour effect of pullulan-doxorubicin micelles and CMP-doxorubicin (CMP-Dox) conjugates was investigated by Nogusa and co-workers [151–154] in rats bearing Walker 256 carcinosarcoma and Yoshida sarcoma. In his survey, Mehvar [155] presented this work as an example of applications of polysaccharides as drug delivery carriers. On the one hand, the authors demonstrated that pullulan-doxorubicin micelles had higher affinities towards tumour cells than doxorubicin alone, providing for a direct release of the drug at a high level. On the other hand, they showed a control of the drug release from the CMP-Dox by adjusting the length and the composition of the oligopeptidic spacer between the carboxyl group of CMP and the amino group of doxorubicin [151–153]. More recently, Lu and co-workers [128, 156] investigated the cytotoxicity of CMP-Dox pH-sensitive nanoparticles to 4T1 mouse breast cancer cells. Conjugates were obtained by the formation of hydrazone bond between the amine group of doxorubicin and hydrazide-activated CMP. *In vitro*, the release of doxorubicin from nanoparticles occurred faster at pH 5.0 than at pH 7.4. Eventually CMP-Dox nanoparticles showed comparable cytotoxicity effect to that of free doxorubicin in 4T1 cells.

Masuda and co-workers [157] demonstrated the potential of CMP as a carrier for targeting immune tissues in a rat model of adjuvant arthritis, allowing one to design novel CMP immunoconjugates. Wotschadlo and co-workers [158] have investigated the interactions between magnetic nanoparticles coated with carboxymethylcellulose (CMC), carboxymethyldextran (CMD) and CMP, and tumour cells and leucocytes. CMC and CMP showed a constant interaction rate with both cell types, whereas CMD exhibited a rapid interaction kinetics with tumour cells but a reduced one with leucocytes as target. They concluded that the interactions with the cells were depending on the backbone of the carboxymethylated polysaccharides used as shell material.

4.3.1.2 Sulfation

Pullulan was sulfated with the aim of developing new alternatives to heparin, which suffers from its heterogeneous and variable structure, the animal origin and multiple *in vivo* effects. In 2001, two independent studies were published about the sulfation process on pullulan as compared to dextran, and optimised conditions to reach DS in sulfate groups up to 2.00 were established [123, 159]. The final characteristics of the sulfated pullulans (PS) depend to a great extent on the temperature, the solvent, the duration of the reaction and the sulfation reagent used. The DS of sulfated pullulans was higher than that of dextran sulfates. This was attributed to the presence of a higher amount of primary hydroxyl groups in the structure of pullulan. With ^{13}C-NMR spectroscopy Mähner and co-workers [159] showed a homogeneous distribution of the sulfate along both polysaccharidic backbones. Later, Alban and co-workers [124] confirmed these results by determining that the sulfation of the OH groups occurred in the order $C_6 > C_2 > C_3 > C_4$ irrespective of the molecular weight of pullulan, the procedure adopted and the DS. They obtained sulfated pullulans by stepwise sulfation of pullulans with SO_3-pyridine complex in DMF at 75 °C and 95 °C for 3–8 hours; pullulans with molecular weights of 50 kDa (soluble in DMF) and 200 kDa (insoluble in DMF) were used. DS in sulfate groups up to 1.99 were reached after activation of the polymer solution/suspension by ultrasound or with an alkali complex. The anticoagulant activity of the sulfated pullulans determined by the coagulation assays, such as prothrombin time, activated partial thromboplastin time (APTT), Heptest® and thrombin time (TT), increased with increasing DS and molecular weight. The action profile of sulfated pullulans changed in accordance with the overall DS in sulfate groups, but also with increasing sulfate groups in positions 2, 3 and 4 (defined as SS) as reflected by the ratio of the TT to the APTT activity (**Figure 4.2**). The authors observed that sulfated pullulans showed the best effect in the TT followed by that in the APTT, compared to unfractionated heparin, which had the same specific activity in both assays. Moreover, sulfated pullulans did not inhibit factor Xa either directly or when mediated by antithrombin as reflected by nonsignificant effect in the Hepest® assay.

The anti-inflammatory activities of heparin have partly been related to its capacity to inhibit the binding of leucocytes to endothelial P-selectin. The capacity of pullulan sulfate to bind to P-selectin was investigated in a study that compared 6 heparins with 15 structurally defined semisynthetic sulfated glucans derived from phycarin, curdlan and pullulan [160]. The inhibitory capacity was analysed in a parallel-plate

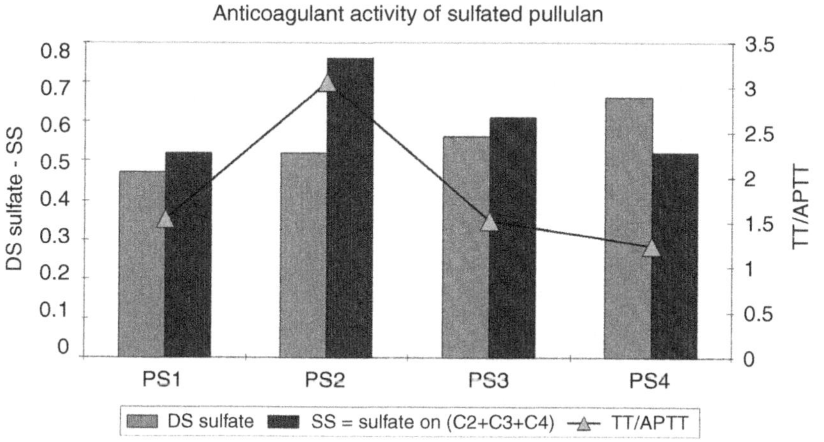

Figure 4.2 Degree of sulfation (DS sulfate) and sulfation pattern (SS) dependence of the TT/APTT ratio as demonstrated by the pullulan sulfates: sulfated pullulans 1 (DS = 0.47; SS = 0.52), sulfated pullulans 2 (DS = 0.52; SS = 0.76), sulfated pullulans 3 (DS = 0.56; SS = 0.61) and sulfated pullulans 4 (DS = 0.66; SS = 0.52). As a comparison, the TT/APTT ratio of unfractionated heparin was 1.0. Adapted from S. Alban, A. Schauerte and G. Franz, *Carbohydrate Polymers*, 2002, **47**, 3, 267 [124]

flow chamber, by observing the rolling of U937 cells on P-selectin layers. In the comparison of glucan sulfates, charge density was found to be the most important parameter for P-selectin binding, with highly sulfated glucans showing excellent inhibitory properties; variation in molecular weight was of minor importance while glycosidic backbone linkage had significant effects.

4.4 Pullulan-Based Hydrogels as New Biomedical Materials

In the last 30 years, polysaccharide-based hydrogels have demonstrated tremendous potential applications in the medical and pharmaceutical fields as drug carriers, in the development of biomimetics, biosensors, artificial muscles and environmentally responsive or smart materials and in chemical separations [30]. As pullulan is by nature not a gelling polysaccharide, an appropriate chemical modification of its backbone is necessary to get macromolecular systems capable of forming hydrogels.

Interactions between pullulan and other macromolecules have been investigated to obtain hydrogels endowed with new physicochemical properties. Without any crosslinking agent, stable hydrogels were obtained by mixing cationic aminated pullulan with anionic CMC. Unlike scleroglucan, aminated pullulan promoted CMC gelation, due to its flexibility [100]. Gradwell and co-workers [161] considered the adsorption of pullulan abietate on regenerated cellulose thin films as a way of permanent surface modification of cellulose fibres to mimic wood composites. Another study mentions the use of carboxylated pullulan to regulate protein (namely β-lactoglobulin) adsorption to air-water interfaces through the formation of a protein/polysaccharide complex [162]. A parallel was evidenced between charge density of carboxylated pullulan and adsorption kinetics as well as rheological behaviour of protein/polysaccharide complex.

Solution properties and gelation of chemically modified and hydrophobised pullulan have been extensively reported for the preparation of gels as microparticles. The use of chemical crosslinkers remains the most widespread method to get pullulan-based macroscopic hydrogels.

4.4.1 From Nanogels ...

Numerous papers deal with pullulan hydrogels as drug delivery systems, particularly in the form of microgels and nanogels. Obvious therapeutic benefits can be achieved by slow release of drugs into the plasma, and thus altering the concentration profiles of the drugs. Hydrogel nanoparticles of crosslinked pullulan with glutaraldehyde have been prepared in order to develop a DNA carrier system, improving gene loading efficiency, controlled-release properties, biocompatibility and enhanced stability [163]. Siloxanic units were used to cross link CMP leading to porous microparticles [164]. Electrostatic and/or hydrophobic interactions between the microparticles and lysozyme followed a Langmuir isotherm adsorption behaviour. Propanolol and quinidine were considered as potential drugs that could be delivered by such systems [165]. CMP or sulfopropylpullulan has been co-crosslinked with cyclodextrin using the same siloxanic units as previously reported [164] to obtain hydrogel microparticles that can sequester organic molecules such as water pollutants, drugs and enzymes [105].

However, pullulan hydrophobised by a modification of the backbone constituted the main type of derivative for forming gels through low-energy interactions of macromolecular structures with hydrophobic associative domains [108, 166, 167]. Without crosslinkers, these domains self-associate into stable nanocolloids with an inner hydrophobic core [168, 169]. The self-assembly method based on associating polymers is an efficient and versatile technique for the preparation of functional nanogels and hydrogels. In particular, amphiphilic pullulans obtained from cholesteryl, acetyl or chloroacetyl graftings onto the hydroxyl groups form nanogels that are able to trap hydrophobic molecules, proteins or peptides, and nucleic acids. Moreover, hydrophobised pullulan-based nanogels interact also with various molecular assemblies such as liposomes and oil-water emulsions [170]. They are stable and their size does not change even after a week at room temperature [35]. As a consequence, hydrophobised pullulan conjugates were used as drug-targeting carriers for bioactive substances, such as metronidazole, nicotinic acid, sulfathiazole, mitoxantrone or epirubicin [112, 115, 171], through their binding to various hydrophobic substances and soluble proteins [108, 166, 172–175], as well as in biotechnology as artificial molecular chaperones in the presence of β-cyclodextrin [133, 176–179] or for the formation of hybrid nanogels [180, 181]. They can be used as polymeric nanocarriers in cancer chemotherapy [169, 170, 172, 182–189], protein stabilisation [190, 191] and artificial vaccines [183, 192–194]. Stimuli-responsive nanogels such as pH-responsive [174, 175, 193, 195], thermoresponsive [138, 193, 196, 197] and photoresponsive [180, 198] nanogels were also designed using a similar self-assembly method.

The most cited paper in the field of hydrophobised pullulan reports the early works of Akiyoshi's research group about the self-assembly of cholesteryl-bearing pullulan (CHP) as stable hydrogel nanoparticles [121]. CHP complexes consisting of polysaccharide fractions of different molecular weights and cholesteryl moiety of different DS were synthesised providing self-aggregates in water (which were regarded as hydrogel nanoparticles) in which pullulan chains were noncovalently crosslinked by associating cholesteryl moieties (**Figure 4.3**).

First experiments showed the formation of a complex with various globular and soluble proteins such as haemoglobin, peroxidase, myoglobin and cytochrome *c* [200, 201]. The nanoparticles also showed an excellent colloidal stability and almost no dissociation of the protein from the complex. Solution properties in water of CHP containing 1.6 cholesterol groups per 100 glucose units were investigated. Relatively monodispersive spherical particles upon ultrasonication with a diameter of about 25 nm and a hydrodynamic radius of 13 nm were observed and characterised by light scattering, electron microscopy, ^{1}H-NMR and fluorescence spectroscopy. Existence of microdomains consisting of both the rigid core of hydrophobic cholesterol and the hydrophilic polysaccharide shell was suggested from spectroscopic data and the incorporation of several hydrophobic fluorescent probes in the particles. The CHP self-aggregates are strongly complexed with hydrophobic and less hydrophilic fluorescent probes [121]. When the DS of the cholesteryl moiety was increased, they observed a decrease in the size of the self-aggregates without any change in the aggregation number of CHP in one nanoparticle, and a decrease in the temperature induced a structural change in the nanoparticles. Insulin subsequently complexed onto the nanoparticles was significantly protected from aggregation and also from thermal denaturation and enzymatic degradation [188]. Moreover, the physiological activity of complexed insulin was preserved *in vivo* after intravenous injection [169]. In further studies, the irreversible aggregation of carbonic anhydrase B upon heating was completely prevented by complexation between the heat-denatured enzyme and the nanoparticles. The complexed carbonic anhydrase B was released by the

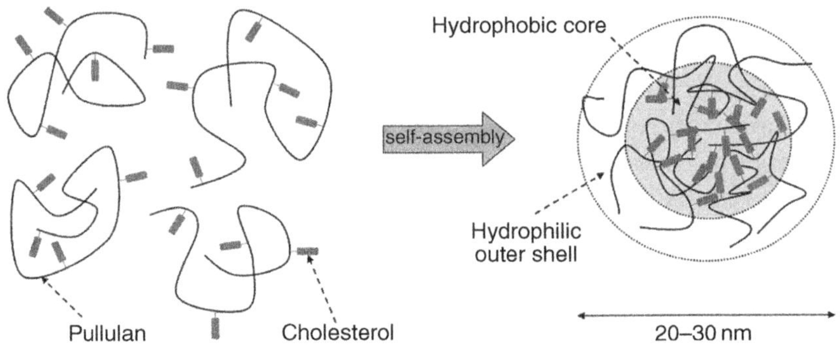

Figure 4.3 Schematic representation of the formation of cholesterol-pullulan conjugate-based nanoparticles by self-aggregation in aqueous solution. There are about 100 glucose units between cholesterol moieties on the pullulan backbone. Reproduced with permission from I. Lee and K. Akiyoshi, *Biomaterials*, 2004, 25, 15, 2911. ©2004, Elsevier

dissociation of the self-aggregates and it refolded to its native form upon the addition of β-cyclodextrin according to a mechanism similar to that of a molecular chaperone. The activity of the released carbonic anhydrase B was maintained [167]. Similar results have been obtained more recently where molecular chaperone-like activity of CHP was evidenced for the stabilisation and the refolding of citrate synthase [190]. Thermoresponsive nanoparticles were prepared by self-assembly of two different hydrophobically modified polymers, CHP and a copolymer of *N*-isopropylacrylamide and *N*-[4-(1-pyrenyl)butyl]-*N*-*n*-octadecylacrylamide, via their hydrophobic moieties [197], as well as hexadecyl group-bearing pullulan self-assembly nanoparticles [202]. Galactoside-conjugated CHP nanoparticles were specifically internalised by rat hepatocytes and HepG2 cells. Tissue distribution of the CHP self-aggregates was highly dependent on the chemical conjugation of the galactose moiety [203]. Interestingly, galactoside-bearing nanoparticles were specifically accumulated in rat hepatocytes compared to unmodified CHP, indicating that in this situation the modified pullulan moiety was no more able to interact with asialoglycoprotein receptors as was native pullulan. The effect of the chemical modification of pullulan on its affinity for liver cells was also observed by Masuda and co-workers [157] in rats treated with intravenous injection of CMP. Carboxymethylation modified the selectivity of the macromolecule from liver (pullulan) to spleen and blood (CMP). Moreover, 24 hours after the injection, accumulation of CMP in lymph nodes was significantly greater than that of the native pullulan.

CHP nanogels also have applications in the prevention of bone resorption when complexed to the W9-peptide, a tumour necrosis factor-α and receptor activator of nuclear factor-κB ligand antagonist that prevents the reduction in bone mineral density [204]. They have been used to ensure the slow release of interleukin-12 *in vivo*, evidencing their usefulness for a cytokine immunotherapy of malignancies [205]. Eventually, recent studies have shown that cholesterol-bearing pullulan would be of great interest in the fight against Alzheimer's disease [206, 207]. The formation of fibrils by amyloid P-protein (Aβ protein) is considered as a key step in this pathology. CHP nanogels as artificial chaperones inhibit the formation of Aβ-(1–42) fibrils with marked amyloidogenic activity and cytotoxicity. The CHP nanogels incorporated up to 6–8 Aβ-(1–42) molecules per particle and induced a change in the conformation of Aβ from a random coil to α-helix- or β-sheet-rich structure. This structure was stable and the aggregation of Aβ-(1–42) was suppressed. Furthermore, the dissociation of the nanogels caused by the addition of methyl-β-cyclodextrin released monomeric Aβ molecules. Besides, nanogels composed of amino-group-modified CHP with positive charges under physiological conditions had a greater inhibitory effect than CHP nanogels, suggesting the importance of electrostatic interactions between primary ammonium and Aβ for inhibiting the formation of fibrils. The interaction of CHP nanogels with oligomeric forms of Aβ-(1–42) led to a reduction of the cytotoxicity of Aβ-(1–42) in primary cortical cells and microglial cells. These results suggest that, more widely, CHP nanogels could provide a valid complementary approach to antibody immunotherapy in neurological disorders characterised by the formation of soluble toxic aggregates, such as Alzheimer's disease.

More recently, CHP nanogels were used as a new vehicle for oro-digestive vaccine therapy, which is a developing area of interest in needle-free vaccine. Intranasal injection of antigen alone does not induce a high level of antigen, and adjuvants are always needed. The adjuvants are generally poorly tolerated and the objective is to develop an antigenic protein delivery system for adjuvant-free vaccines by intranasal route [208]. A nontoxic subunit fragment of *Clostridium botulinum* type A neurotoxin BoHc/A administered intranasally with cationic type of cationic CHP nanogel was taken up by mucosal dendritic cells and induced antibody responses without co-administration of mucosal adjuvant [208]. In the same way, a chemically modified pullulan (palmitoyl derivative (O-palmitoylpullulan))-entrapped bovine serum albumin (BSA) as an antigen model was used for liposome coating. The polysaccharide could produce better IgG and IgA titre levels as compared to plain alum-adsorbed BSA. When administered by oral route, the plain liposomes containing BSA produced significantly higher IgG and IgA levels as compared to control [209].

Self-assembling nanoparticles of pullulan acetate (PA) were conjugated to vitamin H (biotin) to improve their cancer-targeting activity and internalisation [175]. Three samples of biotinylated PA, comprising 7, 20 and 39 vitamin H groups per 100 anhydroglucose units, were synthesised, and adriamycin was loaded into the nanoparticles as a model drug. The rhodamine B isothiocyanate-labelled nanoparticles exhibited very strong adsorption to HepG2 cells, while the fluorolabelled acetylated pullulan nanoparticles did not show any significant interaction. The degree of the interaction as well as the internalisation of the nanoparticles into the cancer cells increased with increasing vitamin H content. Zhang and co-workers [115] have described the preparation of folic acid-conjugated PA nanoparticles (FPA), with improved cancer-targeting activity. Epirubicin-loaded FPA nanoparticles (FPA/EPI) exhibited faster drug release than epirubicin-loaded PA nanoparticles (PA/EPI) *in vitro*. Confocal image analysis and flow cytometry revealed that FPA/EPI nanoparticles exhibited a greater extent of cellular uptake than PA/EPI nanoparticles against KB cells overexpressing folate receptors. The cytotoxicity of PA/EPI nanoparticles to KB cells was inhibited by an excess amount of folic acid, suggesting that the binding and/or uptake were mediated by the folate receptor. Polysaccharide microspheres from acetylated pullulan(s) (PAM) have also been designed for the long-term delivery of peptide/protein drugs, as an alternative to a polylactic-polyglycolic acid (PLGA) depot system [210]. The microspheres with different acetyl content were obtained via a water-in-oil-in-water emulsion method with diameters ranging from 35 to 100 μm and loaded with exenatide, a drug used for the treatment of type II diabetes. Although the release of exenatide from the PLGA microspheres showed a fast and high-burst behaviour, a sustained-release profile was observed with PAM for 21 days. Inflammation of the tissue was minimal with PAM compared to PLGA microspheres.

4.4.2 ... To Macrogels

Crosslinkers such as epichlorohydrin, sodium trimetaphosphate (STMP) and glutaraldehyde are widely used to prepare polysaccharide-based macroscopic hydrogels. Lack and co-workers [211] obtained crosslinked hydrogels by treating pullulan with STMP in alkaline aqueous medium. The crosslinking rate increased with an increase in the concentration of any of the reagents (pullulan, STMP, NaOH). A stronger gel was obtained with increasing polymer concentration whereas increasing STMP concentration led to a plateau. The crosslinking mechanism was elucidated using high-resolution NMR to follow the reaction between STMP and a model molecule, namely methyl α-D-glucopyranoside, as pullulan characteristics did not allow it (long correlation and short relaxation times) [212]. Although STMP could give phosphodiester or phosphotriester bridges between pullulan chains, the authors showed single phosphoester links in pullulan hydrogels. At the same time, pullulan hydrogel discs crosslinked with STMP were developed for vascular cell culture [213]. Most hydrogels are less adhesive than cell culture-specific plastic surfaces, and cellular adhesion on hydrophilic materials such as hyaluronic acid hydrogels was reported to be rather low, as a result of the smooth surface and the highly anionic nature of hyaluronic acid [214]. Following functionalisation of dextrin with hydroxyethyl methacrylate ester, mouse fibroblasts adhered to dextrin hydrogels, and most of the initial cells remained associated to the hydrogels and proliferated [215]. In some cases, cellular adhesion and growth of smooth muscle cells were observed on nonmodified pullulan hydrogels [213]. When pullulan was partially substituted with diethylaminoethyl groups, STMP-crosslinked hydrogels were able to trap and release nucleic acid, allowing cell transfections *in vitro* and *in vivo* [101, 107, 216]. More recently, pullulan-based hydrogels that could mimic extracellular matrix found in tissues were designed for use as tissue engineering and tissue repair scaffolds. In most tissue engineering approaches, cells need to be combined with a scaffold for generating a new tissue or repairing a wide variety of tissues and organs [217]. Hydrogels comprised of naturally derived macromolecules have potential advantages of biocompatibility, biodegradability and intrinsic cellular interaction [218]. With this purpose, pullulan hydrogels containing pores into which live cells could infiltrate and proliferate were prepared either using a combined crosslinking and freeze-drying process [219, 220] or using a salt leaching technique [221]. Cell ingrowth was affected by scaffold porosity, with cells infiltrating deeper into highly porous scaffolds containing large pores (>200 μm). Although the weak mechanical properties of highly porous hydrogels could impair their use in tissue engineering, cell proliferation and extracellular matrix deposition would enhance the mechanical properties of the construct and improve the performance of the porous scaffolds overtime. Co-crosslinking of unmodified pullulan with various other components is scantily reported in the literature. Hydrogels obtained from a pullulan/dextran/fucoidan mixture with STMP as a crosslinking agent led to a novel biomaterial that could support the attachment of human endothelial progenitor cells of different origins [222]. Human CD34+ human umbilical cord blood cells and CD133+ bone marrow cells were differentiated towards the endothelial lineage when cultured on the hydrogel, as well as mature endothelial cells isolated from human saphenous veins. Incorporation of heparin into carboxylated PA hydrogels effectively increased endothelial cell growth, suggesting interactions between heparin and heparin receptors on cell membranes [223]. These heparin-loaded materials could ultimately be used in the fabrication of vascular prostheses that require a nonthrombogenic surface. Addition of vascular endothelial growth factor, a key regulator in new blood vessel formation, to alginate hydrogel has been shown to improve cell migration, thus creating a three-dimensional material niche for vascular progenitor cell populations [224]. This endothelial progenitor cell-based therapy may be applicable to pullulan/dextran/fucoidan hydrogels for the treatment of ischaemic diseases. Vascular progenitor cell populations would be delivered on a bioactive material that provides a microenvironment enhancing cell survival and the sustained release and repopulation of the surrounding tissue by outwardly migrating cells [224]. In addition to light and confocal microscopy performed on transparent hydrogels, high-resolution magnetic resonance imaging (MRI) can be used to assess three-dimensional structures of pullulan porous scaffolds and to image tissue-engineered constructs [225]. To validate cell-seeding procedures, human mesenchymal stem cells (MSC) were magnetically labelled with anionic citrate-coated iron oxide nanoparticles. Cell organisation was strongly dependent on the scaffold structure. Most importantly, high-resolution MRI allowed accurate detection of pullulan scaffolds and seeded cells under *in vivo* conditions. In the future, this noninvasive imaging method would be useful in clinical applications to distinguish the implanted scaffold from the surrounding tissue and to follow cell migration throughout the scaffold and/or outside the scaffold towards neighbouring areas.

Pullulan and cyclodextrins were co-crosslinked using epichlorohydrin to yield microspheres. Once packed in a glass column, these microspheres behaved like an affinity chromatography stationary phase [226]. When pullulan and gelatin are mixed, a phase separation is observed, with the gelation point depending on the temperature and rheology of both phases [227]. These gelatin/pullulan mixtures may lead to gelled viscoelastic microparticles [228]. Hydrogels could also be obtained in water by crosslinking CMP with either adipic acid dihydrazide or dimethylaminopropylamine [143]. The ionic and/or amphiphilic properties of the former hydrogel were demonstrated by its capacity to entrap methylene blue and Coomassie brilliant blue [143] as they were depending on the pH of the solution for the latter [106]. Pullulan-based hydrogels and CMP microparticles were prepared by Mocanu and co-workers [164] using 3-(glycidoxypropyl)trimethoxysilane as a crosslinker. The porosity of the gel was assessed by electronic transmission microscopy and also through the absorption of poly(ethylene glycol) of various molecular weights. Hydrogels have also been prepared from methacrylated pullulan derivatives via a 'living' free radical process, namely the reversible addition-fragmentation chain transfer technique, the network formation being monitored by rheological measurements [132].

In some studies, both self-associative properties of hydrophobised pullulan and the type of chemical crosslinker used have been considered to give birth to ultrastructured macrogels endowed with finely tuned drug/protein entrapment and release properties [133, 189, 229–231]. Associative microparticles of crosslinked pullulan and CMP amidified by akylated amine were obtained by using STMP and compared with those crosslinked with epichlorohydrin [189]. The property of the studied supports to retain great amounts of lysozyme, which is released in NaCl solutions, could be useful in separation/purification processes of the enzyme from production media and to develop new crosslinked pullulan-based drug delivery systems. Moreover, the immobilised lysozyme retained its enzymatic activity. The lysozyme/hydrophilic support complex could also be used for healing of the infected wounds, where it would act both as a fluid adsorbent and as a topical antibacterial agent.

As another example, Morimoto and co-workers [232] synthesised poly (*N*-isopropylacrylamide) (PNIPAM) hydrogels by crosslinking with self-assembled nanogels consisting of cholesteryl- and methacryloyl-substituted pullulan as a crosslinker for a hydrophilic nanodomain. The shrinking half-time of the new hydrogel was approximately 2 minutes, which is about 3400 times faster than that of a PNIPAM hydrogel crosslinked by classical methylene(bisacrylamide).

New hybrid macrogels with well-defined nanostructures can be obtained by utilising self-assembled nanogels as building blocks [30, 133, 229]. Morimoto and co-workers [133] prepared methacryloyl group-bearing CHP nanogels to form a macrogel after polymerisation within which they were well dispersed as evidenced by transmission electron microscopy (TEM). The hybrid hydrogels showed two well-defined networks such as a nanogel intranetwork structure of less than 10 nm (physically crosslinking) and an internetwork structure of several hundred nanometres (chemical crosslinking). The immobilised nanogels retained their ability to trap and release insulin used as a model protein by host-guest interaction of the cholesteryl group and cyclodextrin, and also showed a chaperone-like activity for refolding the denatured protein. Based on pullulan hydrogel properties, we have developed a new hybrid mesh composed of clinical-grade polypropylene mesh embedded in a pullulan polysaccharide hydrogel [233]. Evaluation after *in vivo* implantation and comparison with two clinically used materials, namely porcine decellularised small intestinal submucosa and plain polypropylene mesh, showed that the new hybrid mesh did not have any adverse effects and was associated with better tissue organisation [233]. These prosthetic materials could be used in surgery and tissue engineering since many postoperative complications are due to poor integration of the materials, which delays the healing process.

Other hydrophobised pullulan conjugates such as acetylated pullulan have also been used to prepare films with a tunable biodegradability or to deliver anticancer drugs. Teramoto and Shibata [109] have prepared and characterised biodegradable films from PA with DS ranging from 1.0 to 3.0. Tensile modulus of the films was comparable to that of a widely used cellulose acetate film. The biodegradation rate of PA, estimated by measuring the biochemical oxygen demand under aerobic conditions, decreased with the increasing degree of acetylation.

The use of pullulan microspheres for cell preservation and transport was patented recently. Usually living cells are sent by Express Mail in plates sealed with film or in flasks containing liquid transport medium. However, this mode of transport suffers from drawbacks such as mechanical stress exerted by the medium during transport, increased contamination rate of samples when cells come in contact with air bubbles and liquid leakage. Therefore, it is essential to develop storage and transport systems that could maintain viability and function of the tissue. Based on the biocompatible and hydrophilic properties of pullulan hydrogels, a cell transport system was developed. By adding the pullulan hydrogel to the culture medium, a semisolid stopper was formed and cell viability and function were maintained for 4 days at room temperature [234].

4.4.3 Pullulan-based Core-shell Nanoparticles for Imaging and Therapy

The use of pullulan or some of its derivatives as a carrier for the radioisotope ^{99m}Tc, to target cells with quantum dots (QD) or as a coating of iron oxide particles dedicated to medical imaging or hyperthermia treatments is the subject of the more recent works. In a first study published in 2003, Song and co-workers [235] used pH-responsive nanoparticles of PA derivatives to target colon cancer cells in mice. They obtained labelling efficiencies above 90% and per cent retention rates significantly higher than that of ^{99m}Tc-pertechnetate. Monodisperse hybrid nanoparticles (38 nm in diameter) containing QD were obtained by simple mixing with nanogels of CHP modified with amino groups ($CHPNH_2$) [236, 237]. The authors investigated the capacity of $CHPNH_2$ to deliver QD into HeLa cells and rabbit MSC in comparison to conventional cationic liposome. $CHPNH_2$-QD hybrid nanoparticles remained detectable inside MSC for at least 2 weeks of culture and had little effect on their *in vitro* chondrogenic ability.

More recently, the cellular uptake and the cytotoxicity of acetylated pullulan-coated superparamagnetic iron oxide nanoparticles (PAMN) have been examined by Gao and co-workers [238] and their hyperthermic effect on tumour cells (KB) has been evaluated. Hyperthermia using magnetic nanoparticles is a very promising treatment for cancer based on the hypothesis that cancerous cells are more sensitive to an increase of temperature than normal cells. TEM and dynamic light scattering showed that PAMN had spherical morphology with an average diameter of 25.8 ± 6.1 nm. The presence of the adsorbed layer of PA on the magnetite surface was confirmed by Fourier transform infrared spectroscopy. The PAMN exhibited good heating properties in an alternating magnetic field as well as a low cytotoxicity against L929 cells. After internalisation of PAMN into KB cells, showed by TEM, a significant death of tumour cells (80%) was obtained *in vitro* by magnetic field-induced hyperthermia, as compared to magnetic field alone (no cell death). Hence, this magnetic pullulan-derived material constitutes a very promising nanosystem for hyperthermic treatment of tumours.

References

1. B.J. Catley and P.J. Kelly, *Biochemical Society Transactions*, 1975, **3**, 6, 1079.

2. R. Bauer, *Zentralblatt fur Bakteriologie, Parasitenkunde Infektionskrankheiten und Hygiene (Reihe A)*, 1938, **98**, 5.

3. H. Bender, J. Lehmann and K. Wallenfels, *Biochimica et Biophysica Acta*, 1959, **36**, 309.

4. K. Wallenfels, G. Keilich, G. Bechtler and D. Freudenberger, *Biochemische Zeitschrift*, 1965, **341**, 433.

5. Y. Tsujisaka and M. Mitsuhachi in *Industrial Gums: Polysaccharides and Their Derivatives*, Eds., J.N. BeMiller and R. Whistler, Academic Press, San Diego, CA, USA, 1993, p.447.

6. T. Kimoto, T. Shibuya and S. Shiobara, *Food and Chemical Toxicology*, 1997, **35**, 3/4, 323.

7. R.S. Singh, G.K. Saini and J.F. Kennedy, *Carbohydrate Polymers*, 2008, **73**, 4, 515.

8. B.W. Wolf, inventors; Abbot Laboratories, assignee; US 6916796, 2005.

9. R.F. Childers, P.L. Oren and W.M.K. Seidler, inventors; Eli Lilly and Company, assignee; US 5015480, 1991.

10. Y. Izutsu, K. Sogo, S. Okamoto and T. Tanaka, inventors; Dainippon Pharmaceutical Co., Ltd., assignee; US 4650666, 1987.

11. Y. Miyamoto, H. Goto, H. Sato, H. Okano and M. Ijima, inventors; Zeria Shinyaku Kogyo Kabushiki Kaisha, assignee; US 4610891, 1986.

12. S. Nakashio, K. Tsuji, N. Toyota and F. Fujita, inventors; Sumitomo Chemical Company, Ltd., assignee; US 3972997, 1976.

13. W. Xu and S.W. Rhee, *Organic Electronics*, 2010, **11**, 6, 996.

14. S. Savic, D. Pantelic and D. Jakovijevic, *Applied Optics*, 2002, **41**, 22, 4484.

15. Y. Onda, H. Muto and H. Suzuki, inventors; Shin-Etsu Chemical Co., Ltd., assignee; US 4322524, 1977.

16. A.M. Stephen in *Food Polysaccharides and Their Applications*, Ed., A.M. Stephen, Marcel Dekker, New York, NY, USA, 1995, p.341.

17. G.H. Ma, T. Fukutomi and S. Nozaki, *Journal of Applied Polymer Science*, 1993, **47**, 7, 1243.

18. H. Fredriksson, R. Andersson, K. Koch and P. Aman, *Journal of Chromatography A*, 1997, **768**, 2, 325.

19. A. Mizutani, K. Nagase, A. Kikuchi, H. Kanazawa, Y. Akiyama, J. Kobayashi, M. Annaka and T. Okano, *Journal of Chromatography A*, 2010, **1217**, 4, 522.

20. J.M. Gaddy and P.A. Patton, inventors; Tate & Lyle Ingredients Americas, Inc., assignee; US 7022838, 2006.

21. T.J. Boyd, G. Xu, A. Gaffar and D.B. Visco, inventors; Micron Technology, Inc., assignee; US 0193791, 2006.

22. S.H. Leung, R.S. Leone, L.D. Kumar, N. Kulkarni and A.F. Sorg, inventors; Warner-Lambert Company LLC, assignee; US 7025983, 2006.

23. N. Ikewaki, N. Fuji and T. Onaka, inventors; Onaka, Yasushi, assignee; US 6956120, 2005.

24. R. Scott, D. Cade and X. He, inventors; Hyperion Catalysis International, Inc., assignee; US 6887307, 2005.

25. D. Cade, R. Scott and X. He, inventors; Warner-Lambert Co., assignee; US 6517865, 2003.

26. W.M. Kulicke and T. Heinze, *Macromolecular Symposia*, 2006, **231**, 47.

27. T. Igarashi, K. Nomura, K. Naito and M. Yoshida, inventors; Hayashibara Biochemical Laboratories, Inc., assignee; US 4370472, 1983.

28. H. Seibutsu and M. Kenkyujo, inventors; Hayashibara Biochemical Laboratories, assignee; GB 2109391, 1983.

29. S. Yuen, *Advances in Chemistry Series*, 1974, **4**, 172.

30. T. Coviello, P. Matricardi, C. Marianecci and F. Alhaique, *Journal of Controlled Release*, 2007, **119**, 1, 5.

31. Y. Sakata and M. Otsuka, *International Journal of Pharmaceutics*, 2009, **374**, 1/2, 33.

32. M. Kawahara, K. Mizutani, S. Suzuki, S. Kitamura, H. Fukada, T. Yui and K. Ogawa, *Bioscience, Biotechnology, and Biochemistry*, 2003, **67**, 4, 893.

33. J.W. Rhim, *Food Science and Biotechnology*, 2003, **12**, 2, 161.

34. European Food and Safety Authority, *EFSA Journal*, 2004, **85**, 1.

35. K.I. Shingel, *Carbohydrate Research*, 2004, **339**, 3, 447.

36. T.D. Leathers, *Applied Microbiology and Biotechnology*, 2003, **62**, 5/6, 468.

37. T.D. Leathers in *Polysaccharides II: Polysaccharides from Eukaryotes*, Eds., E.J. Vandamme, S. De Baets and A. Steinbüchel, Wiley-VCH, Weinheim, Germany, 2002, p.1.

38. C. Israilides, A. Smith, B. Scanlon and C. Barnett in *Biotechnology and Genetic Engineering Reviews*, Volume 16, Intercept Ltd. Scientific, Technical & Medical Publishers, Andover, UK, 1999, p.309.

39. P.A. Gibbs and R.J. Seviour in *Polysaccharides in Medicinal Applications*, Ed., S. Dimitriu, Marcel Dekker, New York, NY, USA, 1996, p.59.

40. B.J. Catley, A. Ramsay and C. Servis, *Carbohydrate Research*, 1986, **153**, 1, 79.

41. B.J. Wiley, D.H. Ball, S.M. Arcidiacono, S. Sousa and J.M. Mayer, *Journal of Environmental Polymer Degradation*, 1993, **1**, 3.

42. T. Roukas and F. Mantzouridou, *Journal of Chemical Technology and Biotechnology*, 2001, **76**, 4, 371.

43. K. Nishinari, K. Kohyama, P.A. Williams, G.O. Phillips, W. Burchard and K. Ogino, *Macromolecules*, 1991, **24**, 20, 5590.

44. M. Yalpani in *Polysaccharides: Synthesis, Modification and Structure/Property Relations*, Ed., M. Yalpani, Elsevier, Amsterdam, The Netherlands, 1998, p.321.

45. C.G. Fraser and H.J. Jennings, *Canadian Journal of Chemistry*, 1971, **49**, 1804.

46. N. Waksman, R.M. de Lederkremer and A.S. Cerezo, *Carbohydrate Research*, 1977, **59**, 2, 505.

47. M.M. Corsaro, C. De Castro, A. Evidente, R. Lanzetta, A. Molinaro, M. Parrilli and L. Sparapano, *Carbohydrate Polymers*, 1998, **37**, 2, 167.

48. R.A. Reis, C.A. Tischer, P.A.J. Gorin and M. Iacomini, *FEMS Microbiology Letters*, 2002, **210**, 1, 1.

49. Z.M. Chi and S.Z. Zhao, *Enzyme and Microbial Technology*, 2003, **33**, 2/3, 206.

50. H.J. Jennings and I.C.P. Smith, *Journal of the American Chemical Society*, 1973, **95**, 2, 606.

51. C. Arnosti and D.J. Repeta, *Starch-Starke*, 1995, **47**, 2, 73.

52. M. Doman-Pytka and J. Bardowski, *Critical Reviews in Microbiology*, 2004, **30**, 2, 107.

53. H. Aoki, Yopi, A. Padmajanti and Y. Sakano, *Bioscience, Biotechnology, and Biochemistry*, 1996, **60**, 11, 1795.

54. F. Niehaus, A. Peters, T. Groudieva and G. Antranikian, *FEMS Microbiology Letters*, 2000, **190**, 2, 223.

55. H. Bender and K. Wallenfels, *Biochemische Zeitschrift*, 1961, **334**, 79.

56. K. Wallenfels, H. Bender and J.R. Rached, *Biochemical and Biophysical Research Communications*, 1966, **22**, 3, 254.

57. R. Ohba and S. Ueda, *Biotechnology and Bioengineering*, 1980, **22**, 10, 2137.

58. D. Bruneel and E. Schacht, *Journal of Bioactive and Compatible Polymers*, 1995, **10**, 4, 299.

59. D. Bruneel and E. Schacht, *Polymer*, 1993, **34**, 12, 2628.

60. D. Bruneel and E. Schacht, *Polymer*, 1993, **34**, 12, 2633.

61. D. Bruneel and E. Schacht, *Polymer*, 1994, **35**, 12, 2656.

62. M. Vihinen and P. Mäntsälä, *Critical Reviews in Biochemistry and Molecular Biology*, 1989, **24**, 4, 329.

63. D.H. Ball, B.J. Wiley and E.T. Reese, *Canadian Journal of Microbiology*, 1992, **38**, 4, 324.

64. Y. Ohe, K. Ohtani, Y. Sone and A. Misaki, *Bioscience, Biotechnology, and Biochemistry*, 1993, **57**, 2, 227.

65. T. Roukas, *Food Biotechnology*, 1999, **13**, 3, 255.

66. G.Q. Li, H.W. Qiu, Z.M. Zheng, Z.L. Cai and S.Z. Yang, *Journal of Chemical Technology and Biotechnology*, 1995, **62**, 4, 385.

67. A. Richard and A. Margaritis, *Critical Reviews in Biotechnology*, 2002, **22**, 4, 355.

68. B.A. Nemet, Y. Shabtai and M. Cronin-Golomb, *Optics Letters*, 2002, **27**, 4, 264.

69. J. Audet, H. Gagnon, M. Lounes and J. Thibault, *Bioprocess Engineering*, 1998, **19**, 1, 45.

70. Y. Hemar and D.N. Pinder, *Biomacromolecules*, 2006, **7**, 3, 674.

71. K.I. Shingel and P.T. Petrov, *Colloid and Polymer Science*, 2002, **280**, 2, 176.

72. T. Harada, Y. Kanzawa, K. Kanenaga, A. Koreeda and A. Harada, *Food Structure*, 1991, **10**, 1, 1.

73. A.C. Kshirsagar, V.B. Yenge, A. Sarkar and R.S. Singhal, *Food Chemistry*, 2009, **113**, 4, 1139.

74. E.C. Lopez, D. Champion, G. Blond and M. Le Meste, *Carbohydrate Polymers*, 2005, **59**, 1, 83.

75. E. Tsaliki, S. Pegiadou and G. Doxastakis, *Food Hydrocolloids*, 2004, **18**, 4, 631.

76. F. Spyropoulos, W.J. Frith, I.T. Norton, B. Wolf and A.W. Pacek, *Journal of Rheology*, 2007, **51**, 5, 867.

77. F. Spyropoulos, W.J. Frith, I.T. Norton, B. Wolf and A.W. Pacek, *Food Hydrocolloids*, 2008, **22**, 1, 121.

78. A. Lazaridou, C.G. Biliaderis and V. Kontogiorgos, *Carbohydrate Polymers*, 2003, **52**, 2, 151.

79. N. Gontard, R. Thibault, B. Cuq and S. Guilbert, *Journal of Agricultural and Food Chemistry*, 1996, **44**, 4, 1064.

80. E. Kristo, C.G. Biliaderis and A. Zampraka, *Food Chemistry*, 2007, **101**, 2, 753.

81. L. Lebrun, J. Blanco and M. Metayer, *Carbohydrate Polymers*, 2005, **61**, 1, 1.

82. W.J. Lee and S.H. Kim, *Macromolecular Research*, 2008, **16**, 3, 247.

83. N. Teramoto, M. Saitoh, J. Kuroiwa, M. Shibata and R. Yosomiya, *Journal of Applied Polymer Science*, 2001, **82**, 9, 2273.

84. P. Prasad, G.S. Guru, H.R. Shivakumar and K.S. Rai, *Journal of Applied Polymer Science*, 2008, **110**, 1, 444.

85. K.I. Shingel, V.M. Tsarenkov and P.T. Petrov, *Carbohydrate Research*, 2000, **324**, 4, 283.

86. K.I. Shingel and P.T. Petrov, *Polymer Science Series B*, 2001, **43**, 3/4, 81.

87. K.L. Xi, Y. Tabata, K. Uno, M. Yoshimoto, T. Kishida, Y. Sokawa and Y. Ikada, *Pharmaceutical Research*, 1996, **13**, 12, 1846.

88. J. Jo, T. Kai, A. Okazaki, M. Yamamoto, Y. Hirano and Y. Tabata, *Journal of Controlled Release*, 2007, **118**, 3, 389.

89. J.I. Jo, T. Ikai, A. Okazaki, K. Nagane, M. Yamamoto, Y. Hirano and Y. Tabata, *Journal of Biomaterials Science: Polymer Edition*, 2007, **18**, 7, 883.

90. T. Tanaka, S. Hamano, Y. Fujishima and Y. Kaneo, *Biological & Pharmaceutical Bulletin*, 2005, **28**, 3, 560.

91. Y. Tabata, Y. Matsui, K. Uno, Y. Sokawa and Y. Ikada, *Journal of Interferon and Cytokine Research*, 1999, **19**, 3, 287.

92. T. Yamaoka, Y. Tabata and I. Yoshito, *Drug Delivery*, 1993, **1**, 1, 75.

93. Y. Kaneo, T. Tanaka, T. Nakano and Y. Yamaguchi, *Journal of Controlled Release*, 2001, **70**, 3, 365.

94. T. Tanaka, Y. Fujishima, S. Hanano and Y. Kaneo, *International Journal of Pharmaceutics*, 2004, **286**, 1/2, 9.

95. G. Keilich, P. Salminen and E. Husemann, *Die Makromolekulare Chemie*, 1971, **141**, 117.

96. M. Shibata, R. Nozawa, N. Teramoto and R. Yosomiya, *European Polymer Journal*, 2002, **38**, 3, 497.

97. F. Fujita, K. Fukami and M. Fujimoto, inventors; Sumitomo Chemical Company, Hayashibara Biochemical Laboratories, Inc., assignee; US 4167623, 1979.

98. Y. Nishijima, H. Niwase and M. Fujimoto, inventors; Japan Kokai Tokkyo Koho, assignee; JP 7911108, 1979.

99. K. Glinel, J.P. Sauvage, H. Oulyadi and J. Huguet, *Carbohydrate Research*, 2000, **328**, 3, 343.

100. M. Miani, R. Gianni, G. Liut, R. Rizzo, R. Toffanin and F. Delben, *Carbohydrate Polymers*, 2004, **55**, 2, 163.

101. A. San Juan, H. Hlawaty, F. Chaubet, D. Letourneur and L.J. Feldman, *European Heart Journal*, 2006, **27**, 967.

102. I. Murase, F. Fujita, T. Ohnishi and T. Tamura, inventors; Sumitomo Chemical Co., Ltd., assignee; EP 76698, 1983.

103. G. Mocanu, D. Vizitiu, D. Mihai and A. Carpov, *Carbohydrate Polymers*, 1999, **39**, 3, 283.

104. K. Imai, T. Shiomi and Y. Tesuka, inventors; Covalent Associates Inc., assignee; JP 0321602 (9121602), 1991.

105. G. Mocanu, D. Mihai, D. LeCerf, L. Picton and M. Moscovici, *Journal of Applied Polymer Science*, 2009, **112**, 3, 1175.

106. Z. Souguir, S. Roudesli, L. Picton, D. Le Cerf and E. About-Jaudet, *European Polymer Journal*, 2007, **43**, 12, 4940.

107. A. San Juan, G. Ducrocq, H. Hlawaty, I. Bataille, E. Guenin, D. Letourneur and L.J. Feldman, *Journal of Biomedical Materials Research Part A*, 2007, **83A**, 3, 819.

108. K. Na, E.S. Lee and Y.H. Bae, *Journal of Controlled Release*, 2003, **87**, 1–3, 3.

109. N. Teramoto and M. Shibata, *Carbohydrate Polymers*, 2006, **63**, 4, 476

110. Y. Tezuka, *Carbohydrate Research*, 1997, **305**, 2, 155.

111. W. Henni-Silhadi, M. Deyme, M.R. de Hoyos, D. Le Cerf, L. Picton and V. Rosilio, *Colloid and Polymer Science*, 2008, **286**, 11, 1299.

112. W.Z. Yang, H.L. Chen, F.P. Gao, M.M. Chen, X.M. Li, M.M. Zhang, Q.Q. Zhang, L.R. Liu, Q. Jiang and Y.S. Wang, *Current Nanoscience*, 2010, **6**, 3, 298.

113. M.A. Hussain and T. Heinze, *Polymer Bulletin*, 2008, **60**, 6, 775.

114. M.A. Hussain, D. Shahwar, M.N. Hassan, M.N. Tahir, M.S. Iqbal and M. Sher, *Collection of Czechoslovak Chemical Communications*, 2010, **75**, 1, 133.

115. H.Z. Zhang, X.M. Li, F.P. Gao, L.R. Liu, Z.M. Zhou and Q.Q. Zhang, *Drug Delivery*, 2010, **17**, 1, 48.

116. A. Kaya, X.S. Du, Z.L. Liu, J.W. Lu, J.R. Morris, W.G. Glasser, T. Heinze and A.R. Esker, *Biomacromolecules*, 2009, **10**, 9, 2451.

117. Z. Liu, Y. Jiao, Y. Wang, C. Zhou and Z. Zhang, *Advanced Drug Delivery Reviews*, 2008, **60**, 15, 1650.

118. G. Mocanu, M. Constantin and A. Carpov, *Angewandte Makromolekulare Chemie*, 1996, **241**, 1.

119. M. Shibata, M. Asahina, N. Teramoto and R. Yosomiya, *Polymer*, 2001, **42**, 1, 59.

120. Y. Muroga, K. Hayashi, M. Fukunaga, T. Kato, S. Shimizu and K. Kurita, *Biophysical Chemistry*, 2006, **121**, 2, 96.

121. K. Akiyoshi, S. Deguchi, N. Moriguchi, S. Yamaguchi and J. Sunamoto, *Macromolecules*, 1993, **26**, 12, 3062.

122. J.M. Mayer, M. Greenberger, D.H. Ball and D.L. Kaplan, *Polymer Materials: Science and Engineering*, 1990, **63**, 732.

123. D. Mihai, G. Mocanu and A. Carpov, *European Polymer Journal*, 2001, **37**, 3, 541.

124. S. Alban, A. Schauerte and G. Franz, *Carbohydrate Polymers*, 2002, **47**, 3, 267.

125. G.C. Mocanu, G. Stanciulescu, A. Carpov, D. Mihai, R.P. Ghiocel and M. Moscovici, inventors; Institute of Chimii Macromolecular, Romania, assignee; Romanian Patent 88034, 1985.

126. F.L. Bragd, A.C. Besemer and H. van Bekkum, *Journal of Molecular Catalysis A: Chemical*, 2001, **170**, 1/2, 35.

127. A.E.J. Denooy, A.C. Besemer and H. Vanbekkum, *Carbohydrate Research*, 1995, **269**, 1, 89.

128. D.X. Lu, X.T. Wen, J. Liang, Z.W. Gu, X.D. Zhang and Y.J. Fan, *Journal of Biomedical Materials Research Part B - Applied Biomaterials*, 2009, **89B**, 1, 177.

129. M. Shinkai, M. Suzuki, S. Iijima and T. Kobayashi, *Biotechnology and Applied Biochemistry*, 1995, **21**, 125.

130. I. Bataille, J. Huguet, G. Muller, G. Mocanu and A. Carpov, *International Journal of Biological Macromolecules*, 1997, **20**, 3, 179.

131. D. Bontempo, G. Masci, P. De Leonardis, L. Mannina, D. Capitani and V. Crescenzi, *Biomacromolecules*, 2006, 7, 7, 2154.

132. V. Crescenzi, M. Dentini, D. Bontempo and G. Masci, *Macromolecular Chemistry and Physics*, 2002, **203**, 10/11, 1285.

133. N. Morimoto, T. Endo, Y. Iwasaki and K. Akiyoshi, *Biomacromolecules*, 2005, **6**, 4, 1829.

134. S.J. Wu, Z.Y. Jin, J.M. Kim, Q.Y. Tong and H.Q. Chen, *Carbohydrate Polymers*, 2009, **76**, 1, 129.

135. R.C. Tian, J.P. Gao, J.G. Yu and M.L. Duan, *Journal of Applied Polymer Science*, 1992, **45**, 4, 591.

136. G. Masci, D. Bontempo and V. Crescenzi, *Polymer*, 2002, **43**, 20, 5587.

137. Y. Ohya, S. Maruhashi and T. Ouchi, *Macromolecules*, 1998, **31**, 14, 4662.

138. N. Morimoto, R. Obeid, S. Yamane, F.M. Winnik and K. Akiyoshi, *Soft Matter*, 2009, **5**, 8, 1597.

139. M. Legros, P. Cardinael, V. Dulong, L. Picton and D. Le Cerf, *Polymer Journal*, 2008, **40**, 3, 233.

140. M. Legros, V. Dulong, L. Picton and D. Le Cerf, *Polymer Journal*, 2008, **40**, 12, 1132.

141. W. Henni-Silhadi, M. Deyme, M.M. Boissonnade, M. Appel, D. Le Cerf, L. Picton and V. Rosilio, *Pharmaceutical Research*, 2007, **24**, 12, 2317.

142. V. Dulong, G. Mocanu and D. Le Cerf, *Colloid and Polymer Science*, 2007, **285**, 10, 1085.

143. V. Dulong, D. Le Cerf, L. Picton and G. Muller, *Colloids and Surfaces A: Physicochemical and Engineering Aspects*, 2006, **274**, 1–3, 163.

144. K. Glinel, J. Huguet and G. Muller, *Polymer*, 1999, **40**, 25, 7071.

145. Z. Souguir, S. Roudesli, E. About-Jaudet, L. Picton and D. Le Cerf, *Carbohydrate Polymers*, 2010, **80**, 1, 123.

146. A. Guyomard, B. Nysten, G. Muller and K. Glinel, *Langmuir*, 2006, **22**, 5, 2281.

147. D. Magnin, V. Callegari, S. Matefi-Tempfli, M. Matefi-Tempfli, K. Glinel, A.M. Jonas and S. Demoustier-Champagne, *Biomacromolecules*, 2008, **9**, 9, 2517.

148. K. Horie, M. Sakagami, K. Kuramochi, K. Hanasaki, H. Hamana and T. Ito, *Pharmaceutical Research*, 1999, **16**, 2, 314.

149. K. Horie and P.A. Insel, *Journal of Biological Chemistry*, 2000, **275**, 38, 29433.

150. K. Horie, M. Sakagami, K. Masuda, M. Notoya, H. Hamana, T. Yoshikawa and K. Hirano, *Biological & Pharmaceutical Bulletin*, 2004, **27**, 8, 1275.

151. H. Nogusa, K. Yamamoto, T. Yano, M. Kajiki, H. Hamana and S. Okuno, *Biological & Pharmaceutical Bulletin*, 2000, **23**, 5, 621.

152. H. Nogusa, T. Yano, N. Kashima, K. Yamamoto, S. Okuno and H. Hamana, *Bioorganic & Medicinal Chemistry Letters*, 2000, **10**, 3, 227.

153. H. Nogusa, H. Hamana, N. Uchida, R. Maekawa and T. Yoshioka, *Japanese Journal of Cancer Research*, 2000, **91**, 12, 1333.

154. H. Nogusa, T. Yano, S. Okuno, H. Hamana and K. Inoue, *Chemical & Pharmaceutical Bulletin (Tokyo)*, 1995, **43**, 11, 1931.

155. R. Mehvar, *Current Pharmaceutical Biotechnology*, 2003, **4**, 5, 283.

156. D.X. Lu, X.T. Wen, J. Liang, X.D. Zhang, Z.W. Gu and Y.J. Fan, *Chinese Journal of Polymer Science*, 2008, **26**, 3, 369.

157. K. Masuda, M. Sakagami, K. Horie, H. Nogusa, H. Hamana and K. Hirano, *Pharmaceutical Research*, 2001, **18**, 2, 217.

158. J. Wotschadlo, T. Liebert, T. Heinze, K. Wagner, M. Schnabelrauch, S. Dutz, R. Muller, F. Steiniger, M. Schwalbe, T.C. Kroll, K. Hoffken, N. Buske and J.H. Clement, *Journal of Magnetism and Magnetic Materials*, 2009, **321**, 10, 1469.

159. C. Mähner, M.D. Lechner and E. Nordmeier, *Carbohydrate Research*, 2001, **331**, 2, 203.

160. J. Fritzsche, S. Alban, R.J. Ludwig, S. Rubant, W.H. Boehncke, G. Schumacher and G. Bendas, *Biochemical Pharmacology*, 2006, **72**, 4, 474.

161. S.E. Gradwell, S. Renneckar, A.R. Esker, T. Heinze, P. Gatenholm, C. Vaca-Garcia and W. Glasser, *Comptes Rendus Biologies*, 2004, **327**, 9/10, 945.

162. R.A. Ganzevles, H. Kosters, T. van Vliet, M.A.C. Stuart and H.H.J. de Jongh, *Journal of Physical Chemistry B*, 2007, **111**, 45, 12969.

163. M. Gupta and A.K. Gupta, *Journal of Controlled Release*, 2004, **99**, 1, 157.

164. G. Mocanu, D. Mihai, D. LeCerf, L. Picton and V. Dulong, *Reactive & Functional Polymers*, 2007, **67**, 1, 60.

165. G. Mocanu, D. Mihai, M. Legros, L. Picton and D. LeCerf, *Journal of Bioactive and Compatible Polymers*, 2008, **23**, 1, 82.

166. Y.I. Jeong, J.W. Nah, H.K. Na, K. Na, I.S. Kim, C.S. Cho and S.H. Kim, *Drug Development and Industrial Pharmacy*, 1999, **25**, 8, 917.

167. K. Akiyoshi, Y. Sasaki and J. Sunamoto, *Bioconjugate Chemistry*, 1999, **10**, 3, 321.

168. K. Akiyoshi, I. Taniguchi, H. Fukui and J. Sunamoto, *European Journal of Pharmaceutics and Biopharmaceutics*, 1996, **42**, 4, 286.

169. K. Akiyoshi, S. Kobayashi, S. Shichibe, D. Mix, M. Baudys, S.W. Kim and J. Sunamoto, *Journal of Controlled Release*, 1998, **54**, 3, 313.

170. I. Taniguchi, K. Akiyoshi, J. Sunamoto, Y. Suda, M. Yamamoto and K. Ichinose, *Journal of Bioactive and Compatible Polymers*, 1999, **14**, 3, 195.

171. G. Mocanu, M. Constantin, G. Fundueanu and A. Carpov, *STP Pharma Sciences*, 2000, **10**, 6, 439.

172. M. Yamamoto, K. Ichinose, N. Ishii, T. Khoji, K. Akiyoshi, N. Moriguchi, J. Sunamoto and T. Kanematsu, *Oncology Reports*, 2000, **7**, 1, 107.

173. X.G. Gu, M. Schmitt, A. Hiasa, Y. Nagata, H. Ikeda, Y. Sasaki, K. Akiyoshi, J. Sunamoto, H. Nakamura, K. Kuribayashi and H. Shiku, *Cancer Research*, 1998, **58**, 15, 3385.

174. K. Na and Y.H. Bae, *Pharmaceutical Research*, 2002, **19**, 5, 681.

175. K. Na, T.B. Lee, K.H. Park, E.K. Shin, Y.B. Lee and H.K. Cho, *European Journal of Pharmaceutical Sciences*, 2003, **18**, 2, 165.

176. Y. Sasaki and K. Akiyoshi, *Current Pharmaceutical Biotechnology*, 2010, **11**, 3, 300.

177. Y. Nomura, M. Ikeda, N. Yamaguchi, Y. Aoyama and K. Akiyoshi, *FEBS Letters*, 2003, **553**, 3, 271.

178. Y. Nomura, Y. Sasaki, M. Takagi, T. Narita, Y. Aoyama and K. Akiyoshi, *Biomacromolecules*, 2005, **6**, 1, 447.

179. W. Asayama, S.I. Sawada, H. Taguchi and K. Akiyoshi, *International Journal of Biological Macromolecules*, 2008, **42**, 3, 241.

180. T. Hirakura, Y. Nomura, Y. Aoyama and K. Akiyoshi, *Biomacromolecules*, 2004, **5**, 5, 1804.

181. T. Hirakura, K. Yasugi, T. Nemoto, M. Sato, T. Shimoboji, Y. Aso, N. Morimoto and K. Akiyoshi, *Journal of Controlled Release*, 2010, **142**, 3, 483.

182. K. Satoh, F. Chen, A. Aoyama, H. Date and K. Akiyoshi, *European Journal of Cancer Supplements*, 2008, **6**, 9, 139.

183. H. Shiku, L.J. Wang, Y. Ikuta, T. Okugawa, M. Schmitt, X.G. Gu, K. Akiyoshi, J. Sunamoto and H. Nakamura, *Cancer Chemotherapy and Pharmacology*, 2000, **46**, S77.

184. S. Matsukawa, M. Yamamoto, K. Ichinose, N. Ohata, N. Ishii, T. Kohji, K. Akiyoshi, J. Sunamoto and T. Kanematsu, *Anticancer Research*, 2000, **20**, 4, 2339.

185. L. Wang, H. Ikeda, Y. Ikuta, M. Schmitt, Y. Miyahara, Y. Takahashi, X.G. Gu, Y. Nagata, Y. Sasaki, K. Akiyoshi, J. Sunamoto, H. Nakamura, K. Kuribayashi and H. Shiku, *International Journal of Oncology*, 1999, **14**, 4, 695.

186. Y. Ikuta, N. Katayama, L. Wang, T. Okugawa, Y. Takahashi, M. Schmitt, X. Gu, M. Watanabe, K. Akiyoshi, H. Nakamura, K. Kuribayashi, J. Sunamoto and H. Shiku, *Blood*, 2002, **99**, 10, 3717.

187. T. Nishikawa, K. Akiyoshi and J. Sunamoto, *Macromolecules*, 1994, **27**, 26, 7654.

188. K. Akiyoshi, T. Nishikawa, S. Shichibe and J. Sunamoto, *Chemistry Letters*, 1995, **8**, 707.

189. G. Mocanu, D. Mihai, L. Picton, D. LeCerf and G. Muller, *Journal of Controlled Release*, 2002, **83**, 1, 41.

190. S. Sawada, Y. Nomura, Y. Aoyama and K. Akiyoshi, *Journal of Bioactive and Compatible Polymers*, 2006, **21**, 6, 487.

191. S. Sawada and K. Akiyoshi, *Macromolecular Bioscience*, 2010, **10**, 4, 353.

192. M. Aoki, S. Ueda, H. Nishikawa, S. Kitano, M. Hirayama, H. Ikeda, H. Toyoda, K. Tanaka, M. Kanai, A. Takabayashi, H. Imai, T. Shiraishi, E. Sato, H. Wada, E. Nakayama, Y. Takei, N. Katayama, H. Shiku and S. Kageyama, *Vaccine*, 2009, **27**, 49, 6854.

193. S. Kageyama, S. Kitano, M. Hirayama, Y. Nagata, H. Imai, T. Shiraishi, K. Akiyoshi, A.M. Scott, R. Murphy, E.W. Hoffman, L.J. Old, N. Katayama and H. Shiku, *Cancer Science*, 2008, **99**, 3, 601.

194. S.A. Ferreira and F.M. Gama in Proceedings of the Boston, USA, *Vaccine Publishers*, Supplement S, 2008.

195. K. Na, K.H. Lee and Y.H. Bae, *Journal of Controlled Release*, 2004, **97**, 3, 513.

196. N. Morimoto, F.M. Winnik and K. Akiyoshi, *Langmuir*, 2007, **23**, 1, 217.

197. K. Akiyoshi, E.C. Kang, S. Kurumada, J. Sunamoto, T. Principi and F.M. Winnik, *Macromolecules*, 2000, **33**, 9, 3244.

198. H. Hasuda, O.H. Kwon, I.K. Kang and Y. Ito, *Biomaterials*, 2005, **26**, 15, 2401.

199. I. Lee and K. Akiyoshi, *Biomaterials*, 2004, **25**, 15, 2911.

200. K. Akiyoshi, S. Yamaguchi and J. Sunamoto, *Chemistry Letters*, 1991, **7**, 1263.

201. K. Akiyoshi, K. Nagai, T. Nishikawa and J. Sunamoto, *Chemistry Letters*, 1992, **9**, 1727.

202. K. Kuroda, K. Fujimoto, J. Sunamoto and K. Akiyoshi, *Langmuir*, 2002, **18**, 10, 3780.

203. I. Taniguchi, K. Akiyoshi and J. Sunamoto, *Macromolecular Chemistry and Physics*, 1999, **200**, 6, 1554.

204. N. Alles, N.S. Soysa, M.D.A. Hussain, N. Tomomatsu, H. Saito, R. Baron, N. Morimoto, K. Aoki, K. Akiyoshi and K. Ohya, *European Journal of Pharmaceutical Sciences*, 2009, **37**, 2, 83.

205. T. Shimizu, T. Kishida, U. Hasegawa, Y. Ueda, J. Imanishi, H. Yamagishi, K. Akiyoshi, E. Otsuji and O. Mazda, *Biochemical and Biophysical Research Communications*, 2008, **367**, 2, 330.

206. K. Ikeda, T. Okada, S. Sawada, K. Akiyoshi and K. Matsuzaki, *FEBS Letters*, 2006, **580**, 28/29, 6587.

207. S. Boridy, H. Takahashi, K. Akiyoshi and D. Maysinger, *Biomaterials*, 2009, **30**, 29, 5583.

208. T. Nochi, Y. Yuki, H. Takahashi, S. Sawada, M. Mejima, T. Kohda, N. Harada, I.G. Kong, A. Sato, N. Kataoka, D. Tokuhara, S. Kurokawa, Y. Takahashi, H. Tsukada, S. Kozaki, K. Akiyoshi and H. Kiyono, *Nature Materials*, 2010, **9**, 7, 572.

209. Z.R. Cui and R.J. Mumper, *Pharmaceutical Research*, 2002, **19**, 7, 939.

210. H.J. Yang, I.S. Park and K. Na, *Colloids and Surfaces A: Physicochemical and Engineering Aspects*, 2009, **340**, 1–3, 115.

211. S. Lack, V. Dulong, D. Le Cerf, L. Picton, J.F. Argillier and G. Muller, *Polymer Bulletin*, 2004, **52**, 6, 429.

212. S. Lack, V. Dulong, L. Picton, D. Le Cerf and E. Condamine, *Carbohydrate Research*, 2007, **342**, 7, 943.

213. A. Autissier, D. Letourneur and C. Le Visage, *Journal of Biomedical Materials Research Part A*, 2007, **82A**, 2, 336.

214. A. Ramamurthi and I. Vesely, *Journal of Biomedical Materials Research Part A*, 2002, **60**, 1, 195.

215. C. Joana, M. Susana, M. João and M.G. Francisco, *Journal of Biomedical Materials Research Part A*, 2009, **93A**, 1, 389.

216. A. San Juan, H. Hlawaty, F. Chaubet, D. Letourneur and L.J. Feldman, *Journal of Biomedical Materials Research Part A*, 2007, **82A**, 2, 354.

217. A. Khademhosseini and R. Langer, *Biomaterials*, 2007, **28**, 34, 5087.

218. K.Y. Lee and D.J. Mooney, *Chemical Reviews*, 2001, **101**, 7, 1869.

219. A. Autissier, C. Le Visage, C. Pouzet, F. Chaubet and D. Letourneur, *Acta Biomaterialia*, 2010, **6**, 9, 3640.

220. C. Le Visage, F. Chaubet, A. Autissier and D. Letourneur, inventors; INSERM, assignee; WO/2009/047347, 2009.

221. C. Le Visage and D. Letourneur, inventors; INSERM, assignee; WO/2009/047346, 2009.

222. N.B. Thebaud, D. Pierron, R. Bareille, C. Le Visage, D. Letourneur and L. Bordenave, *Journal of Materials Science - Materials in Medicine*, 2007, **18**, 2, 339.

223. K. Na, D. Shin, K. Yun, K-H. Park and K.C. Lee, *Biotechnology Letters*, 2003, **25**, 5, 381.

224. E.A. Silva, E-S. Kim, H.J. Kong and D.J. Mooney, *Proceedings of the National Academy of Sciences of the United States of America*, 2008, **105**, 38, 14347.

225. M. Poirier-Quinot, G. Frasca, C. Wilhelm, N. Luciani, J.C. Ginefri, L. Darrasse, D. Letourneur, C. Le Visage and F. Gazeau, *Tissue Engineering, Part C: Methods*, 2010, **16**, 2, 185.

226. G. Fundueanu, M. Constantin, D. Mihai, F. Bortolotti, R. Cortesi, P. Ascenzi and E. Menegatti, *Journal of Chromatography B: Analytical Technologies in the Biomedical and Life Sciences*, 2003, **791**, 1/2, 407.

227. P. Ding, A.W. Pacek, W.J. Frith, I.T. Norton and B. Wolf, *Food Hydrocolloids*, 2005, **19**, 3, 567.

228. P. Ding, I.T. Norton, Z. Zhang and A.W. Pacek, *Journal of Food Engineering*, 2008, **86**, 3, 307.

229. N. Morimoto, T. Endo, M. Ohtomi, Y. Iwasaki and K. Akiyoshi, *Macromolecular Bioscience*, 2005, **5**, 8, 710.

230. G. Fundueanu, M. Constantin and P. Ascenzi, *Biomaterials*, 2008, **29**, 18, 2767.

231. S. Yamane, Y. Sasaki and K. Akiyoshi, *Chemistry Letters*, 2008, **37**, 12, 1282.

232. N. Morimoto, T. Ohki, K. Kurita and K. Akiyoshi, *Macromolecular Rapid Communications*, 2008, **29**, 8, 672.

233. A. Abed, B. Deval, N. Assoul, I. Bataille, P. Portes, L. Louedec, D. Henin, D. Letourneur and A. Meddahi-Pelle, *Tissue Engineering Part A*, 2008, **14**, 4, 519.

234. D. Letourneur, F. Chaubet, A. Meddahi-Pelle, I. Bataille and C. Chesne, inventors; INSERM, assignee; WO/2005/053396, 2005.

235. H.C. Song, H.S. Bom, K. Na, K.Y. Lee, Y.J. Heo and S.M. Kim, *Journal of Nuclear Medicine*, 2003, **44**, 5, 1087.

236. U. Hasegawa, S.I.M. Nomura, S.C. Kaul, T. Hirano and K. Akiyoshi, *Biochemical and Biophysical Research Communications*, 2005, **331**, 4, 917.

237. S. Toita, U. Hasegawa, H. Koga, I. Sekiya, T. Muneta and K. Akiyoshi, *Journal of Nanoscience and Nanotechnology*, 2008, **8**, 5, 2279.

238. F.P. Gao, Y.Y. Cai, J. Zhou, X.X. Xie, W.W. Ouyang, Y.H. Zhang, X.F. Wang, X.D. Zhang, X.W. Wang, L.Y. Zhao and J.T. Tang, *Nano Research*, 2010, **3**, 1, 23.

5 Cellulose and Its Use for Blood Purification

Nicholas Andrew Hoenich

5.1 Introduction

Cellulose is a natural polymer with unique chemical physical and mechanical characteristics and is used in a variety of industrial applications. It is also used clinically in blood purification processes as well as in a number of evolving clinical applications, for example, as a scaffold material in tissue engineering, a temporary skin substitute, a haemostatic agent and a postoperative adhesion barrier. In this chapter, the clinical use of cellulose in blood purification processes is discussed.

5.2 Concepts of Blood Purification

Blood purification within the body is performed by different organs - carbon dioxide is removed by the lungs; excess water and products arising from the dietary breakdown of ingested protein and cellular metabolism are removed by the kidney. The liver has a wide range of functions, including detoxification, protein synthesis and production of biochemicals necessary for digestion. Extracorporeal circulatory procedures may be used as a technique to replace or augment a number of these functions. Such techniques rely upon the removal of compounds from the blood by diffusion, convection or adsorption. The most commonly used extracorporeal circulatory process is the replacement or augmentation of kidney function, with around 20,000 patients in the United Kingdom being treated currently. In the United States in 2004 (the most recent year for which complete data are available), 104,364 patients (approximately 0.03% of the US population) began renal replacement therapy; however, considerable inequalities in the number of patients treated are known to exist, in both emerging economies and Western countries [1, 2].

The primary function of the kidney is to maintain homeostasis in the body. It does so by removing excess fluid, and metabolites resulting from metabolic activity and the intake of food, by a complex process involving filtration, selective reabsorption and excretion. In addition, it has a number of metabolic functions related to the control of blood pressure, red cell production and the conversion of vitamin D. Damage or injury to the kidney necessitates augmentation or replacement of function by artificial means. Such augmentation or replacement focuses on fluid and metabolite removal but does not at present involve any augmentation of the metabolic functions of the organ. The augmentation or replacement of kidney function is generally initiated when a substantial or total loss of the ability of the human kidneys to remove water, excrete metabolic waste products or maintain body homeostasis occurs. The most commonly used augmentation or replacement method is known as haemodialysis, and involves continuous passage of patient's blood through an artificial kidney or a haemodialyser containing a semipermeable membrane and returning it back to the patient. Blood flows on one side of the membrane while a dilute electrolyte solution (dialysis fluid) flows on the other side of the membrane. The processes that occur within the haemodialyser can be summarised as follows:

- Equilibration of the electrolyte composition of blood and dialysis fluid.

- Elimination of metabolites elevated as a consequence of renal insufficiency by diffusion into the dialysis fluid. Some convective mass transport due to ultrafiltration or fluid removal also takes place and contributes to the overall solute removal.

- In both reversible and end-stage renal disease, there is a need to remove fluid from the body, and this removal is via a process known as ultrafiltration, which is governed by the hydrostatic pressure difference between the blood and the dialysis fluid. The hydrostatic pressure gradient is generally supplemented by an osmotic pressure gradient induced by the inclusion of glucose in the dialysis fluid.

Haemodialysis is the most widely used procedure for the treatment of end-stage renal disease. The process of haemodialysis can be divided further into low and high flux dialysis. Low flux dialysis utilises membranes with a low hydraulic permeability, blood and dialysis fluid flow rates of 250 and 500 ml/min, respectively, and a treatment duration of 4 hours or more. If the dialyser surface area is increased and the blood and dialysis fluid rates are increased up to 400 and 800 ml/min, respectively, the treatment is termed high flux dialysis.

Haemodialysis is a process in which the solute transport from the blood to the dialysis fluid occurs by diffusion. This tends to favour the removal of low-molecular-weight solutes. To eliminate high-molecular-weight compounds whose presence in nonphysiological levels is associated with long-term complications in dialysis patients, such as β2-microglobulin (~11 kDa), which aggregates into amyloid fibres that deposit in joint spaces, resulting in dialysis-related amyloidosis, recent approaches to the treatment of kidney disease by extracorporeal circulatory approaches have focused on the use of high flux membranes and a combination of convection and diffusion or convection alone to enhance the range of metabolites retained that have to be removed [3].

As the clinical requirements for the treatment of acute or reversible kidney failure differ, the treatments used are primarily those in which transmembrane transport of metabolites occurs by convection.

Irreversible renal failure is treated intermittently. Historically, this was because of problems of gaining access to the circulation as well as the complexity of the equipment used. Currently, the accepted standard for the treatment of irreversible or end-stage renal disease is three times a week treatment. Since such treatment is uneven in its spacing, there is considerable interest in moving to more frequent schedules to avoid the large changes in patient biochemistry that are associated with the traditional approach [4].

Acute or reversible renal failure arising from an injury or damage to the kidney is treated continuously over a period of several days. Furthermore, as acute renal failure may frequently be associated with sepsis, the dialytic process in such instances needs to be supplemented by extracorporeal adsorption techniques to remove endotoxins. Such removal can be attained by the use of cellulose-coated microspheres, as they demonstrate a high adsorptive capacity for endotoxins [5, 6]. When ingested, poisons, with or without renal failure, can also be similarly removed [7].

End-stage liver disease is generally treated by liver transplantation. However, transplantation is often delayed due to the unavailability of a donor organ, and temporary liver support may be used in this interim period. Such temporary support is by the use of bioartificial liver systems, which function as bridging devices. Such devices utilise liver cells or hepatocytes derived from human liver tumours, such as the hepatoma cell line HepG2, or *in vitro* immortalised cell lines, such as the NKNT-3 cell line. Tumour-derived cell lines such as HepG2 and C3A express a variety of liver functions, but some specific liver functions, such as ammonia detoxification and ureagenesis, are poorly expressed and are insufficient to sustain life. More recently, a novel, clonal, immortalised human fetal liver cell line, cBAL111, that displays hepatocyte-specific functions has been described in the literature [8].

There is considerable interest in the use of alternative approaches such as the use of livers of transgenic pigs; however, in such an approach, the possibility exists that porcine endogenous retrovirus may infect human tissues, and clinical application of this approach remains in abeyance.

Culture-based extracorporeal liver assist devices may be simple hollow fibre devices in which the area around the fibres is used to grow cells to a high density with the culture remaining stable for an extended period. In such devices, the blood from the patient flows through the lumen of the fibres and is separated from the liver cells by the membrane wall. Early approaches used a cellulose acetate membrane whose nominal molecular cut-off was 70 kDa thereby permitting the diffusion of not only small-molecular metabolites but also low-molecular-weight proteins. This type of approach was used clinically by Sussman and co-workers [9].

Current culture-based extracorporeal liver assist devices use synthetic membranes and are more complex in their design. Poyck and co-workers in their recent studies used a device that externally retained the structure of a hollow fibre dialyser, but differed internally. It contained a three-dimensional nonwoven hydrophilic polyester matrix for high-density hepatocyte culture. The matrix was circularly wound around a polycarbonate core, and hydrophobic polypropylene gas capillaries situated between the matrix layers in a parallel fashion supplied oxygen and removed CO_2. This combination of the polyester matrix and the oxygenation capillaries creates a third compartment used to perfuse the plasma or medium through the bioreactor. Within this design, the plasma or medium has direct access to the hepatocytes in the polyester fabric. Plasma is perfused through the bioreactor via the side ports. The integrated oxygenation capillaries are embedded in polyurethane resin and fitted with gas inlet and outlet caps. The homogeneous distribution of the oxygenation capillaries throughout the bioreactor compartment ensures that every hepatocyte has an oxygen source within its immediate surroundings [8].

Liver failure following ingestion of poisons may also be treated using such an approach, but more frequently, hybrid systems, which combine dialytic concepts with approaches intended to remove protein-bound metabolites, are used such as the molecular adsorbents recirculation system [10], single-pass albumin dialysis [11] and the Prometheus system [12].

Other blood purification methods include plasmapheresis, a therapeutic treatment of blood in which the cellular component is separated from the plasma and which is used primarily in the treatment of immune-mediated diseases such as Guillain-Barré syndrome, lupus and thrombotic thrombocytopenic purpura. There are four main types of membrane plasmapheresis: plasma exchange (PE), double-filtration plasmapheresis (DFPP), plasma adsorption and immunoadsorption (IA). In PE, plasma separated with a plasma separator is discarded and replaced with the same volume of fresh frozen plasma or albumin solution. In DFPP, plasma separated with a plasma separator is allowed to pass through the plasma component separator with a small pore size. High-molecular-weight proteins are discarded and low-molecular-weight substances including valuable albumin are returned to the patient. A small amount of substitution fluid such as albumin may be added. A variant of this technique is rheopheresis, which is used to treat microcirculatory disorders or impaired microcirculation; in this process, fibrinogen and high-molecular-weight substances responsible for microcirculatory disorders (e.g., age-related macular degeneration) are removed. In plasma adsorption, plasma separated with a plasma separator is allowed to flow into a plasma adsorption column. Pathogenic substances are adsorbed and removed due to affinity between ligands and pathogenic substances. In contrast to the above approaches, no substitution fluid is used. IA is a subcategory of plasma adsorption in which the adsorption column selectively adsorbs immune complexes and autoantibodies. Current clinical practice relies on the use of hollow fibres for PE and DFPP; however, the membranes used (cellulose acetate or synthetic polymers) have a larger pore size (0.1–0.5 μm).

Haemoperfusion, used for removing drugs or poisons from the blood in emergency situations, is a technique that involves passing the patient's blood over an adsorbent substance and then returning it back to the circulation. The adsorbent substances most commonly used are resins or activated carbon. When activated carbon is used, to minimise the shedding of microparticles during use and to eliminate damage to blood cells, it is encapsulated in cellulose acetate [13].

For many years, the clinical application of haemoperfusion was limited to treatment of acute poisoning. Since the 1990s, interest in the use of adsorbents in extracorporeal medical devices has been increasing and a number of new sorbent materials have become available [14, 15].

Hollow fibre devices may also be used for the removal of viruses from the blood and are primarily used for the enhancement of safety in the production of biotherapeutic drug products such as biopharmaceuticals and plasma derivatives. Historically, cuprammonium rayon was used for this application, but today it has been replaced by a hydrophilised polyvinylidene fluoride membrane.

Lung function can also be replaced or augmented by artificial means. Oxygen and carbon dioxide transfer can be accomplished by diffusion through semipermeable membranes interposed between the blood and gas phases without the injurious effects of the direct blood-oxygen interface, which characterised the earlier forms of oxygenators. Early forms of this approach used ethyl cellulose and cellophane membranes [16]. Such membranes were subsequently replaced by synthetic materials, which offered a markedly improved gas transport. Today, in addition to forming part of the heart-lung machine used in cardiopulmonary bypass operations, to take over the function of the heart and lungs and to maintain the circulation of blood and the oxygen content of the body during surgery, extracorporeal membrane oxygenation is used to provide respiratory support to patients whose heart and lungs are severely diseased or damaged. The technique is most commonly used in neonatal intensive care units, for newborns in pulmonary distress, but can also be used for adults. One of the recent uses of this approach has been the treatment of adults and children with respiratory distress following infection by the H1N1 flu virus [17].

5.3 Cellulose: Derivation and Structure

Cellulose in its native form can be found in plants. Cellulose from plants, forest trees and cotton is assembled from glucose, which is produced in the living plant cell by photosynthesis; about 33% of all plant matter is cellulose, the cellulose content of cotton is 90% and that of wood is 50%. It is also present in the cell walls of green algae (e.g., *Valonia ventricosa* and *Chaetomorpha melagonium*) and fungi.

Cellulose never occurs in a pure form, and plant-derived cellulose usually contains hemicellulose, lignin and pectin; however, current cellulose extraction and purification processes are able to yield materials of high purity. Commercial production of cellulose uses easily harvested sources such as wood and cotton; in addition, cellulose may also be derived from bacteria. The common vinegar bacterium *Acetobacter xylinum* is used to produce bacterial cellulose. This nonphotosynthetic organism can procure glucose, sugar, glycerol or other organic substrates and convert them into pure cellulose. *A. xylinum* is one of the most prolific cellulose-producing bacteria, and a typical single cell can convert up to 108 glucose molecules per hour into microbial cellulose. Microbial cellulose differs in a number of respects from cellulose obtained from plants, namely, purity, as it does not contain lignin or hemicellulose, mechanical strength and high absorbency in the hydrated state.

The chemical structure of cellulose is well established. It is a polydisperse linear homopolymer, consisting of repeating β-1,4-glycosidic linked D-glucopyranose units (or anhydroglucose units) (**Figure 5.1**). It has been shown by ^{1}H-NMR spectroscopy that the β-D-glucopyranose adopts the $^4C_1$ chain conformation, which is the lowest free energy conformation of the molecule. As a consequence, the hydroxyl groups are positioned in the ring plane (equatorial), while the hydrogen atoms are in the vertical position (axial). The polymer contains free hydroxyl groups at the C_2, C_3 and C_6 atoms. Because of the β-configuration of the intermonomer links, the size of the cellulose molecule can vary depending on the degree of polymerisation or chain length, which in turn is dependent upon the source of the cellulose. Several different crystalline structures of cellulose are known, corresponding to the location

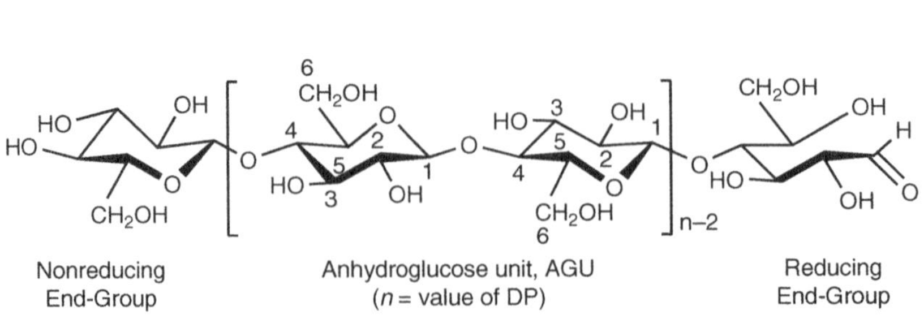

Figure 5.1 The molecular structure of cellulose

of hydrogen bonds between and within strands. Natural cellulose is cellulose I, with structures Iα and Iβ. Cellulose produced by bacteria and algae is enriched in Iα, while cellulose of higher plants mainly consists of Iβ. Cellulose in regenerated cellulose fibres is cellulose II. The conversion of cellulose I to cellulose II is not reversible, suggesting that cellulose I is metastable and cellulose II is stable. With various chemical treatments it is possible to produce the structures cellulose III and cellulose IV.

As early as 1838, Payen recognised cellulose as a definitive substance and coined the name. Cellulose nitrate (frequently referred to as nitrocellulose) was discovered in 1846; the preparation of Schweizer's reagent, that is, a cuprammonium hydroxide solution, represented the first cellulose solvent in 1857; and the synthesis of an organosoluble cellulose acetate occurred a few years later in 1865.

Partially functionalised cellulose nitrate mixed with camphor as softener was one of the first polymeric materials used as a plastic and is well known under the trade name of Celluloid. Cellulose nitrates of higher nitrogen content have been used extensively for military purposes.

Cellulose was first isolated from wood in 1885 by Charles F. Cross and Edward Bevan at the Jodrell Laboratory of the Royal Botanic Gardens, Kew, London. They also showed that when cellulose fibres are dissolved in an alkali and carbon disulfide, a viscose solution is formed. Subsequently, Cross and Bevan, in collaboration with Clayton Beadle, went on to manufacture a cellulose film from viscose by passing the viscose solution through a bath of dilute sulfuric acid and sodium sulfate. The commercial production of cellulose in sheet form ('Cellophane') began some years later in 1913 when Dr. Jacques Brandenberger began its production at Bezons in France. Later, the material was also produced in the United States by the E.I. duPont de Nemours and Company, who modified it by coating it with a nitrocellulose lacquer to make it moisture proof, thereby widening its suitability for commercial applications. Today, the word 'Cellophane' has become genericised but in the United Kingdom and in many other countries it remains a registered trademark and the property of Innovia Films Ltd.

5.4 Cellulose and Its Use in Blood Purification

The term dialysis was coined by Thomas Graham, Professor of Chemistry at Anderson's University in Glasgow. In 1861, he demonstrated through his experiments that crystalloids were able to diffuse through vegetable parchment coated with albumin. Although Graham predicted that his findings might be applicable to medicine, he did not venture into this area. Dialysis may therefore be considered as a rate-governed membrane process in which microsolutes are driven across a semi-permeable membrane by means of a concentration gradient. The microsolutes diffuse across the membrane into the dialysis fluid, which in chemical engineering terms is the receiving fluid. If this solution is not continuously renewed, the solute concentrations on both sides of the membrane will equilibrate, negating the driving force.

The first membranes used in blood purification were handmade tubular membranes derived from a cellulose nitrate derivative made from a mixture of cotton, sulfuric acid and nitric acid dissolved in alcohol. This material was difficult to produce and to sterilise; it was subject to leakage and suffered from a lack of consistency with respect to pore size. All these contributed to its limited application. The tubes, which were connected at either end to a glass manifold to receive and return the blood, were approximately 8 cm in length and were encased in a glass cylinder through which the dialysis fluid passed [18]. The design of this early device was in many respects similar to the hollow fibre designs in use today.

The first blood purification in humans was performed in Germany by Hans Georg Haas (1886–1971), who developed his system for the dialysis of blood (haemodialysis) in 1923. He also used collodion tubes. Progress in the development of dialysis membranes and dialysers was not made until 1938, when Thalheimer experimented with dog blood using cellulose-based tubular membranes and heparin, which was used as an anticoagulant [19].

The next milestone was the development of an artificial kidney by Kolff in the Netherlands, which used cellulose tubing wrapped around a rotating drum [20]. In this device, a blood-filled tube was connected to the patient's circulatory system and was wrapped concentrically around a rotating wooden drum placed horizontally in a trough containing an electrolyte solution. Blood flow in the circuit was facilitated by a pump and as the membrane passed through the bath, the uraemic toxins would pass into the electrolyte solution. Examples of the Kolff rotating drum kidney were sent after World War II to the Peter Brent Brigham Hospital in Boston, where they were modified, and the modified machines became known as the Kolff-Brigham kidney. Between 1954 and 1962, many such devices were shipped from Boston to other hospitals worldwide and formed the basis of treatment of acute renal failure. Although such devices could remove waste products or toxins, whose levels had increased as a consequence of renal failure, the removal of fluid that would normally be passed as urine was difficult. This was solved by Nils Alwall in Sweden, who developed a variant of the Kolff-Brigham kidney, which combined both dialysis and fluid removal (ultrafiltration), by placing the membrane between two concentric wire drums [21].

With the demonstration that uraemic patients could be successfully treated using artificial kidneys, considerable activity around the world followed, focusing upon the development of improved and more effective dialysers. Although many experimental designs were produced, the design originating from Kiil was the most significant [22]. In this device, pairs of membrane sheets were supported between an assembly of grooved plates. The blood flowed within the envelope formed by the pairs of membranes when the assembly was put together. The dialysis fluid flowed on the outer side of the membrane envelope in grooves, which also supported the membranes. In contrast to earlier designs, due to its low pressure resistance, the parallel-plate dialyser could be used without the need to use an external pump.

With the availability of reliable methods to gain repeated access to the patient's circulatory system, blood purification extended beyond the treatment of acute or reversible renal insufficiency to the treatment of end-stage renal disease. Such treatments used exclusively cellulose membranes produced from cotton by either the xanthase or cuoxam process. In the former, a solution of purified cellulose is dissolved in sodium hydroxide, followed by an ageing process, which lowers the molecular weight of cellulose. Carbon disulfide is added to convert the material to cellulose xanthate. This is then dissolved in excess sodium hydroxide and extruded into an acid bath. This process was used for the production of Visking tubing in the United States, and for the production of Cellophane in Europe. In the cuoxam process, cellulose is solubilised in an ammonia solution of cupric oxide. The cuprammonium cellulose complex is extruded into an acid film to yield the regenerated cellulose, which in Europe was manufactured by Membrana GmbH and known by the trade name Cuprophan®. Japanese manufacturers of cellulose membranes used the term cuprammonium rayon, when produced by similar methods. Membranes thus produced were macroscopically homogeneous; they were hydrophilic, absorbed water and formed a hydrogel in which diffusion of solutes takes place through water-filled regions rather than pores. By the late 1960s, the first hollow fibre device was manufactured utilising cellulose membranes produced using a gel spinning process [23, 24]. European production of hollow fibres based on cellulose and produced by a wet spinning process based on the cuprammonium process followed some years later in 1974. Originally such membranes had a wall thickness of 16 μm, which from the mid-1970s gradually decreased to the current thickness of 8 μm.

Today, treatment of renal failure, be it acute or end stage, relies on the hollow fibre device (**Figure 5.2**). In this, the membrane is fixed at both ends in a Perspex jacket and blood enters and leaves the assembly via headers, which distribute the blood into the fibres. The dialysis fluid enters and leaves the assembly via the Perspex outer casing in a countercurrent flow configuration to that of the blood and is in contact with the outer side of the fibres during its passage through the device. The manufacture of membranes used in such devices based on cellulose is either by solution spinning or by melt spinning. In solution spinning, the polymer is first dissolved in a solvent or a mixture of solvents. The viscous material thus formed is forced through a spinneret, with a precision orifice containing a coaxial tube; a core liquid being forced through

Figure 5.2 A modern hollow fibre haemodialyser

the coaxial tube stabilises the emerging thread. The fibre thus formed then passes through a coagulation bath, in which precipitation of the polymer occurs. The fibre then passes through an extraction bath for the removal of residual solvents or additives. To retain morphology and ensure constant transport properties, a plasticiser is added, and the speed of precipitation is controlled. A variety of techniques may be used for this, including the addition of plasticisers and pore formers to the polymer, the addition of solvents to the core fluid and variation of the composition of the coagulation bath or the adjustment of the temperature and extrusion rate.

In melt spinning, the polymer is melted at high temperature and extruded through the spinneret and then cooled in air. The membrane formation occurs as a result of solidification of the polymer material as it changes from the liquid to the solid phase. Generally, an inert gas is used as the core fluid. This process can only be used in the production of membranes that have a high thermal stability. As in solution spinning, a variety of different approaches may be used to control the morphology of the material, namely, variation of the amount of plasticiser in the polymer and variation of the spinning temperature.

The composition of polymer dope, the coagulation bath, the extraction bath and the core liquid plasticising or wetting agent for a number of cellulose membranes is summarised in **Table 5.1**. The production processes described have been collated from published information, and the information presented is far from complete since it includes only information that is in the public domain. Manufacturers are reticent about the exact conditions and compositions used in the production of their product and much of the information is available only in the form of patents. Many of the membranes produced are known by specific names and are detailed in **Table 5.2**. This table contains not only the currently produced materials, but also those that were produced specifically for blood purification historically. Cuprophan® and Hemophan®, both manufactured in Germany by Membrana GmbH, were withdrawn from clinical use at the beginning of the current decade owing to a gradual global switch to synthetic-based membranes and the purchase of the company from the Accordis Group by Polypore International Inc. (Charlotte, North Carolina). Currently, only a single membrane based on cellulose, known by the name SMC®, remains in production; this membrane is a synthetically modified membrane with a unique surface structure for enhanced haemocompatibility, and when incorporated in a dialyser it is used with a novel spacer yarn (Performance Enhancing Technology®) to optimise solute removal performance.

As indicated in **Figure 5.1**, cellulose contains OH groups (~31.48% by weight). In blood purification processes, the presence of such groups is considered a disadvantage since their presence governs the material's blood-contact behaviour. To optimise such behaviour, a variety of different approaches that reduce the number of hydroxyl groups are used. The group of membranes thus produced are known as modified cellulose membranes. Commonly used approaches include etherification, esterification, crosslinking or graft copolymerisation.

In cellulose ethers, the hydroxyl groups are replaced by either alkyl or hydroxyl alkyl groups. Cellulose esters include cellulose acetate and cellulose triacetate prepared by reacting cellulose with acetic anhydrite, utilising acetic acid as the solvent and sulfuric acid as a catalyst. Graft polymerisation of cellulose is complex and outside the scope of this chapter, but the approaches used have been recently reviewed [25].

In the context of membranes for blood purification, in cellulose acetate membranes the three hydroxyl groups of the glucose monomer are replaced by acetyl groups; a differing degree of substitution can be used (2, 2.4 and 3) and this substitution yields cellulose diacetate and cellulose triacetate. The first modification of the classical cellulose membranes was in 1987, where there was a substitution of the OH groups by *N,N*-diethylaminoethyl (*N,N*,-DEAE). This material was produced by Membrana GmbH (Wuppertal, Germany) and sold under the trade name of Hemophan®. In this material, the tertiary amino groups replaced only a small number (1.5%) of hydroxyl groups, and resulted in the presence of hydrophobic regions on a hydrophilic surface, thereby hindering the interaction of the surface with complement fractions. The presence of DEAE groups was however implicated in the increased consumption of anticoagulant when the material was used clinically. A similar approach was used in the case of synthetically modified cellulose (SMC), in which a small proportion (~1%) of the OH groups were replaced by hydrophobic benzyl groups through ether bonds. Surface modifications of cellulose membranes

Table 5.1 Materials used in the manufacture of hollow fibres based on cellulose

	Polymer dope	Core fluid	Coagulation	Plasticiser	Morphology
Membranes produced by solution spinning					
Cellulose	$Cu(OH)_2$ NH_4OH $NaOH+CS_2$	Isopropyl	H_2SO_4	Glycerine	Homogeneous
Cellulose acetate	Dimethyl formamine Poly(ethylene glycol) (PEG) *N*-methylpyrrolidone	Liquid paraffin Glycerine	Water	Glycerine	Homogeneous
Cellulose	*N*-methylpyrrolidone PEG	*N*-methylpyrrolidone PEG Water	Water *N*-methylpyrrolidone	Glycerine	Homogeneous
Membranes manufactured by melt spinning					
Cellulose acetate	PEG with glycerine or diglycerol	Nitrogen	Air	Glycerine	Homogeneous
Saponified cellulose ester	PEG Glycerine	Nitrogen	Air	Glycerine	Homogeneous

Table 5.2 Cellulose-based haemodialysis membranes	
Cuprammonium rayon	Regenerated cellulose (unmodified)
Cuprophan®	Regenerated cellulose (unmodified)
Hemophan®	Cellulose etherified with diethylaminoethyl (DEAE)
Excerbane®	Regenerated cellulose coated with vitamin E
PEG-modified cellulose	Cuprammonium rayon coated with PEG
SMC®	Synthetically modified cellulose in which there is benzyl group substitution of the hydroxyl groups
Cellulose acetate	Regenerated cellulose esterified with acetate
Cellulose triacetate	Regenerated cellulose esterified with 3.0 acetate
Cellulose 2.5 acetate	Regenerated cellulose esterified with 2.5 acetate

produced in Japan focused on the production of a bioreactive membrane in which vitamin E (D-α-tocopherol) is used. The surface modification was carried out during the fibre spinning process: the modifying solution contained a hydrophilic acrylic polymer with reactive epoxy groups, a fluororesin polymer and an oleyl alcohol chain dissolved in the core solution. The regenerated cellulose and the modified core solution are coextruded in a conventional manner through a spinneret into a coagulation bath where two phases are formed. The outer circumference of the fibre is composed of cellulose and the inner surface is covered by the hydrophobic modifier linked to the outer surface by covalent bonding of the reactive epoxy group of the modifier with the OH group of the cellulose membrane with about 150 mg/m^2 of vitamin E immobilised. Grafting has also been used to modify cellulose membrane; to the cellulosic backbone of the current Japanese produced cuprammonium rayon, a polyethylene glycol (PEG) is grafted (Asahi Kasei Kuraray Medical Biomembrane™). In this approach, alkylethylene carboxylic acid is esterified with its terminal hydroxyl group to the hydroxyl group of the cellulose polymer. The PEG chains thus grafted form a hydrogel layer (~2.4 nm thick) on the cellulose surface acting as a buffer zone between the cellulose and blood, hindering the direct contact of plasma proteins with the membrane surface.

5.5 Membrane Performance with Respect to Blood Purification

In modelling the membrane transport properties, two different approaches are possible: first modelling homogeneous membranes, which in this context would relate to polymer liquid gels, and modelling porous membranes, which are viewed as an impervious polymer phase containing liquid-filled pores. Transport phenomena in homogeneous membranes are well matched by a phenomenological approach and a solution diffusion theory. This approach focuses upon the relationship between the driving force and the resulting flux. In dialysis, the primary flux arises from diffusion, which in turn is driven by a concentration gradient across the membrane and can be quantified by consideration of Fick's first law of diffusion. Mathematically, diffusive solute transport across the membrane is governed by the relationship:

$$J_D = -DA\frac{dC}{dx}$$

where J_D is the diffusive flux across the membrane, D is the solute diffusion coefficient, A is the area available for transport and dC/dx is the concentration gradient.

Thus, the transport clearly decreases as solute size increases because of two effects: first an increased molecular size is associated with a reduced solute diffusion coefficient, and an increased solute size produces more collisions with the pore walls.

The transport properties of a porous membrane can be described by a hydrodynamic approach, which considers the structure of the membrane. Linear nonequilibrium or irreversible thermodynamics is a more general approach that is used and may be applied to either class of membrane. In describing transport across either type of membrane, the problem may be simplified to a single question, namely, how to describe the size distribution of the channels through which the solutes pass and the interaction between the solute and the membrane material. In the transport of solutes across a membrane, however, there is a simultaneous pressure- and concentration-driven transport, resulting in a mixed diffusive and convective flux. Furthermore, if such fluxes are large, then their interaction must also be considered. Irreversible thermodynamics provides a framework whereby these fluxes and interactions may be analysed. This approach owes its origin to the work undertaken by Kedem and Katchalsky [26].

The membranes hydraulic permeability (L_P) is a parameter that reflects the relationship between the volumetric flux and the transmembrane pressure difference (ΔP). In the absence of an osmotic pressure difference, the volumetric flux (J_V) can be expressed mathematically as:

$$J_V = L_P \Delta P$$

which in the presence of an osmotic pressure difference becomes:

$$J_V = L_P[\Delta P - \sigma \Delta \pi_S]$$

where $\Delta\pi_S$ is the osmotic pressure difference across the membrane and σ the Staverman reflection coefficient.

As the volumetric flow is associated with a degree of solute transport, such transport can be expressed as:

$$J_s = C_s(1 - \sigma)J_v + \omega\Delta\pi_s$$

where ω is the solute permeability coefficient and C_S the solute concentration.

In addition to the flux across the membrane, concentration changes between the two liquid phases separated by the membrane may also arise from adsorption to the membrane, which may be considered as a special case of membrane fouling. Adsorption reflects the interaction at a molecular level between the material surface and the molecule and is determined by the hydrostatic forces and the nature of the compound. It can be quantified by different mathematical relationships namely the Temkin, the Hill, the Freundlich or the Langmuir equations. The Langmuir equation is frequently used since it describes the relationship between the concentration of a compound adsorbing to binding sites and the fractional occupancy of the binding sites. The shape of the equation is a gradual positive curve that flattens to a constant value. This shape arises from the fact that adsorption can occur only at a fixed number of definite localised sites, and that each site can hold only one molecule, and all sites are equivalent with no interaction between adsorbed molecules.

In addition to the transport processes, when used clinically, the biocompatibility of the membrane is also of paramount importance.

Contact of blood with membrane induces interactions and subsequent alterations of the blood elements. In general, the topic 'membrane biocompatibility' is concerned with such interactions. The general sequences of events following contact of the blood with a foreign surface are shown schematically in **Figure 5.3**. The basic sequence of events that follow blood-material contact are an initial rapid absorption of plasma proteins onto the surfaces of the foreign material, the activation of the coagulation cascade and of platelets, leading to thrombin generation and thrombus formation and the activation of the immune system. Such activation is principally via the alternative pathway. It is initiated by the deposition of C3b on the material surface which with Factor B forms C3 and C5 convertases. These enzymes cleave the anaphylatoxins C3a and C5a from C3 and C5 by an autocatalytic process. Once in the circulation, the C-terminal arginine is removed and C3a des Arg and C5a des Arg are formed. These fractions are generally measured when membrane-induced complement activation is studied. The cleavage of C5 by C5 convertase results not only in the production of C5a but also in the production of C5b, which initiates a macromolecular complex of proteins, the membrane attack complex, formed from C5b, C6, C7, C8 and C9 (C5b-9).

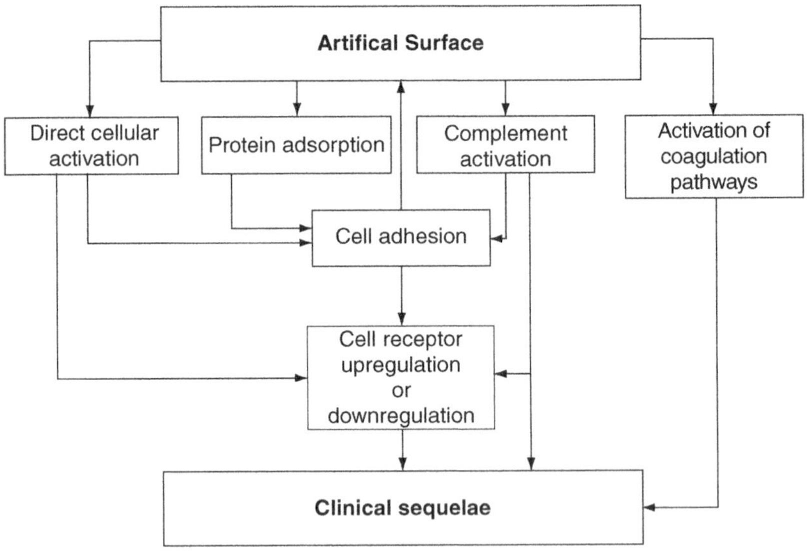

Figure 5.3 Blood pathway activation following contact with a nonphysiological surface

Complement products are the principal mediators of the inflammatory response produced by blood contact with a foreign surface and are important markers for the biocompatibility of such materials. A wide range of clinical sequelae are associated with the generation of C3a, C5a and C5b-9 and virtually every blood cell type responds either directly via receptors or indirectly via secondary mediators to these complement fractions.

These responses are in part a consequence of blood recognising the material as non-self, and in part a consequence of the interaction between the material and cell surfaces, although other factors such as underlying disease condition, drugs administered, as well as biochemical abnormalities also play a part. There is no uniform agreement on the definition of biocompatibility of membranes. Some investigators define 'biocompatible' as 'beneficial' and others 'least is best'. Furthermore, despite numerous studies addressing the differences between membrane types as well as studies investigating the role of membrane type on the outcomes associated with treatment, the precise clinical impact of each of these bioincompatible events is difficult to determine. This is because the uraemic state '*per se*' is associated with numerous physiological and metabolic disorders that can be confused with the effects of membrane biocompatibility. Furthermore, the clinical status and the underlying illness of the patient population in which these aspects are of interest are often highly heterogeneous and variable, posing difficulties for studies aimed at investigating clinical outcomes.

5.6 Clinical Concepts of Membrane Performance

The previous expressions relate to events occurring at the liquid or blood membrane interface. While such approaches are of interest to material scientists, they are of limited interest in the clinical setting since in such setting the membrane is incorporated into a device that has been sterilised prior to use, and through which there is a flow of fluid on both sides of the membrane. Thus, it is the performance of the complete unit that is of clinical interest.

5.6.1 Solute Transport

Solute transport in the clinical setting is generally defined as clearance (K), a term analogous to the clearance concept of the human kidney. It focuses on the solute removal by the device and can be defined as the amount of solute removed from the blood per unit of time, divided by incoming blood concentration, that is, the volumetric rate of removal by the device and is expressed mathematically as:

$$K = Q_B\left(\frac{C_{BI} - C_{Bo}}{C_{BI}}\right)$$

where Q is the volumetric flow rate and C the concentration; subscript B refers to blood, and i and o represent inlet and outlet.

As the blood flow entering and leaving the device is not identical due to fluid removal (Q_F) during passage through the device, it is necessary to account for this, which modifies the relationship to:

$$K = Q_B\left(\frac{C_{BI} - C_{Bo}}{C_{BI}}\right) + Q_F\left(\frac{C_{Bo}}{C_{BI}}\right)$$

It should be noted that this correction does not provide a quantification of solute transport via convection but merely corrects the diffusion equation for differences in the flow rates entering and leaving the device. These relationships relate to the loss of a metabolite from the blood and equivalent relations exist for the gain in the dialysis fluid. Both relationships can be adapted to take into consideration the special case in which there is metabolite of interest in the dialysis fluid, or if the metabolite of interest is electrically charged.

For the quantification of solute transport across the membrane in the presence of convection, it is possible to present clearance as:

$$K = K_0 + T_r\, Q_F$$

where K_0 is a pure diffusive clearance, Q_F is the ultrafiltration flow rate and T_r is the transmittance coefficient (the ratio of the transmitted flux to the incident flux).

In the presence of simultaneous convective and diffusive mass transport, the combined solute transport is not the sum of the individual components, due to an interaction between the convective and diffusive components, and several models have been proposed to explain this mathematically, of which the simplest is:

$$K = K_0 + Q_F T$$

where K_0 is the clearance at zero ultrafiltration, Q_F the ultrafiltration rate and T the transmittance coefficient, a parameter that is a function of flow conditions and membrane properties.

An expression for the transmittance coefficient that is universal for all solutes has been proposed by Jaffrin [27], namely:

$$K = K_0 + 0.46Q_F$$

for ultrafiltration rates below 70 ml/min, which for fluid flux rates above 70 ml/min modifies to:

$$K = K_0 + 0.43Q_F + 0.00083Q_{F^2}$$

In blood purification, the relative contributions to the total solute transport from diffusion and convection are determined by the treatment modality. For example, in conventional haemodialysis, the dominant mode of solute transport is diffusive and thus the contribution from convection is small. On the other hand, in newer treatment modalities such as haemodiafiltration, in which focus is on the removal of high-molecular-weight solutes elevated as a consequence of renal insufficiency, the dominant solute mechanism is convective rather than diffusive. In the treatment of acute renal failure, by continuous therapies, solute removal is entirely by convection.

These formulae refer to the solute removal from whole blood. In practice, solutes are removed from the plasma water, and consequently plasma water clearance is generally used to provide an accurate estimate of the solute removal during clinical treatment.

Equivalent approaches to therapies utilising convection alone which are used in the treatment of acute or reversible renal failure require the use of the solute sieving coefficient (S), the ratio of the solute concentration in the ultrafiltrate (C_F) to the solute concentration of bulk plasma water (C_W), such that:

$$S = \frac{C_F}{C_W}$$

In the absence of protein binding, solute concentration in the plasma water is related to the plasma concentration by

$$\frac{C_P}{C_W} = 1 - \phi$$

where ϕ is the volume fraction of hydrated proteins ~$0.0107C_P$.

The concentration in the blood is related to that in plasma by:

$$\frac{C_B}{C_P} = 1 - \text{HCT} + \lambda\text{HCT}$$

where λ is the distribution coefficient between red cells and plasma and HCT is the haematocrit. During convective therapies, a diluting fluid may be added to the blood either before it enters the filter (predilution) or after it leaves the filter (postdilution) and the solute removal is more complex and the mathematical relationships are outside the scope of this review and may be found elsewhere [28].

5.6.2 Fluid Removal

During the clinical application of membrane separation devices, not only solutes but also fluid needs to be removed. Transmembrane passage of fluid is governed by the membrane hydraulic permeability, the pressure gradient across the membrane and the plasma oncotic pressure. While the earlier mathematical relationship applies, there is a dependence of fluid removal rate on the fluid dynamic conditions within the device. Also, there is a relationship between the rate of fluid removal across the membrane and time, which is a consequence of membrane fouling by plasma proteins.

5.6.3 Biocompatibility

Clinical biocompatibility studies may be divided into two groups of studies. First, studies comparing membranes over a short period of clinical use, to establish differences in functionality, and second, studies aimed at establishing a relationship between outcomes in terms of mortality and morbidity to the membrane material, and thus, indirectly to its functionality. Craddock and co-workers [29] convincingly demonstrated the activation of complement during haemodialysis using cellulosic membranes [29], and Cheung [30] subsequently demonstrated the production of complement activation products C3a and C5a during dialysis with Cuprophan membranes. Because of the potent biological activities of the anaphylatoxins and the frequently large magnitude of increase during haemodialysis that is easily detectable using radioimmunoassay, plasma C3a (and its stable des-arginine derivative) levels have been widely used as an index of membrane biocompatibility.

The studies by Craddock and co-workers [29] showed that during haemodialysis using Cuprophan® membranes, a 15- to 30-fold increase in dialyser venous plasma C3a concentrations can be detected [29]. Ivanovich and co-workers [31] compared Cuprophan® to cellulose acetate membranes, and found that haemodialysis using the latter was associated with lower plasma C3a antigen levels.

Chenoweth and co-workers [32] and Hakim and co-workers [33] both demonstrated an attenuation of these effects when the dialyser was reprocessed for use in the same patient.

In addition to the changes in the circulating levels of plasma complements, another widely used measure of biocompatibility has been the quantification of temporal changes in the white cell numbers during treatment. When using cellulose-based membranes, a profound neutropenia or loss of white cells from the circulation is observed. This loss is rapid, and occurs within minutes of blood coming into contact with the membrane. A nadir is generally reached around 15 minutes following the initial blood-membrane contact, and thereafter there is a gradual return to pretreatment levels by the end of treatment or around 2–4 hours. This temporal profile of white cells is a consequence of the overexpression of receptors on the cells (CD11b/CD18, CD15s) leading to an increased cellular adhesiveness, and the aggregation and subsequent sequestration in the lung vasculature. The reversal of neutropenia is a consequence of the downregulation of responsiveness to C5a or to the internalisation of C5a receptors on the cells. The rebound results from the return of sequestrated neutrophils to the systemic circulation, and the recruitment from the marginated pool or bone marrow stores. Regenerated or unmodified cellulose membranes exhibit the strongest changes, while changes noted when using modified cellulose membranes are less marked and depend on the type of modification employed. Because the temporal profile of neutrophils during haemodialysis is complement mediated, complement activation and neutropenia correlate with each other.

Blood contact with the membrane surface also induces activation of the coagulation cascade as well as activation of platelets. These effects in a clinical setting are minimised by the administration of an anticoagulant. Traditionally, unfractionated heparin is the agent of choice, although today there is a trend to use low-molecular-weight fractions of this compound. The presence of an anticoagulant during clinical use, while advantageous to the patient, makes it difficult to quantify the *in vivo* thrombus-generating potential of the material. Red cells are not influenced by contact with the membrane, but may be damaged by shear stresses within the extracorporeal circuit as well as the blood pump.

Historically, measurements relating to complement activation (and the associated neutropenia due to the activation of complement receptors on the cells) were used widely for comparing membranes. Such comparisons demonstrated that for cellulose-based membranes, the manufacturing or regenerating processes can modify the transport and biocompatibility properties such that they are significantly different from those of Cuprophan®. Furthermore, when compared with membranes based on synthetic polymers, the magnitude of changes observed overlaps [34, 35].

Since many of the events associated with membrane-blood contact have clinical sequelae, research emphasis has shifted away from the quantification of the magnitude of changes observed to understanding the molecular mechanisms involved, for example, the release of oxygen radicals (oxidative stress) and intragranular proteases arising from the membrane-induced neutrophil activation.

Historic data indicate that compared to unmodified cellulose membranes, the use of modified cellulose and synthetic dialysis membranes was associated with an approximate 20% lower risk for all-cause mortality [36]. A subsequent study by Bloembergen and co-workers [37] additionally found that clinical use of unmodified cellulose membranes also increases the risk of infection and atherogenesis.

5.7 Contemporary Issues Relating to the Use of Cellulose in Blood Purification

Concerns began in the 1970s regarding the narrow range of molecules that could be removed by dialysis. Recall that at this time, the membranes of choice were primarily based on cellulose. These concerns led to the development of membranes with an opened pore structure and their use with new treatment techniques such as haemofiltration, or haemodiafiltration to enhance the removal of a wider spectrum of molecular weights than possible with conventional haemodialysis. Since there were no cellulose-based membranes commercially available at that time that could meet these requirements, the manufacture and the use of synthetic polymer blends such as polysulfone began. Such membranes offered enhanced solute transport compared to cellulose-based membranes, and their biocompatibility profile was also enhanced.

To narrow the difference between cellulose and synthetic membranes, a number of modifications to the manufacturing process of cellulose-based membranes were made, and today cellulose membranes are also available for use in these therapies.

In the artificial kidney, the membrane acts as a semipermeable barrier, separating the sterile blood side from the nonsterile dialysis fluid. The dialysis fluid is produced by the blending of water with an electrolyte concentrate. Tap water is unsuitable for use and requires further treatment to reduce its chemical contaminant levels to those recommended in International Standards (ISO 13959:2009 *Water for haemodialysis and related therapies*). Such treatment removes the chlorine added to drinking water and paves the way for the proliferation of bacteria and the formation of biofilm in the distribution network unless there is rigorous attention to the sterilisation and design of the water distribution network. Bacterial proliferation leads to the formation of biofilm, which once formed is difficult to eradicate.

Such films are dynamic and release bacteria and endotoxins with cytokine-inducing activity. Cytokines are polypeptides with molecular weights ranging from 10 to 45 kDa and are generated by immunocompetent cells in response to infection, inflammation or trauma and have a wide range of biological effects. The membrane in the artificial kidney may act as a barrier to the transport of intact bacteria into the circulation, but lipopolysaccharide (LPS) fragments have the capacity to pass across the membrane and this has raised concerns that membranes with a large pore size may confer a risk to the patient by contributing to the microinflammatory state of patients undergoing regular dialysis [38].

Both cellulose and synthetic membranes are able to adsorb LPS fragments, although the adsorption is minimal in the case of cellulose [39]. In contrast to cellulose membranes which may be considered as a symmetric gel, the structure of synthetic membranes is more complex, and experimental studies have demonstrated that endotoxin adsorption in such membranes occurs in the outer layer of the membrane and is governed by not only the structure of the membrane but also its hydrophobicity and polymer composition.

As previously discussed, biomaterial surfaces in contact with blood activate the complement system. The desire to minimise these responses was the driver in the development of modified cellulose membranes in which the number of OH groups on the membrane surface are reduced. Cardiac events are a major cause of death in dialysed patients. This is due, at least in part, to the high prevalence of atherosclerotic coronary heart disease. To a large extent, coronary lesions are acquired in the predialytic phase of chronic renal failure, but factors related to the dialysis procedure itself may also influence early or late events in atherogenesis. The second most common cause of death in dialysis patients is infection. Activated proinflammatory pathways have the potential to produce a diminished immune response limiting patients' ability to fight infection. Although this has been demonstrated in now historic studies for Cuprophan®, the role of modified cellulose membranes has not been studied; furthermore, the sensitivity of dialysis patient response to infection may also be influenced by other factors such as iron supplementation [40].

Definitive evidence relating to the role of biocompatibility and outcomes is lacking, as the results of prospective randomised studies are conflicting. In addition, there may be a difference between the membranes in terms of flux, and it has been difficult practically to separate the effects of biocompatibility and flux. When treating patients with acute or reversible renal failure, it has also been speculated that membrane biocompatibility may have a role in patient survival and the recovery of renal function. A recent meta-analysis by Subramanian and co-workers [41], however, showed that while synthetic membranes appeared to confer a significant survival advantage over cellulose-based membranes (cumulative odds ratio (OR) for survival, 1.37; 95% confidence interval (CI), 1.02–1.83; $p = 0.03$). There was no association between membrane type and recovery of renal function (cumulative OR, 1.23; 95% CI, 0.90–1.68; $p = 0.18$).

Membranes used in blood purification are exposed to protein-containing solution for a period of several hours. Such extended contact can give rise to leachables from the membrane resulting in allergic or more severe reactions. Leachables originating from the haemodialyser membrane include residues of chemicals used in the production of hollow fibres or from the testing of membrane integrity following assembly into the dialyser. One such residue, PF 5070, a perfluorocarbon, used during the integrity testing of dialysers incorporating modified cellulose membranes was found to be implicated in a series of fatalities in Europe in 2002 [42, 43].

Cellulose acetate degrades with time and a fragmentation of the material into acetylated carbohydrate derivatives occurs. Such degradation has been described in cellulose acetate-containing dialysers when there was a failure to adhere to the recommended storage time. Patient exposure to the degradation products results in adverse reactions which have been described in the literature. Oba and co-workers [44] described the migration of acetylated hemicellulose from a hollow fibre dialyser as being responsible for scleritis or iritis. Ocular symptoms were also noted in 2001 in a group of patients dialysed with a new batch of cellulose acetate [45]. Subsequent investigations implicated an incorrect storage of dialysers [46]. To minimise the reoccurrence of this problem, mathematical modelling techniques have been developed to allow the safe shelf life of cellulose-containing dialysers to be established [47].

5.8 The Future Role of Cellulose in Blood Purification Processes

The most widely used application of cellulose in medicine has been as a membrane in the treatment of renal failure. However, from a position of universal use in the early 1960s, there has been a gradual decline in the use of this material as worldwide trends indicate [48].

An important question whether patients treated with cellulose membranes are at a disadvantage, compared to patients receiving treatment using synthetic membranes, in terms of outcomes and well-being remains unanswered. On the one hand, historic studies suggested that a difference may have existed; however, other aspects of the dialytic treatment may have also contributed. A more recent Cochrane review on the other hand found no evidence of benefit when synthetic membranes were compared with cellulose membranes or modified cellulose membranes in terms of reduced mortality or the reduction in dialysis-related adverse symptoms [49]. Despite the inability to demonstrate a disadvantage, the European Best Practice guidelines for haemodialysis suggest that dialyser membranes with the lowest degree of complement and leucocyte activation should be used in the treatment of renal failure and membranes that induce strong complement and leucocyte activation, inflammatory reactions and/or a blunting of the response of leucocytes to stimuli should be avoided. Clearly this precludes the continuing use of unmodified cellulose membranes and indeed the manufacture of such membranes was discontinued earlier this decade.

The discontinuation of the use of modified cellulose membranes in favour of synthetic membranes is however less compelling, since such membranes overlap in terms of their biocompatibility with those manufactured from synthetic materials. Furthermore, there may be other factors, such as cost, favouring the continuing use of modified cellulose membranes, particularly in emerging economies.

As the clinical application of cellulose in the form of membranes for dialysis declines, new applications for the material are emerging. Cellulose acetate butyrate microcapsules, as well as cellulose-based microspheres, have been used for the delivery of drugs [50]. Surgicel® (Ethicon Inc., USA) (oxidised regenerated cellulose) is used widely to control haemorrhage, as this material, when saturated with blood, swells rapidly. Viscose cellulose sponges have also been used as a scaffold for cartilage tissue engineering [51].

Cellulose may also be derived from different types of microorganisms. Of these, cellulose synthesised by *A. xylinum* is characterised by a three-dimensional structure consisting of an ultrafine network of cellulose nanofibres (3–8 nm) that are largely uniaxially oriented. This unique nanomorphology results in a large surface area, which can hold a large amount of water and display great elasticity and high wet strength. These features favour its use in a wound healing system. Furthermore, since the material does not dry out, it also allows potential transfer of antibiotics or other medicines into the wound, while at the same time serving as an efficient physical barrier against any external infection [52].

References

1. R.N. Foley and A.J. Collins, *Journal of American Society of Nephrology*, 2007, **18**, 10, 2644.

2. K. Hommel, S. Rasmussen, A.L. Kamper and M. Madsen, *Nephrology Dialysis and Transplantation*, 2010, **25**, 8, 2624.

3. P.J. Blankestijn, I. Ledebo and B. Canaud, *Kidney International*, 2010, **77**, 7, 581.

4. A. Pierratos, *Nature Clinical Practice Nephrology*, 2008, **4**, 11, 602.

5. C. Weber, B. Henne, F. Loth, M. Schoenhofen and D. Falkenhagen, *Journal of the American Society for Artificial Internal Organs*, 1995, **41**, 3, M430.

6. T. Taniguchi, A. Kurita, K. Yamamoto and H. Inaba, *Transfusion and Apheresis Science*, 2009, **40**, 1, 55.

7. H.W. de Koning, R.A. Chamuleau and A. Bantjes, *Artificial Organs*, 1982, **6**, 3, 280.

8. P.P. Poyck, A.C. van Wijk, T.V. van der Hoeven, D.R. de Waart, R.A. Chamuleau, T.M. van Gulik, R.P. Oude Elferink and R. Hoekstra, *Journal of Hepatology*, 2008, **48**, 2, 266.

9. N.L. Sussman, M.G. Chong, T. Koussayer, D.E. He, T.A. Shang, H.H. Whisennand and J.H. Kelly, *Hepatology*, 1992, **16**, 1, 60.

10. S.R. Mitzner, J. Stange, S. Klammt, S. Koball, H. Hickstein and E.C. Reisinger, *Journal of the American Society for Artificial Internal Organs*, 2009, **55**, 5, 498.

11. U. Boonsrirat, K. Tiranathanagul, N. Srisawat, P. Susantitaphong, P. Komolmit, K. Praditpornsilpa, K. Tungsanga and S. Eiam-Ong, *Artificial Organs*, 2009, **33**, 8, 648.

12. A. Santoro, S. Faenza, E. Mancini, E. Ferramosca, F. Grammatico, A. Zucchelli, M.G. Facchini and A.D. Pinna, *Transplantation Proceedings*, 2006, **38**, 4, 1078.

13. J. Tijssen, A. Bantjes, A.W. van Doorn, J. Feijen, B.van Dijk, C.R. Vonk and I.C. Dijkhuis, *Artificial Organs*, 1979, **3**, 1, 11.

14. S.V. Mikhalovsky, *Perfusion*, 2003, **18**, 1, 47.

15. D.J. Malik, G.L. Warwick, M. Venturi, M. Streat, K. Hellgardt, N. Hoenich and J.A. Dale, *Biomaterials*, 2004, **25**, 15, 2933.

16. C.N. Firme, R. Anthone, S. Antone and A.E. Macneill, *Journal of Thoracic and Cardiovascular Surgery*, 1960, **40**, August, 253.

17. E. Buckley, D. Sidebotham, A. McGeorge, S. Roberts, S.J. Allen and J. Beca, *British Journal of Anaesthesia*, 2010, **104**, 3, 326.

18. J.J Abel, L.G. Rowntree and B.B. Turner, *Transfusion Science*, 1990, **11**, 2, 164.

19. J.S. Cameron, *Nephrology Dialysis and Transplantation*, 2000, **15**, 7, 1086.

20. P. Fagette, *Journal of the American Society for Artificial Internal Organs*, 1999, **45**, 4, 238.

21. N. Alwall, *Artificial Organs*, 1986, **10**, 2, 86.

22. F. Kiil, *Acta Chirugica Scandinavica Supplement*, 1960, **253**, 142.

23. J.A. Sargent, *International Journal of Artificial Organs*, 2007, **30**, 11, 953.

24. F. Gotch, B. Lipps, J. Weaver, Jr., J. Brandes, J. Rosin, J. Sargent and P. Oja, *Transactions of the American Society for Artificial Internal Organs*, 1969, **15**, 87.

25. D. Roy, M. Semsarilar, J.T. Guthrie and S. Perrier, *Chemical Society Review*, 2009, **38**, 7, 2046.

26. A. Katchalsky and O. Kedem, *Biophysics Journal*, 1962, **2**, 2, 53.

27. M.Y. Jaffrin, L.H. Ding and J.M. Laurent, *Journal of Biomechanical Engineering*, 1990, **112**, 2, 212.

28. L.W. Henderson in *Replacement of Renal Function by Dialysis: a Textbook of Dialysis,* 2nd edition, Eds., W. Drukker, F.M. Parsons and J.F. Maher, Martinus Nijhoff Publishers, Boston, MA, USA, 1983, p.242.

29. P.R. Craddock, J. Fehr, A.P. Dalmasso, K.L. Brigham and H.S. Jacob, *Journal of Clinical Investigation*, 1977, **59**, 5, 879.

30. A.K. Cheung, *Nephrology Dialysis and Transplantation*, 1994, **9**, 12, 96.

31. P. Ivanovich. D.E. Chenoweth, R. Schmidt, H. Klinkmann, L.A. Boxer, H.S. Jacob and D.E. Hammerschmidt, *Kidney International*, 1983, **24**, 6, 758.

32. D.E. Chenoweth, A.K. Cheung, D.M. Ward and L.W. Henderson, *Kidney International*, 1983, **24**, 6, 770.

33. R.M. Hakim, D.T. Fearon and J.M. Lazarus, *Kidney International*, 1984, **26**, 2, 194.

34. N.A. Hoenich, C. Woffindin, S. Stamp, S.J. Roberts and J. Turnbull, *Biomaterials*, 1997, **18**, 19, 1299.

35. N.A. Hoenich, C. Woffindin, J.N. Mathews and J. Vienken, *Biomaterials*, 1995, **16**, 8, 587.

36. R.M. Hakim, P.J. Held, D. Stannard, R.A. Wolfe, F.K. Port, J.T. Daugirdas and L. Agodoa, *Kidney International*, 1996, **50**, 2, 566.

37. W.E. Bloembergen, R.M. Hakim, D.C. Stannard, P.J. Held, R.A. Wolfe, L.Y. Agodoa and F.K. Port, *American Journal of Kidney Diseases*, 1999, **33**, 1, 1.

38. A. Merino, J. Portolés, R. Selgas, R. Ojeda, P. Buendia, J. Ocaña, M.A. Bajo, G. Del Peso, J. Carracedo, R. Ramírez, A. Martín-Malo and P. Aljama, *Clinical Journal of the American Society for Nephrology*, 2010, **5**, 2, 227.

39. Y. Takemoto, T. Nakatani, K. Sugimura, R. Yoshimura and K. Tsuchida, *Artificial Organs*, 2003, **27**, 12, 1134.

40. G. Sunder-Plassmann, S.I. Patruta and W.H. Horl, *American Journal of Kidney Diseases*, 1999, **34**, 4, Supplement 2, S25.

41. S. Subramanian, R. Venkataraman and J.A. Kellum, *Kidney International*, 2002, **62**, 5, 1819.

42. B. Canaud, *Nephrology Dialysis and Transplantation*, 2002, **17**, 4, 545.

43. B. Canaud, P. Aljama, C. Tielemans, V. Gasparovic, A. Gutierrez and F. Locatelli, *Journal of the American Society for Nephrology*, 2005, **16**, 6, 1819.

44. T. Oba, K. Tsuji, A. Nakamura, H. Shintani, S. Mizumachi, H. Kikuchi, M. Kaniwa, S. Kojima, K. Kanohta and Y. Kawasaki, *Artificial Organs*, 1984, **8**, 4, 429.

45. Z. Averbukh, D. Modai, J. Sandbank, S. Berman, M. Cohn, E.Z. Galperin, N. Cohen, V. Dishi and J. Weissgarten, *Artificial Organs*, 2001, **25**, 6, 437.

46. J.C. Hutter, M.J. Kuehnert, R.R. Wallis, A.D. Lucas, S. Sen and W.R. Jarvis, *Journal of the American Medical Association*, 2000, **283**, 16, 2128.

47. J.C. Hutter, M.C. Long, H.M. Luu and L.W. Schroeder, *Journal of the American Society for Artificial Internal Organs*, 2001, **47**, 5, 522.

48. A. Grassmann, S. Gioberge, S. Moeller and G. Brown, *Artificial Organs*, 2006, **30**, 12, 895.

49. A. MacLeod, C. Daly, I. Khan, L. Vale, M. Campbell, S. Wallace, J. Cody, C. Donaldson and A. Grant, *Cochrane Database Systematic Review*, 2001, **3**, CD003234.

50. G. Fundueanu, M. Constantin, E. Esposito, R. Cortesi, C. Nastruzzi and E. Menegatti, *Biomaterials*, 2005, **26**, 20, 4337.

51. H. Pulkkinen, V. Tiitu, E. Lammintausta, M.S. Laasanen, E.R. Hämäläinen, I. Kiviranta and M.J. Lammi, *Biomedical Materials and Engineering*, 2006, **16**, 4, S29.

52. W.K. Czaja, D.J. Young, M. Kawecki and R.M. Brown, Jr., *Biomacromolecules*, 2007, **8**, 1, 1.

6 Immunomodulatory Effects of Botanical Polysaccharides

Igor A. Schepetkin and Mark T. Quinn

6.1 Introduction

Over the past several decades, research on the medicinal properties of natural products has increased significantly, and a large body of evidence suggests that plant extracts may represent a source of novel compounds with medicinal properties (reviewed in [1–3]). Polysaccharides (PS) isolated from botanical sources such as mushrooms, algae, lichens and higher plants have also attracted a great deal of attention in the biomedical arena because of their broad spectrum of therapeutic properties and relatively low toxicity [1, 4, 5] (**Figure 6.1**). Native and modified polysaccharides derived from mushrooms, algae and plants have been found to exhibit various beneficial pharmacological properties, including immunomodulatory, anticoagulant, anticancer, wound-healing, antihyperglycaemic and lipid-reducing properties, and have also been used for drug delivery [6–10] (**Figure 6.1**). The ability of plant-derived polysaccharides to boost certain immune responses may explain, in part, some of the beneficial effects of medicinal plants. Although the mechanism of action of these substances is still under investigation, it appears that one of the primary effects is nonspecific induction of the immune system [3, 4]. Indeed, the immunostimulatory, antitumour, bactericidal and other therapeutic effects of botanical polysaccharides are thought to occur via stimulation of macrophages and lymphocytes [3, 11, 12]. Much evidence indicates that immune responsive cells, such as lymphocytes and macrophages, can be activated by the binding of polysaccharides to cell-specific surface receptors. For this reason, polysaccharides have recently become attractive as beneficial therapeutics for the treatment of infectious diseases. Clearly, modulation of innate immunity can have a significant impact on the host's ability to respond rapidly and potently to a diverse array of pathogens [13]. In this chapter, current understanding of the immunomodulatory properties of polysaccharides derived from mushrooms, algae, lichens and higher plants, including their antioxidant activity, mitogenic activity, antiviral activity, adjuvant effects, antitumour properties and effects on leucocyte function (**Figure 6.1**) is described.

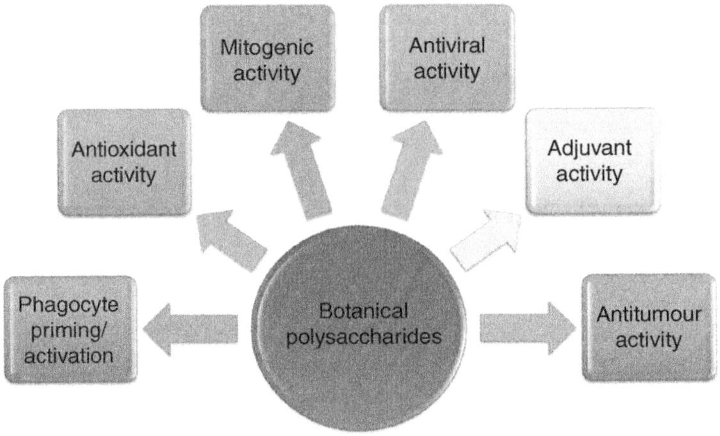

Figure 6.1 Therapeutic properties of botanical polysaccharides. Botanical polysaccharides exhibit a wide range of immunomodulatory and therapeutic properties. Each of these properties is discussed in this chapter

6.2 Immunomodulatory Activity of Fungal-, Algae- and Lichen-derived Polysaccharides

Among the group of botanical polysaccharides, fungal-derived polysaccharides are the best-known and one of the most powerful immune stimulants. The reason may be that these types of polysaccharides are also constituents of cell walls of certain pathogenic microorganisms such as *Pneumocystis carinii*, *Cryptococcus neoformans*, *Aspergillus fumigatus*, *Histoplasma capsulatum* and *Candida albicans*. It has recently been recognised that the initial interaction between fungi and the innate immune system is via binding between fungal-specific chemical signatures and pattern recognition receptors on mononuclear phagocytes. Fungal pattern-associated molecular patterns are restricted to complex carbohydrates in the cell wall, including phospholipomannan, β-glucans and chitin. These patterns bind specifically to two classes of pattern recognition receptors in phagocyte membranes, Toll-like receptors (TLR) and C-lectin-like receptors, through which they initiate signalling responses that culminate in the release of pro- and anti-inflammatory cytokines, link the innate immune response with the adaptive immune response and initiate phagocytosis and intracellular killing [14].

The healing and immunostimulating properties of mushrooms have been known for thousands of years. Polysaccharides isolated from mushrooms predominantly enhance or activate macrophage host defence responses (**Table 6.1** and **Figure 6.2**). A wide range of immunostimulatory polysaccharides with different chemical structures have been isolated from mushrooms (basidiomycetes) [5, 51]. The majority of immune-active polysaccharides derived from mushrooms are β-glucans, including lentinan

Table 6.1 Effects of polysaccharides from mushrooms, fungi and lichen on macrophage function

Common name and family	PS features (name, source, structure)	Cell type	Effects on macrophages[a]	References
Coriolus versicolor (Polyporaceae)	Intramycelial and extramycelial PS	Murine peritoneal MØ	↑ NO	[15]
	Polysaccharopeptide	Murine peritoneal MØ	↑ NO, ROS and TNF-α (oral)	[16]
	Protein-bound PS (Krestin, PSK)	Murine peritoneal MØ	↑ NO, GSHPx and SOD activity (ip)	[17]
Cordyceps sinensis (fungus) (Clavicipitaceae)	-	Human U937 MØ	↑ MØ differentiation	[18]
	Cordysinocan (M_r 82 kDa)	Rat alveolar NR8383 MØ	↑ Phagocytosis	[19]
Flammulina velutipes (Physalacriaceae)	*F. velutipes* PS (FVP)	Murine peritoneal MØ	↑ TNF-α	[20]
Ganoderma lucidum (Ganodermataceae)	β-Glucan	Murine peritoneal MØ and J774.1 MØ, human MØ	↑ NO and IL-1	[21]
Grifola frondosa (Schizophyllaceae)	Maitake D-fraction [MD-fraction, β-(1,3;1,6)- glucan]	Murine peritoneal MØ, murine RAW 264.7 MØ	↑ IL-1 (ip) and oral); ↑ iNOS message, NO and IL-12	[22, 23]
	Grifolan [GRN, β-(1,3;1,6)- glucan]	Murine peritoneal MØ, murine RAW 264.7 MØ, Kupffer cells	↑ TNF-α, IL-6 and NO; ↑ CD11b, cytokines and NO in Kupffer cells (*in vivo*)	[24–29]
Hericium erinaceus (Hericiaceae)	β-(1,3;1,6)-glucan	Murine peritoneal MØ	↑ NO, TNF-β and IL-1β	[30]

Table 6.1 Continued

Common name and family	PS features (name, source, structure)	Cell type	Effects on macrophages[a]	References
Lentinus edodes (Lentinaceae)	Lentinan [β-(1,3;1,6)-glucan, M_r 500 kDa]	Murine peritoneal MØ, murine C4M MØ, human monocytes	↑ NO, ROS, TNF-α, IL-1, phagocytosis and cytotoxicity	[31–36]
	Arabinogalactan (M_r 576 kDa)	Murine RAW 264.7 MØ	↑ NO ; ↑ TNF-α, IL-1β, IL-6 and IL-10 message	[37]
Lentinus lepideus (Lentinaceae)	Soluble heteroglycan (PG101)	Human U937 MØ	↑ TNF-α, IL-1β, IL-10, IL-12 and GM-CSF	[38]
Morchella esculenta (Morchellaceae)	Galactomannan (M_r 1000 kDa)	Human THP-1 MØ	↑ NF-κB activation	[39]
Phellinus linteus (Hymenochaetaceae)	Acidic PS	Murine peritoneal MØ	↑ NO production and cytotoxicity	[40]
Pleurotus florida (Tricholomataceae)	Glucan	Murine peritoneal MØ	↑ NO	[41]
Poria cocos Wolf (Coriolaceae)	Heteromannan (M_r 8 kDa)	Murine RAW 264.7 MØ	↑ NF-κB/Rel, p38 kinase activation, iNOS and NO	[42, 43]
Sarcodon aspratus (Thelephoraceae)	Fucogalactan	Murine peritoneal MØ	↑ TNF-α and NO	[36]
Schizophyllum commune (Schizophyllaceae)	Schizophyllan/sonifilan [SPG, β-(1,3;1,6)-D-glucan]	Murine peritoneal MØ, human THP-1 and U937 MØ	↑ TNF-α, IL-6, IL-8, IL-12 and cytotoxicity	[28, 44, 45]
Sclerotium rolfsii (Typhulaceae)	Scleroglucan [β-(1,3)-glucan, M_r 1560 kDa]	Murine peritoneal MØ	↑ Phagocytosis and cytotoxicity	[46]

Ramalina celastri (Ramalinaceae)	Galactomannan polymer (GMPOLY)	Murine peritoneal MØ	↑ NO	[47]
	α-Glucan	Murine peritoneal MØ	↑ ROS and phagocytosis (ip)	[48]
Thamnolia vermicularis var. *subuliformis* (lichen)	Thamnolia heteroglycans (Ths-4, Ths-5)	Rat peritoneal MØ	↑ TNF-α	[49]
Tremella fuciformis Berk (Tremellaceae)	Tremella acidic heteroglycans (T2a-T2d, M_r 20–410 kDa)	Human monocytes	↑ IL-1, IL-6 and TNF-α	[50]

GM-CSF: Granulocyte-macrophage colony-stimulating factor
GSHPx: Glutathione peroxidase
IL: Interleukin
iNOS: Inducible nitric oxide synthase
MØ: Macrophages
M_r: relative molecular weight
NF-κB: Nuclear factor κB
NO: Nitric oxide
PS: Polysaccharides
ROS: Reactive oxygen species
SOD: Superoxide dismutase
TNF: Tumour necrosis factor
[a]*Where applicable, the route of administration is indicated as oral or intraperitoneal (ip)*

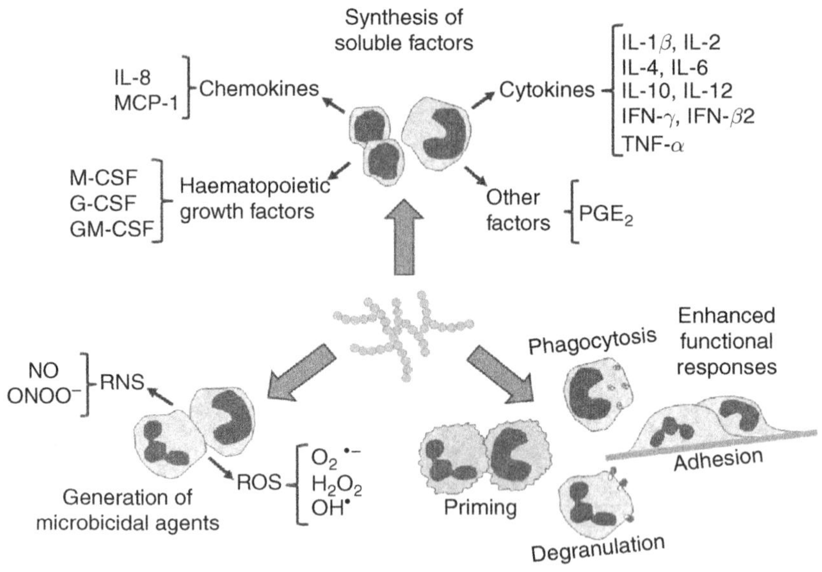

Figure 6.2 Botanical polysaccharides activate a variety of leucocyte responses. Botanical polysaccharides induce secretion of soluble mediators from macrophages (larger cell with kidney-shaped nucleus) and lymphocytes (smaller cell with round nucleus). These mediators include cytokines [interleukin (IL)-1β, IL-2, IL-4, IL-6, IL-10 and IL-12; interferon (IFN)-γ and IFN-β2; and tumour necrosis factor-α (TNF-α)], chemokines [IL-8 and monocyte chemoattractant protein-1 (MCP-1)], haematopoietic growth factors [macrophage colony-stimulating factor (M-CSF), granulocyte colony-stimulating factor (G-CSF) and granulocyte/macrophage colony-stimulating factor (GM-CSF)] and other factors such as prostaglandin E_2 (PGE_2). Polysaccharides also induce macrophages and neutrophils (cell with multilobed nucleus) to produce microbicidal oxidants, including reactive oxygen species [ROS; superoxide anion ($O_2^{\bullet -}$), hydrogen peroxide (H_2O_2) and hydroxyl radical ($OH^{\bullet}$)] and reactive nitrogen species [RNS; nitric oxide (NO) and peroxynitrite ($ONOO^-$)]. Finally, botanical polysaccharides induce/enhance a number of functional responses in phagocytes, including neutrophil and macrophage priming, macrophage phagocytosis and neutrophil degranulation

from *Lentinus edodes* [52], schizophyllan from *Schizophyllum commune* [53], krestin from *Coriolus versicolor* [17], grifolan from *Grifola frondosa* [24] and scleroglucan from *Sclerotinia sclerotiorum* [54]. Similarly, β-glucans and other immunomodulatory polysaccharides have been isolated from lichens, including lichenan and isolichenan from *Cetraria islandica* [55] and thamnolan from *Thamnolia subuliformis* [56].

β-Glucans isolated from algae have also been reported to enhance the immune system function (**Table 6.2**). Cyclic and linear β-glucans with immunomodulatory activities

Table 6.2 Effects of algal polysaccharides on macrophage function

Common name and family	Name of the PS and structure	Cell type	Effects on macrophages[a]	References
Aphanizomenon flos-aquae (Nostocaceae)	Immunon (M_r >10,000 kDa)	Human THP-1 MØ	↑ NF-κB activation; ↑ IL-1β and TNF-α message	[57]
Chlorella pyrenoidosa (Chlorellaceae)	Immurella (M_r >10,000 kDa)	Human THP-1 MØ	↑ NF-κB activation; ↑ IL-1β and TNF-α message	[58]
Cladosiphon okamuranus Tokida (Chordariaceae)	Fucoidan	Murine peritoneal MØ	↑ TNF-α and ROS	[59]
Enteromorpha intestinalis (Ulvaceae)	Dried algae extract PS (DAEB, M_r 46.8 kDa)	Murine peritoneal MØ	↑ NO, TNF-α and phagocytosis	[60]
Euglena gracilis (Euglenaceae)	Paramylon [β-(1,3)-D-glucan]	Murine peritoneal MØ	↑ IL-1 and IL-6	[61]
Gracilaria verrucosa (Gracilariaceae)	Enzyme-treated Gracilaria water-soluble PS (E-GWS)	Murine peritoneal MØ	↑ Phagocytosis and ROS (oral and ip)	[62]
Hizikia fusiformis (Sargassaceae)	Hizikia PS I (HPS-I; M_r 2000 kDa) and II (HPS-II, M_r 70 kDa)	Murine peritoneal MØ	↑ MØ suppression of EL-4 tumour cell growth	[63]
Laminaria digitata (Phaeophyceae)	Phycarine, β-glucan	Murine peritoneal MØ	↑ Phagocytosis	[64]

Table 6.2 Continued

Common name and family	Name of the PS and structure	Cell type	Effects on macrophages[a]	References
Porphyra yezoensis (Bangiaceae)	Porphyra water-soluble fraction (PWSF) and Porphyra acid-soluble fraction (PASF); porphyrans	Murine peritoneal MØ	↑ Glucose consumption, NO, TNF-α and IL-1	[65, 66]
Sargassum thunbergii (Sargassaceae)	Fraction GIV-A (fucoidan, M_r 19 kDa)	Murine peritoneal MØ	↑ ROS and phagocytosis	[67]
Spirulina platensis (Pseudanabaenaceae)	Immulina (M_r > 10,000 kDa)	Human THP-1 MØ	↑ NF-κB activation; ↑ IL-1β and TNF-α message	[58]

[a] *Where applicable, the route of administration is indicated as oral or intraperitoneal (ip)*

have been isolated from *Laminaria digitata* and *Chlorella pyrenoidosa* [64, 68]. Laminarin, a β-glucan from *L. digitata*, has been reported to be a dectin-1 antagonist and has been shown to inhibit the activation of nuclear factor κB (NF-κB)-dependent signalling pathways in human monocytes [69]. In contrast, polysaccharides isolated from *Aphanizomenon flos-aquae* activated NF-κB and enhanced the DNA-binding activity of this transcription factor [57]. Likewise, Hsu and co-workers [70] recently reported that polysaccharides from *C. pyrenoidosa* induced interleukin (IL)-1β secretion in macrophages via TLR-4-mediated protein kinase signalling pathways. Apart from these few studies, little is known about the molecular mechanisms of macrophage activation by algal polysaccharides. However, based on the responses reported, it is likely that algal polysaccharides modulate macrophage immune function in a manner similar to plant- and mushroom-derived polysaccharides.

The immunomodulatory properties of β-glucans have been extensively studied (reviewed in [71, 72]). β-Glucans stimulate a wide range of macrophage responses, such as release of cytokines [73–75], generation of reactive oxygen species (ROS) [76, 77] and nitric oxide (NO) [78] and release of arachidonic acid metabolites, such as prostaglandins [74, 79–81] (**Figure 6.2**). Due to the poor solubility and direct leucocyte-activating properties of β-glucans, their clinical usefulness has been limited. To address these issues, modified forms of β-glucans have been developed, resulting in soluble β-(1,6)-branched β-(1,3)-glucans. Indeed, soluble β-glucans have been shown to prime or enhance microbicidal activities of neutrophils and macrophages without exhibiting direct activating effects [82]. This may be due to differential bioavailability and/or mechanism of action of soluble *versus* insoluble β-glucans. Importantly, soluble β-glucans have been shown to reduce postoperative infection rates and the length of hospitalisation in clinical trials [83, 84], suggesting great promise as immune-enhancing therapeutics.

β-Glucan exerts its effects through direct stimulation of macrophages, neutrophils, dendritic cells and natural killer (NK) cells via β-glucan receptor targets on their cell surface membranes, such as complement receptor 3 (CR3), dectin-1, scavenger receptor and lactosylceramide receptor [85–88] (**Figure 6.3**). One of the principal receptors of β-glucan is dectin-1, which is expressed by macrophages, neutrophils, dendritic cells and vascular endothelial cells [89–91]. Dectin-1 (or Clec7a) is a member of the C-type lectin receptor family and functions as a pattern recognition receptor. Dectin-1 signals to the cell via its cytoplasmic tail, which contains an immunoreceptor tyrosine-based activation motif (ITAM) [92–94]. Dectin-1 binds exclusively to glucan residues with β-(1,3) linkages, which are not present on mammalian cells, whereas carbohydrates with β-(1,4) linkages or α-linkages do not bind to dectin-1 [88, 95]. Indeed, soluble β-(1,3)-glucans or β-(1,3;1,6)-glucans inhibited binding and uptake of zymosan by macrophages [88, 95]. Likewise, Kennedy and co-workers [91] showed that soluble β-glucan and an anti-dectin-1 monoclonal antibody could inhibit binding and phagocytosis of zymosan by human neutrophils. In addition, they reported that soluble β-glucan and the anti-dectin-1 monoclonal antibody also inhibited phagocytosis and killing of *C. albicans* by neutrophils, verifying the importance of dectin-1 in the recognition and killing of fungal pathogens by the innate immune system [91]. Human U937 monocytes have been reported to express non-CR3 glucan receptors, which specifically interact with β-(1,3)-glucans [96]. The rank order of competitive binding affinity of various β-glucans to these receptors was scleroglucan > schizophyllan > laminarin > glucan phosphate > glucan sulfate. Based on the recent work described above, it is likely that at least part of this response is due to dectin-1.

Stimulation of dendritic cells with β-glucan induces translocation of dectin-1 to lipid rafts [97]. In addition, two key signalling molecules, Syk and phospholipase Cγ2, are also recruited to lipid rafts upon activation of dectin-1, suggesting that lipid raft microdomains facilitate dectin-1 signalling [97] (**Figure 6.3**). Downstream signalling pathways lead to activation of NF-κB through a Syk-mediated pathway that involves signalling via nuclear factor of activated T cells (NFAT) (reviewed in [72]). Ultimately, activation by β-glucan induces a cascade of immune defences that protect the organism from various viral [98], bacterial [99] and fungal [100] challenges. However, the mechanisms involved are at least partially dependent on the route of administration. For example, protection after oral administration results primarily from ingestion of small particles of β-glucan by pinocytic M cells located in Peyer's patches of the small intestine. In addition, soluble β-glucans, such as laminarin and scleroglucan, can directly bind to, and be internalised by, intestinal epithelial cells and gut-associated lymphoid tissue (GALT) cells [101]. Once activated, these cells can migrate to the lymph nodes and activate other macrophages, NK cells and T lymphocytes via the release of cytokines [102–104]. As an example, low-molecular-weight scleroglucan hydrolysates (<5 kDa) have been shown to stimulate the activation and maturation of porcine monocyte-derived dendritic cells by upregulation of CD40 and CD80/86 [105].

A second important receptor involved in β-glucan recognition by leucocytes is CR3 (**Figure 6.3**). CR3, also known as CD11b/CD18, is expressed on NK cells, neutrophils and lymphocytes [106]. β-Glucans bind to the lectin domain of CR3, and the I-domain of CD11b has multiple lectin-binding sites that are specific for β-glucans and bind these polysaccharides with high affinity [107]. A number of studies have demonstrated the importance of CR3 in macrophage activation by β-glucan and its therapeutic potential [108]. For example, polysaccharides derived from *Phellinus linteus* have been reported to activate macrophages via CR3 [40]. Likewise, protein-containing heteromannan extracted from *Poria cocos* induced macrophage activation via CR3, as well as via CD14 and TLR-4 [42]. The expression of CR3 can also be modulated by β-glucan exposure. For example, the administration of grifolan increased the expression of CD11b on Kupffer cells [25]. Recently, van Bruggen and co-workers [109] provided additional evidence for the role of CR3 as a β-glucan receptor on human neutrophils.

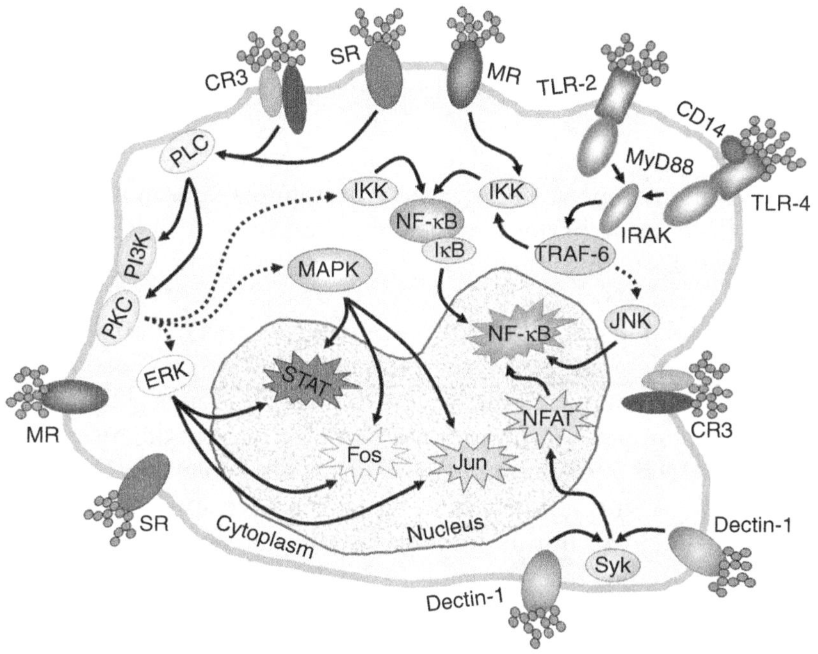

Figure 6.3 Potential receptors and signalling pathways involved in macrophage activation by botanical polysaccharides. Botanical polysaccharides can activate macrophages via a range of different receptors, including complement receptor 3 (CR3), mannose receptor (MR), scavenger receptor (SR), dectin-1 and Toll-like receptors 2 and 4 (TLR-2, TLR-4). It is also likely that several different receptor types cooperate with each other, forming clusters of signalling complexes. SR and CR3 signalling pathways lead to phospholipase C (PLC) activation, which generates intracellular messengers that activate protein kinase C (PKC) and phosphoinositide 3-kinase. Ultimately, these signals lead to activation of the mitogen-activated protein kinase (MAPK), extracellular signal-regulated kinase (ERK) and IκB kinase (IKK). These three pathways induce gene transcription via a number of transcription factors, including signal transducers and activators of transcription (STAT), Fos, Jun and nuclear factor κB (NF-κB). MR activation stimulates macrophage phagocytosis, oxidant production, endocytosis and NF-κB activation. TLR-4/CD14 and/or TLR-2 ligation leads to activation of IL-1R-associated kinase (IRAK) via an adaptor myeloid differentiation protein 88 (MyD88), with subsequent activation of TNF receptor-associated factor 6 (TRAF-6), MAP kinases [e.g., p38 MAPK and Jun N-terminal kinase (JNK)] and finally NF-κB. Activation of dectin-1 leads to clustering in lipid rafts with signalling molecules, such as spleen tyrosine kinase (Syk). Syk activates nuclear factor of activated T cells (NFAT), which ultimately leads to NF-κB activation and gene transcription. Activation of the various transcription pathways outlined above induces expression of proinflammatory cytokines and inducible nitric oxide synthase (iNOS). See text for further details

TLR are also important in mediating responses to mushroom-derived polysaccharide-protein complexes (**Figure 6.3**). For example, TLR-4 was reported to be required for macrophage activation by glycoproteins isolated from *Ganoderma lucidum* and *Poria cocos* [21, 42, 110]. Likewise, Gantner and co-workers [111] found that dectin-1 collaborated with TLR-2 in the inflammatory activation of NF-κB by β-glucans [111]. However, the presence of non-carbohydrate moieties could also account for the reports that the polysaccharide complexes interacted with TLR, and further work is still needed to evaluate this issue.

The role of intracellular targets of β-glucans remains an area of current investigation. After binding to specific surface receptors, soluble β-glucans can become internalised by macrophages, which may facilitate access to internal targets [112]. One reported possibility is that these polysaccharides may bind to and modulate single-stranded RNA and DNA. In a cell-free system, single chains of schizophyllan molecules behaved as complementary polynucleotides, binding to the corresponding single-stranded polynucleotide chain [113]. In this study, incubation of single schizophyllan chains with polynucleotide chains [poly(C), poly(A), poly(dA) or poly(dT)] resulted in the formation of macromolecular complexes.

In addition to the direct macrophage-activating effects of mushroom polysaccharides, effects on haematopoiesis have also been reported. For example, intravenous injection of lentinan increased the absolute number of monocytes in peripheral blood, as well as the number of granulocyte-macrophage progenitor cells in spleen and bone marrow [114]. Likewise, schizophyllan induced the expression of macrophage colony-stimulating factor (M-CSF) in bone marrow cell cultures [115]. Recently, Ito and co-workers [116] reported that the production of granulocyte colony-stimulating factor (G-CSF) was significantly enhanced by β-glucan extracted from the fruiting body of *G. frondosa* in a murine granulocytopaenia model. In addition, this extract induced a biphasic increase in the number of granulocytes in the spleen [116]. The haematopoietic properties of mushroom-derived polysaccharides have been suggested to contribute to radioprotective effects. For example, treatment of mice with a water-soluble fraction isolated from cultured mycelia of *Lentinus lepideus* (called PG101) maintained the number of granulocyte-macrophage colony-forming units in the bone marrow of irradiated mice at a level similar to that seen in nonirradiated control mice [117]. Moreover, monocytes/macrophages, but not T and B lymphocytes, were shown to be the major target cell type responsive to PG101 [117].

6.3 Immunomodulatory Activity of Plant-derived Polysaccharides

Most polysaccharides derived from higher plants are relatively nontoxic and do not cause significant side effects, unlike immunomodulatory bacterial- and fungal-derived polysaccharides. Thus, plant polysaccharides are ideal candidates for therapeutics with immunomodulatory, antitumour and wound-healing action. In studies investigating the effects of plant polysaccharides on macrophage responses, the predominant finding is that polysaccharides derived from higher plants activate macrophages (**Table 6.3** and **Figure 6.2**). Indeed, many polysaccharides derived from different plant families have been shown to enhance macrophage function. It should be noted that some polysaccharides (e.g., BP1002 fractions from *Biophytum petersianum*) can activate macrophages and dendritic cells but not T cells, B cells and NK cells [132]. Plant-derived polysaccharides have been shown to increase macrophage cytotoxic activity against tumour cells and microorganisms, activate phagocytic activity, increase ROS and NO production, and enhance secretion of cytokines and chemokines, such as tumour necrosis factor-α (TNF-α), IL-1β, IL-6, IL-8 and IL-12 and interferon (IFN)-γ and IFN-β2 (reviewed in [3]) (**Figure 6.2**). As an example of this general paradigm, we recently found that arabinogalactan-containing polysaccharides extracted from the cones of *Juniperus scopolorum* (juniper) had potent immunomodulatory effects on human and murine macrophages, as demonstrated by the induction of macrophage-inducible nitric oxide synthase (iNOS) expression and NO production, priming of ROS production and enhanced production of both inflammatory (IL-1, IL-6, IL-12 and TNF-α) and anti-inflammatory (IL-10) cytokines [157]. Likewise, we found that polysaccharides isolated from *Opuntia polyacantha* (prickly pear cactus) [163], *Tanacetum vulgare* L (tansy) [181] and *Artemisia tripartita* [129] also had potent immunomodulating activity and induced a number of host responses in human and murine phagocytes. Furthermore, modulation of macrophage function by these polysaccharides was mediated, at least in part, through activation of NF-κB [163, 181].

Interestingly, different polysaccharides isolated from the same plant species can activate macrophages via distinct signal transduction pathways. For example, enhancement of macrophage myeloperoxidase activity by pectic polysaccharides from *Silene vulgaris* required extracellular Ca^{2+}, whereas activation of this response by callus arabinogalactan from the same species was independent of extracellular Ca^{2+} [180]. In addition to modulation of macrophage responses, plant-derived polysaccharides, including arabinogalactan-containing polysaccharides, can have potent complement-fixing activity, which has been suggested to contribute to their immunomodulatory properties [190, 191]. Indeed, we found that polysaccharides isolated from *T. vulgare* L period [181] and *A. tripartita* [129] exhibited complement-fixing activity. However, complement-fixing activity is not a property common to all arabinogalactan polysaccharides and is not a property of juniper-derived polysaccharides [157]. The absence of anticomplement activity in immunomodulatory polysaccharides may be a therapeutically beneficial property in cases where normal complement activation is desirable.

Table 6.3 Effects of higher plant polysaccharides on macrophage function

Common name and family	PS features (name, source, structure)	Cell type	Effects on macrophages[a]	References
Acanthopanax senticosus (Araliaceae)	PS from seeds	Murine peritoneal MØ	↑ NO, IL-1β, IL-6 and TNF-α	[118]
Aloe barbadensis (Asphodelaceae)	Acemannan	Murine RAW 264.7 MØ	↑ NO, IL-6 and TNF-α	[119, 120]
	Aloeride (M_r 4–7000 kDa)	Human THP-1 MØ	↑ IL-1β and TNF-α message	[121]
Aloe vera L. var. *chinensis* (Haw.) Berg. (Asphodelaceae)	Mannan	Murine peritoneal MØ	↑ NO, TNF-α and phagocytosis	[122]
Anadenanthera colubrina (Fabaceae)	Acid heteropolysaccharide from gum	Murine peritoneal MØ	↑ Phagocytosis and ROS	[123]
Angelica gigas Nakai (Apiaceae)	Angelan	Murine RAW 264.7 MØ	↑ iNOS, IL-1β and TNF-α	[124–126]
Angelica sinensis (Oliv) and Diels (Apiaceae)	Arabinogalactan (M_r 52 kDa)	Murine peritoneal MØ	↑ NO and TLR-4 message	[127]
Arnica montana (Asteraceae)	Arabinogalactan (M_r 95–100 kDa)	Murine peritoneal MØ	↑ TNF-α	[128]
Artemisia tripartita Rydb. (Asteraceae)	Arabinogalactan type II (M_r 251, 126, 78 and 49 kDa)	Murine RAW 264.7 MØ	↑ ROS, NO, IL-6, TNF-α, IL-10, MCP-1	[129]
Astragalus membranaceus (Fabaceae)	Astragalus PS (M_r 1.5–3500 kDa)	Murine peritoneal MØ, murine RAW 264.7 MØ	↑ IL-1β, TNF-α and NO; ↑ iNOS message	[110, 130]

Avena sativa (Poaceae)	Oat β-glucan [ObetaG, β-(1,3;1,4)-glucan]	Murine P338D1 MØ	↑ IL-1 and TNF-α	[99]
Bergenia crassifolia (L.) Fritsch. (Saxifragaceae)	Pectic PS bergenan		↑ ROS	[131]
Biophytum petersianum Klotzsch (Oxalidaceae)	*B. petersianum* PS (BP1002, arabinogalactan, M_r 64 kDa)	Monocyte-derived rat MØ, rat R2 MØ	↑ TNF-α and NO	[132]
Bletilla striata (Orchidaceae)	Glucomannan	Murine RAW 264.7 MØ	↑ NO, IL-1β and TNF-α	[133]
Bupleurum falcatum L. (Apiaceae)	Bupleuran 2IIb (pectic PS)	Murine peritoneal MØ	↑ Fc receptor expression	[134]
Carthamus tinctorius (Asteraceae)	Safflower PS 1 (SF1) and 2 (SF2) (M_r >100 kDa)	Murine peritoneal MØ	↑ TNF-α and NO	[135]
Celosia argentea (Amaranthaceae)	Celosian (acidic PS from seeds)	Murine J774.1 MØ, human monocytes	↑ IL-1β and NO	[136]
Centrosema pubescens (Fabaceae)	*C. pubescens* PS 1 (CP1, M_r 2 kDa) and 2 (CP2, M_r 375 kDa)	Murine phagocytes	↑ Phagocytosis	[137]
Chelidonium majus (Papaveraceae)	*C. majus* protein-bound PS (CM-Ala)[0]	Murine peritoneal MØ	↑ NO	[138]
Crocus sativus (Iridaceae)	Proteoglycan	Murine peritoneal MØ	↑ NO	[139]
Cucurbita moschata Duchesne ex Poiret (Cucurbitaceae)		Murine peritoneal MØ	Antioxidative and cytoprotective effect	[140]

Table 6.3 Continued

Common name and family	PS features (name, source, structure)	Cell type	Effects on macrophages[a]	References
Curcuma zedoaria (Zingiberaceae)	*C. zedoaria* PS (CZ-1-III)	Murine peritoneal MØ, murine RAW 264.7 MØ	↑ ROS, NO and TNF-α	[141]
Curcuma xanthorrhiza Roxb. (Zingiberaceae)	Arabinogalactan (M_r 33 kDa)	Murine RAW 264.7 MØ	↑ ROS, NO, TNF-α, PGE_2 and phagocytosis	[142]
Dioscorea batatas (Dioscoreaceae)	Mucopolysaccharide	Murine peritoneal MØ	↑ MPO activity, ROS, NO, TNF-α and cytotoxicity	[143]
	β-(1,4)-Mannan	Murine J774.1 MØ	↑ ROS and TNF-α	[144]
Dipsacus asperoides (Dipsacaceae)	*D. asperoides* root PS-protein complex (DAP-1)	Murine peritoneal MØ	↓ Phagocytosis (protein moiety)	[145]
Echinacea purpurea (Asteraceae)	Arabinogalactan and xyloglucan	Murine peritoneal MØ, rat alveolar MØ	↑ IL-1, IL-6, TNF-α, IFN-β2, ROS, NO and cytotoxicity; ↑ phagocytosis (oral)	[146–150]
Glycine max (Fabaceae)	Soybean PS-protein complex (SP-MAF1)	Murine peritoneal MØ	↑ Glycolysis and IL-1	[151]
Glycyrrhiza glabra and *Glycyrrhiza uralensis* (Fabaceae)	Glycyrrhiza polysaccharide (GPS)	Murine peritoneal MØ	↑ NO, IL-1 and phagocytosis	[152, 153]
	α-(1-4)-D-Glucan	Murine peritoneal MØ	↑ iNOS, NO, ROS, IL-1, IL-6 and IL-12	[154, 155]
Gynostemma pentaphyllum Makino (Cucurbitaceae)	Non-starch heteropolysaccharide	Murine peritoneal MØ	↑ NO, ROS and TNF-α	[156]

Juniperus scopolorum (Cupressaceae)	Arabinogalactan (M_r 200–680 kDa)	Murine peritoneal MØ, murine J774.1 MØ, human monocytes	↑ iNOS, NO, ROS, IL-1, IL-6, IL-12, TNF-α and IL-10	[157]
Lemna minor (Lemnaceae)	Lemnan (apiogalacturonanic pectin)	Murine peritoneal MØ	↑ ROS and cell adhesion	[158]
Larix occidentalis (Pinaceae)	Arabinogalactan	Human monocytes	↑ TNF-α, IL-1β and IL-6	[159]
Mahonia aquifolium (Berberidaceae)	PS from stem bark	Human THP-1 MØ	↑ IL-8	[160]
Morinda citrifolia (Rubiaceae)	Arabinogalactan	Murine peritoneal MØ	↑ NO, IFN-γ, IL-1β, IL-12 and TNF-α	[161]
Opilia celtidifolia (Opiliaceae)	Arabinogalactan	Rat R2 MØ	↑ NO	[162]
Opuntia polyacantha (Cactaceae)	PS (M_r 733, 550, 310 and 168 kDa)	Human MonoMac6 and THP-1 MØ, murine J774.1 MØ	↑ NO, ROS, IL-6, TNF-α and NF-κB	[163]
Panax ginseng (Araliaceae)	Ginsan and panaxanes	Murine peritoneal MØ	↑ NO, IL-1β, IL-6, IFN-γ, ROS, phagocytosis and cytotoxicity	[164–169]
	Ginsenan S-IIA	Human THP-1 MØ	↑ IL-8	[170]
Panax notoginseng (Araliaceae)	Arabinogalactan (M_r 37–760 kDa)	Murine peritoneal MØ	↑ TNF-α	[171]
Panax quinquefolius (Araliaceae)	Arabinogalactan	Rat alveolar MØ	↑ TNF-α	[172]

Table 6.3 Continued

Common name and family	PS features (name, source, structure)	Cell type	Effects on macrophages[a]	References
Perilla frutescens var. *crispa* (Lamiaceae)	Purified *P. frutescens* PS (PFB-1-0)	Murine peritoneal MØ	↑ Phagocytosis, NO and TNF-α; ↑ GM-CSF and IL-6 (oral)	[173]
Pinus parviflora (Pinaceae)	Acidic PS from pine cones[0]	Murine J774.1 MØ	↑ MØ differentiation	[174]
Plantago ovata (Plantaginaceae)	Mucopolysaccharides	Yorkshire pig MØ	↑ MØ activation	[175]
Platycodon grandiflorus (Campanulaceae)	PS from roots	Murine peritoneal MØ, murine RAW 264.7 MØ	↑ NO, IL-1β, TNF-α and IL-6	[176, 177]
Prunella vulgaris (Lamiaceae)	*P. vulgaris* PS (PV2)	Murine RAW 264.7 MØ	↑ NO and ROS	[178]
Salicornia herbacea (Chenopodiaceae)	S. herbacea extract (SHE)	Murine RAW 264.7 MØ	↑ NO, IL-1β and TNF-α	[179]
Silene vulgaris (Caryophyllaceae)	Callus acidic arabinogalactan (C1) [0]and pectic PS (P1–P3)	Rat peritoneal MØ, human monocytes	↑ Phagocytosis and lysosomal activity	[180]
Tanacetum vulgare (Compositae)	Arabinogalactan (M_r 151, 64 and 9 kDa)	Murine RAW 264.7 MØ, human THP-1 MØ	↑ ROS, NO, TNF-α and NF-κB	[181]
Tinospora cordifolia (Menispermaceae)	Arabinogalactan	Murine RAW 264.7 MØ, human U937 MØ	↑ NO and TNF-α	[182, 183]
Tinospora cordifolia (Menispermaceae)	α-(1,4)-D-glucan	Murine RAW 264.7	↑ TNF-α	[184]

Trigonella foenum-graecum (Fabaceae)	Galactomannan	Rat peritoneal MØ	↑ Phagocytosis	[185]
Tripterygium wilfordii (Celastraceae)	PS extract (PSP22, M_r 22 kDa)	Human THP-1 MØ	↓ LPS stimulatory effects on TNF-α and adhesion molecule expression	[186]
Ribes nigrum (Grossulariaceae)	Arabinogalactan (M_r 80 kDa)	Murine peritoneal MØ	↑ NO, IL-1β, TNF-α and IL-12	[187]
Rubus crataegifolius Bge. (Rosaceae)	α-D-Glucan		↑ NO, TNF-α, phagocytosis and cytotoxicity	[188]
Rheum tanguticum (Polygonaceae)	*R. tanguticum* PS (RTP, M_r 60–80 kDa[0])	Rat peritoneal MØ	↑ IFN-γ	[189]

IL: interleukin
iNOS: inducible nitric oxide synthase
LPS: Lipopolysaccharide
MCP-1: monocyte chemoattractant protein-1
MPO: myeloperoxidase
MØ: macrophages
NF-κB: nuclear factor κB
NO: nitric oxide
PGE_2: prostaglandin E_2
PS: polysaccharides
ROS: reactive oxygen species
TLR: toll-like receptor
TNF: tumour necrosis factor
[a] *Where applicable, the route of administration is indicated as oral or intraperitoneal (ip)*

Some plant-derived polysaccharides may also be useful for drug delivery. For example, polysaccharide isolated from *Bletilla striata* and modified with *N,N′*-carbonyldiimidazole/ethylenediamine in order to acquire nucleic acid-binding affinity was used for targeting delivery of oligonucleotides into macrophages [192]. This modified polysaccharide exhibited significantly higher affinity for macrophages via mannose and β-glucan receptors, and the authors suggested that this polysaccharide could be capable of conveying antisense nucleotides into the cell (e.g., oligodeoxynucleotide and small interfering RNA) for anti-inflammatory therapy [192].

6.4 Effect of Plant Polysaccharides on Neutrophil Function

Neutrophils are the principal effectors of the initial host response to injury or infection and constitute a significant threat to invading bacterial pathogens [193]. Neutrophils play a key role in innate immunity because of their ability to recognise, ingest and destroy many pathogens by oxidative and nonoxidative mechanisms [194]. Thus, the ability to modulate neutrophil functions represents an important immunomodulatory property of botanical polysaccharides. Indeed, botanical polysaccharides have been shown to modulate neutrophil ROS production, phagocytosis, degranulation, adhesion and other host responses (**Figure 6.2**). For example, polysaccharides purified from *Acacia cyanophylla* stimulated ROS production by murine neutrophils [195]. Likewise, polysaccharides extracted from *Panax notoginseng* were found to prime neutrophils, resulting in enhanced ROS production in response to a subsequent stimulus [196]. Furthermore, β-glucans were reported to enhance neutrophil ROS production by binding to the lactosylceramide receptor and activating NF-κB, leading to increased microbicidal activity by human neutrophils [197]. However, not all β-glucans can activate neutrophil ROS production, and specific structural features must be involved in this response. For example, the soluble β-glucans laminarin and scleroglucan failed to activate ROS production by porcine monocytes and neutrophils, whereas particulate β-glucans from algae (*Euglena gracilis*) had a stimulatory effect [198].

Neutrophil granules contain a large array of antimicrobial and potentially cytotoxic substances that are delivered to the phagosome or to the exterior of the cell following degranulation [199] (**Figure 6.2**). Myeloperoxidase is a key enzyme released into the phagosome or to the extracellular space upon neutrophil activation and plays an important role in microbicidal activity by catalysing the formation of various ROS that can attack microorganisms [200]. Thus, the ability to stimulate neutrophil myeloperoxidase release could represent an important immunomodulatory property of plant polysaccharides, and we found that polysaccharides from *T. vulgare* stimulated neutrophil degranulation and myeloperoxidase release [181].

Fucoidan, a sulfated polysaccharide isolated from the brown algae *Fucus evanescens*, stimulated phagocytic and bactericidic activity of murine peritoneal neutrophils [201]. Neutrophil phagocytosis was also activated by polysaccharides isolated from *Panax ginseng* (panaxanes) [164], polysaccharides isolated from *Lemna minor* (lemnan) [202], arabinogalactans from *S. vulgaris* [180], α-1,4-D-galacturonan from *Bergenia crassifolia* (bergenan) [131] and polysaccharides purified from *G. lucidum* [203].

Leucocyte adhesion represents one of the first steps of the inflammatory response and is essential for the accumulation of active immune cells at sites of inflammation [204]. Current evidence suggests that plant polysaccharides can modulate host immune responses by influencing cell adhesion (**Figure 6.2**). For example, bergenan from *B. crassifolia* increased the spontaneous adhesion of neutrophils to plastic surfaces [131]. Polysaccharides isolated from *Astragalus* activated adhesion between neutrophils and endothelial cells by inducing the expression of intercellular adhesion molecule 1 (ICAM-1) on the surface of endothelial cells and contributed to wound healing [205]. Intravenous application of polysaccharides isolated from *Echinacea purpurea* reduced the number of neutrophils in the peripheral blood, suggesting enhanced adhesion to vascular endothelial cells [146]. On the other hand, polysaccharides from *Ginkgo biloba* were reported to inhibit the interaction between P-selectin and neutrophils [206]. Likewise, fragments of galacturonan chains from various pectins (comaruman from *Comarum palustre*, bergenan from *B. crassifolia*, lemnan from *L. minor*, zosteran from *Zostera marina* and citrus pectin) were reported to inhibit neutrophil adhesion to fibronectin [207]. In contrast, the parent pectins, except for comaruman, had no effect on cell adhesion [207]. Thus, plant polysaccharides can modulate neutrophil function by either enhancing or reducing adhesion.

Recently, Graff and co-workers [208] reported that intraperitoneal injection of mice with polysaccharides isolated from the bark of *Funtumia elastica* induced rapid increases in peritoneal neutrophils that were induced by changes in chemokine expression. This response was diminished in CXCR2-deficient mice, suggesting a role of CXCL1 chemokines in this response.

6.5 Antioxidant Properties of Botanical Polysaccharides

Despite the significant role of mammalian antioxidant and repair mechanisms, oxidative damage is an inescapable outcome of aerobic life. When the production of ROS exceeds the ability of living organisms to prevent their accumulation, the result is oxidative stress. In this regard, oxidative stress has been implicated in many different diseases, including the ageing process in general (reviewed in [209–211]). Excessive amounts of ROS, such as superoxide anion ($O_2^{\bullet -}$), hydroxyl radical ($OH^{\bullet}$)

and hydrogen peroxide (H_2O_2), can induce oxidative damage and ultimately impair functions of tissues and organs. In inflammatory diseases, for example, high levels of ROS are produced by phagocytes, resulting in free radical-mediated tissue injury and the formation of lipid peroxidation products [212]. Ultimately, the production of ROS can lead to the alteration of ion gradients, calcium accumulation and eventually central nervous system cell death [213].

Because of the role of ROS in the aetiology of inflammatory disease and tissue injury, considerable efforts have been directed towards the discovery of effective antioxidant compounds that can reduce inflammation and impede oxidant injury to tissues [214, 215]. There are a large number of antioxidant compounds [e.g., vitamin E, probucol, butylated hydroxytoluene (BHT) and anti-inflammatory drugs], and these agents have been shown to be useful in the treatment of diseases involving the production or overproduction of ROS and subsequent lipid peroxidation, such as atherosclerosis, ischaemia-reperfusion injury and arthritis (reviewed in [215, 216]). In addition, newer antioxidant compounds are constantly being developed. In this respect, protection of the host by dietary antioxidants represents an important approach [217].

Recently, many studies have focused on the characterisation of antioxidant properties of immunomodulating polysaccharides from various plant sources, based on their ability to minimise free radical-induced damage. For example, polysaccharides from *Angelica sinensis* (Oliv) Diels and *Aloe vera* var. *chinensis* were shown to protect macrophages from H_2O_2-induced apoptosis [218, 219]. Likewise, polysaccharides from *Opuntia dillenii* Haw protected rat pheochromocytoma PC12 cells against H_2O_2 insult [220]. Pretreatment of macrophages with polysaccharide fractions from *A. sinensis* significantly enhanced cell survival after treatment with tert-butyl hydroperoxide, maintained intracellular glutathione (GSH) content and superoxide dismutase (SOD) activity and also inhibited tert-butyl hydroperoxide-induced lactate dehydrogenase leakage and malondialdehyde (MDA) formation [221]. Additionally, β-(1,3)-glucans reduced the genotoxic effects of ROS in V79 hamster cells [222]. Polysaccharides from *Rheum tanguticum* have been shown to protect intestinal epithelial cells against oxidative stress [223]. Similarly, a significant reduction in mucosal damage was observed when pectin was administered before the perfusion of rat jejunum with peroxyl and hydroxyl radicals [224].

Kardošová and Machová [225] characterised the antioxidant activities of polysaccharides isolated from a variety of sources (*Arctium lappa* var. *herkules*, *Aloe barbadensis*, *Althaea officinalis* var. *robusta*, *Plantago lanceolata* var. *libor*, *Rudbeckia fulgida* var. *sullivantii*, *Mahonia aquifolium* and *Prunus persica*). Analysis of the ability of these polysaccharides to inhibit peroxidation of soybean lecithin liposomes by $OH^•$ showed that glucuronoxylans from *A. officinalis* var. *robusta* and *P. lanceolata* var. *libor* were highly effective, whereas the other polysaccharides had intermediate activity and fructofuranan from *P. lanceolata* var. *libor* roots was essentially inactive [225]. Polysaccharides isolated from *Ficus carica* and *Saposhnikovia divaricata* (Turez.) Schischk demonstrated $O_2^{\bullet-}$ and $OH^{\bullet}$ scavenging activity [226, 227]. Likewise, polysaccharides from *A. vera* (APS-1) and sulfated polysaccharides isolated from *Porphyra haitanensis* were also effective at scavenging $O_2^{\bullet-}$ and exhibited much higher activity than sulfated polysaccharide fractions isolated from the alga *P. haitanensis* [228]. Polysaccharides isolated from *Dendrobium nobile* Lindl and *Lycium barbarum* Linnaeus demonstrated scavenging activity for 2,2′-azino-bis(3-ethylbenzthiazoline-6-sulfonic acid) (ABTS) radicals, as well as for $OH^{\bullet}$ [229, 230]. Similarly, polysaccharides from *Dendrobium chrysotoxum* Lindl and *Dendrobium denneanum* showed dose-dependent scavenging activity for $O_2^{\bullet-}$ and/or $OH^{\bullet}$ [231, 232]. Proteoglycans isolated from rapeseed meal (water-soluble polysaccharide fraction 1) were found to be effective scavengers of $O_2^{\bullet-}$ and $OH^{\bullet}$, but were less effective at scavenging H_2O_2 [233]. Polysaccharides from *Pteridium aquilinum* exhibited moderate scavenging activities against 1,1-diphenyl-2-picrylhydrazyl (DPPH) radicals and $O_2^{\bullet-}$ [234]. On the other hand, arabinoglucogalactans from *P. notoginseng* and pectins exhibited an excellent ability to scavenge DPPH radicals [235, 236].

The oxidation of botanical polysaccharides is not well defined compared to that of monosaccharides and mammalian carbohydrates, such as hyaluronan, chondroitin and heparans. However, similar processes could play a role in the mechanisms of oxidation of plant polysaccharides, including hydrogen atom abstraction and cleavage of the polysaccharide backbone (reviewed in [237]). The presence of carboxyl, sulfate and amino groups can modify the extent of oxidation of particular C–H bonds. It has been proposed that free radicals break glucoside linkages in the galacturonic acids of pectin, resulting in trapping of the radicals [235], which may explain the high radical scavenging ability of pectins. Indeed, both galacturonic acid content and the molecular weight of acidic polysaccharides may play an important role in the antioxidant activity. Among the acidic polysaccharides, a relatively low molecular weight and a high galacturonic acid content appear to increase antioxidant activity [238]. Administration of sulfated polysaccharides from *Fucus vesiculosus* to hyperoxaluric rats decreased lipid peroxidation and increased the activities of antioxidant enzymes [239]. Furthermore, increased nitrosative stress accompanying hyperoxaluria was also normalised upon sulfated polysaccharide treatment [239].

Due to the complexity of polysaccharides and lack of structural data for most botanical polysaccharides reported with antioxidant activity, it is not possible to deduce structure-antioxidant activity relationships. However, several publications have suggested chemical features that may contribute to antioxidant efficacy. For example, chemical modification of ulvan, a heteropolysaccharide extracted from the green alga *Ulva pertusa*, by acetylation or benzoylation, enhanced its antioxidant activity [240]. On the other hand, Liu and co-workers [241] suggested that the $O_2^{\bullet-}$

scavenging activity of polysaccharide extracts appears to depend on the amount of protein present as proteoglycans. Tsiapali and co-workers [242] described the free radical scavenging activity of a variety of polysaccharides and demonstrated that their free radical scavenging activity was due, in part, to their monosaccharide constituents. They observed that phosphorylated and sulfated glucan showed greater antioxidant ability than other glucans and natural polysaccharides [242]. The antioxidant activity of sulfated polysaccharides depends on several structural features, such as the degree of sulfation, molecular weight, substituted groups, types of sugars, glycosidic branching and additional chemical modifications. Indeed, it has been reported that substituted groups, such as sulfate, acetyl and phosphate, can enhance the antioxidant activity of polysaccharides *in vitro* [243, 244]. In support of this idea, phosphorylated and sulfated fucoidan exhibited excellent scavenging activity for DPPH radicals, which was attributed to its strong hydrogen-donating ability [244].

Activation of antioxidant enzymatic and nonenzymatic systems represents an additional mechanism whereby polysaccharides can modulate oxidant stress. Recently, Hu and co-workers [245] reported that polysaccharides from *Potentilla anserina* (PAP) were able to repair oxidative damage in immunological tissues. The authors reported that PAP increased the thymus and spleen indices, GSH levels, SOD activity and total antioxidant capacity in both thymus and spleen. They also found that PAP treatment decreased H_2O_2 in the spleen and decreased NO in both thymus and spleen of mice with dexamethasone-induced oxidative stress [245]. Likewise, *in vivo* administration of *Glycyrrhiza glabra* polysaccharide (GGP) dose-dependently enhanced the activities of antioxidant enzymes, such as SOD, catalase and glutathione peroxidase, in the blood of treated mice and also decreased blood MDA levels [246]. The authors suggested that direct antioxidant activity of GGP could be attributed to the polyphenolic groups (isoflavan) associated with the polysaccharide [246]. However, antioxidant activity has also been observed in polysaccharides free of phenolic groups [222].

Overall, the ability of polysaccharides to act as immunomodulators as well as antioxidants represents a beneficial dual effect that should be exploited. Hence, such polysaccharides would be able to stimulate immunity and also protect tissues from excessive damage caused by ROS generated during host defence. Indeed, recent studies suggest that while the antioxidant activity of polyphenols has been assumed to be the primary therapeutic mechanism, many polyphenols have a direct impact on cellular signalling events independent of their antioxidant activity [247, 248].

6.6 Mitogenic Activity of Botanical Polysaccharides

Mitogenic stimulation of lymphocytes by plant glycoproteins (lectins) was described over 50 years ago by Nowell [249], who observed that phytohaemagglutinin stimulated mitosis in cultures of normal human leucocytes. Subsequently, a number of studies have revealed that various plant-derived polysaccharides and polysaccharide-protein complexes can activate T- and B-lymphocyte proliferation with similar and distinct mechanisms of action (**Figure 6.4**). For example, polysaccharides from sage (*Salvia officinalis*) exhibited very high mitogenic and comitogenic activities, suggesting the potential for use of these polysaccharides as adjuvants [250]. Polysaccharides isolated from the roots of *Dipsacus asperoides* directly stimulated lymphocyte proliferation but had no effect on lectin-induced mitogenic activity [145]. Similarly, protein-bound polysaccharides extracted from *Chelidonium majus* displayed mitogenic activity on spleen cells [138]. An arabinogalactan isolated from *Tinospora cordifolia* exhibited polyclonal mitogenic activity against B cells, although this effect did not require macrophages [251] (**Figure 6.4**). High-molecular-weight polysaccharides from *Thuja occidentalis* proved to be mitogenic for peripheral blood leucocytes, and the mitogenic effect was dependent on IL-1 and IFN-γ, as anti-IL-1 and anti-IFN antibodies blocked the activity [252]. While zwitterionic polysaccharides, such as β-glucans, have been reported to activate $CD4^+$ T cells, this response appears to be mediated by the major histocompatibility complex-II endocytic pathway [253]. The pectic polysaccharide GR-2IIc, isolated from the roots of *Glycyrrhiza uralensis* Fisch, exhibited B-cell mitogenic activity [254], and it has been proposed that the oligosaccharide side chains of the ramified region of GR-2IIc directly contributed to the mitogenic activity of GG-2IIc, whereas side chains attached to the rhamnogalacturonan may amplify this activity [255]. When bupleuran, isolated from the roots of *Bupleurum falcatum*, was administered orally to C3H/HeJ mice, proliferative responses of spleen cells were enhanced, suggesting that TLR-4 was required for this response [256]. Finally, heteroglycans isolated from the lichens *Thamnolia vermicularis* and *Peltigera canina* stimulated rat spleen cell proliferation and secretion of IL-10 [49, 257]. Interestingly, it appears that polysaccharides with high mitogenic activity also exhibit high complement-fixing activity. For example, pectins from *B. falcatum* [256, 258], *G. uralensis* [255] and *Vernonia kotschyana* [259] containing arabinogalactan side chains were highly active in both the complement and mitogenic assays. Whether this correlation is found among other botanical polysaccharides remains to be determined.

Botanical polysaccharides have also been shown to enhance the mitogenic activity of standard mitogens in a synergistic manner (**Figure 6.4**). For example, the polysaccharides lentinan and ganoderan enhanced spleen lymphocyte proliferation induced by concanavalin A [260]. Likewise, astragalan, ganoderan and lentinan synergised with bacterial lipopolysaccharides (LPS) to promote IgG secretion and proliferation of spleen lymphocytes. Additionally, two of these polysaccharides (lichenan and astragalan) also enhanced the activation of peritoneal macrophages and splenic NK cells [260]. Recently, Sun and co-workers [261] reported that

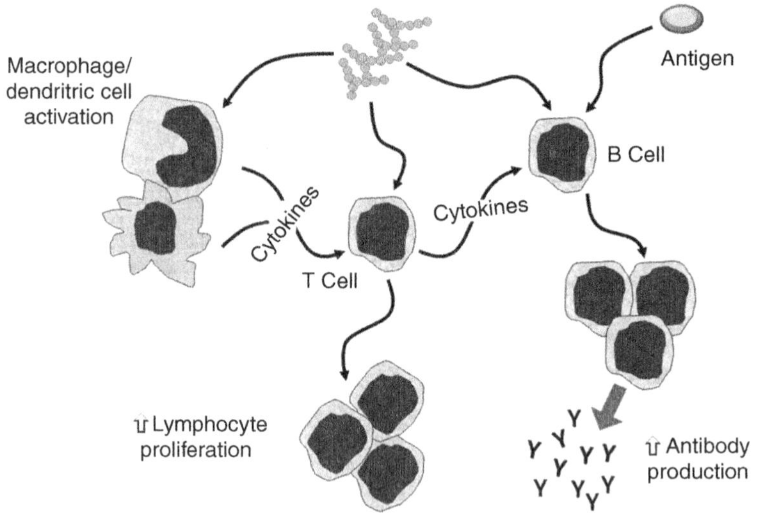

Figure 6.4 Mitogenic properties of botanical polysaccharides. Botanical polysaccharides exhibit a variety of mitogenic properties in T- and B-lymphocytes. Polysaccharides and/or polysaccharide-protein complexes can activate T-lymphocyte proliferation directly or in combination with cytokines (e.g., interleukin (IL)-1 and interferon (IFN)-γ) generated by polysaccharide-activated accessory cells, such as macrophages and/or dendritic cells. Polysaccharides can also enhance the mitogenic activity of other standard mitogens/antigens and induce B-cell proliferation, resulting in increased production of antigen-specific antibodies. This process can be modulated by cytokines (e.g., IL-2, IL-4, IL-10 and IFN-γ) produced concurrently by polysaccharide-activated lymphocytes

polysaccharides from the roots of *Actinidia eriantha* induced splenocyte proliferation and increased serum ovalbumin (OVA)-specific antibody titres in the immunised mice. These polysaccharides also induced the production of Th1 (IL-2 and IFN-γ) and Th2 (IL-10) cytokines and upregulated the message for IL-2, IFN-γ, IL-4 and IL-10 in splenocytes from OVA-immunised mice [261] (**Figure 6.4**).

6.7 Role of Intestinal Immune System in Modulating the Activity of Botanical Polysaccharides

Botanical polysaccharides are acquired orally. However, because of their high molecular weights, polysaccharides are not easily absorbed. Thus, these polysaccharides are first encountered by immune system components at mucosal surfaces, and it is clear that the enteric mucosal immune response plays an important role in the immunomodulatory effects of various botanical polysaccharides.

The intestinal mucosal immune system includes Peyer's patches and other specialised lymphoid tissues (GALT) [262]. Peyer's patches are aggregates of lymphoid tissue and are usually found in the lowest portion of the small intestine; they are unique in that they play important roles as inductive sites for IgA production and regulation of the immune system [262, 263]. Peyer's patch lymphocytes migrate through the mesenteric lymph nodes into the thoracic duct and then to the systemic circulation, where they regulate the local mucosal and systemic immune systems. The epithelium overlying the lymphoid follicles of the Peyer's patches contains specialised M cells, which engulf various substances such as soluble antigens, bacteria and viruses from the intestinal lumen, and transport them to underlying lymphoid cells [264]. Unlike in macrophages, the internalisation of soluble β-glucan by intestinal epithelial cells is not dectin-1 dependent. However, dectin-1 and TLR-2 are responsible for uptake of soluble β-glucan by GALT cells [101].

It has been hypothesised that botanical polysaccharides can interact with both epithelial cells and immunocompetent cells in Peyer's patches, resulting in potentiation of distal mucosal and systemic immune systems [265]. In support of this idea, botanical polysaccharides have also been shown to modulate Peyer's patch cells in the intestinal immune system. For example, the crude polysaccharide fraction isolated from Hachimi-jio-gan, a Japanese and Chinese herbal medicine, was found to augment IgA production in Peyer's patch cells, whereas the low-molecular-weight fraction from this medicine was not active [266]. Likewise, oral administration of polysaccharide complexes isolated from the Japanese herbal medicine Juzen-Taiho-To induced activation of Peyer's patch T cells and secretion of haematopoietic growth factors [267, 268]. Similarly, Yu and co-workers [269] reported that high-molecular-weight polysaccharides from *Atractylodes lancea* contributed to the immunomodulating activity in Peyer's patch cells. Polysaccharides from *G. lucidum* (GLP) stimulated the proliferation of enteric mucosal lymphocytes and enhanced the expression of TNF-α and IL-10 mRNA in intra-epithelial lymphocytes and Peyer's patch lymphocytes, suggesting that the enteric mucosal immune responses may be important for the immunomodulatory activity of GLP [270]. Similarly, bupleuran 2IIb and 2IIc, polysaccharides from *B. falcatum*, induced the proliferation of Peyer's patch cells [256]. Additionally, the *ex vivo* production of IgA and IL-6 from Peyer's patch cells was enhanced two-fold in mice treated with Immulina, a high-molecular-weight (>10,000 kDa) polysaccharide fraction from *Spirulina platensis* [271]. Finally, CVT-E002™ (sold commercially as COLD-fX®) is a poly-furanosyl-pyranosyl polysaccharide-rich extract of the root of North American ginseng (*Panax quinquefolius*) with reported beneficial effects for preventing influenza and the common cold [272]. Recent studies demonstrated that oral administration of CVT-E002™ altered systemic immune responses and affected gut-associated immunity in a manner distinct from that of ginsenoside-containing extracts of North American ginseng [273]. Overall, it is clear that a better understanding of the structures and functions of the carbohydrate structures in polysaccharide molecules will facilitate the development of novel therapeutics that can target immunomodulatory pathways in mucosal tissues of the gut.

6.8 Antiviral Activity of Botanical Polysaccharides

Current medical approaches available for controlling and preventing viral infections, such as influenza, include use of vaccines and antiviral agents [247]. On the other hand, nonspecific enhancement of host immune responses has also been proposed as an effective strategy to provide prophylactic protection against infectious diseases [275]. Indeed, botanical polysaccharides have attracted much attention because of their broad spectrum of therapeutic properties and their relatively low toxicity (reviewed in [3, 4]). Therefore, efforts have focused on the discovery of novel polysaccharides from edible natural resources that either have direct antiviral activity or can act as therapeutic agents to nonspecifically enhance host defence against viral infections.

Polysaccharides from marine algae and cyanobacteria have been extensively studied for their direct antiviral properties (reviewed in [276, 277]). The antiviral effects of algal polysaccharides were first reported in 1958 by Gerber and co-workers [278], who found that seaweed extracts protected chicken embryos against influenza B and mumps virus. Subsequently, the activity spectrum of the sulfated polysaccharides has been found to extend to various enveloped viruses, including opportunistic viral pathogens [276]. For example, Wang and co-workers [279] recently reported that *L. barbarum* polysaccharides inhibited the infectivity of Newcastle disease virus in chicken embryo fibroblasts and that sulfation significantly increased this antiviral activity. Likewise, sulfated polysaccharides have been reported to exhibit anti-human immunodeficiency virus (HIV) activity, and this activity results from the ability of polysaccharides to disrupt the CD4-virion gp120 interaction; however, additional antiviral mechanisms may also be involved [280].

A number of botanical polysaccharides have also been recognised for their ability to enhance host antiviral defences. For example, ingestion of soluble β-glucan derived from oats reduced morbidity and mortality after an intranasal *Herpes simplex* virus type 1 (HSV-1) challenge in mice after strenuous exercise [281]. Likewise, Ohta and co-workers [282] found that intranasal application of acidic polysaccharides isolated from the fungi *Cordyceps militaris* protected mice from lethal influenza infection. In this model, protection was proposed to involve enhancement of host immune function, and the animals treated with *C. militaris* polysaccharides demonstrated increased levels of TNF-α and IFN-γ [282], both being cytokines with antiviral properties [283]. For example, Peng and co-workers [284] found that *Achyranthes bidentata* polysaccharides exhibited anti-HIV-1 activity if they were modified by sulfation. Likewise, oral treatment with exopolysaccharides from halophilous cyanobacterium *Aphanothece halophytica* Fremy significantly inhibited pneumonia in influenza virus A (H1N1)-infected mice [285]. Application of this polysaccharide activated lymphocyte proliferation, lymphocyte IL-2 production, cytolytic activity of NK cells and macrophage phagocytosis in treated mice [285]. Qiu and co-workers [286] showed that polysaccharides isolated from *Astragalus*, Isatis root, *Achyranthes* root and Chinese yam significantly enhanced the antibody titres against Newcastle disease-infectious bronchitis vaccine and activated concanavalin A-induced proliferation of peripheral blood lymphocytes after subcutaneous application of the polysaccharides to chickens. Recently, Chen and co-workers [287] confirmed these observations and showed that *Astragalus* polysaccharide and oxymatrine could synergistically enhance the efficacy of Newcastle disease vaccine in chickens.

6.9 Botanical Polysaccharides as Adjuvants

Recently, many polysaccharides have been shown to possess adjuvant activity, as demonstrated by their ability to enhance specific cellular and humoral immune responses against antigens. Thus, botanical polysaccharides have been proposed to represent excellent candidates to replace alum as the adjuvant of choice for many vaccines [158, 288, 289]. For example, acemannan has been shown to increase lymphocyte responses to alloantigens *in vitro* [290], exhibit adjuvant activity in vaccination against virus [291] or heart-worm antigen [292] and increase survival of virus-infected animals [293, 294]. Selegean and co-workers [289] found that combining vaccines against infectious bursal disease virus with polysaccharides from the edible mushroom *Pleurotus ostreatus* enhanced the production of antibodies against infectious bursal disease virus during the first days of life, which is critical to the broilers' survival. Polysaccharides purified from the roots of *Codonopsis pilosula* and *A. eriantha* (AEPS) significantly enhanced OVA-induced splenocyte proliferation and OVA-specific IgG, IgG1 and IgG2b antibody levels in OVA-immunised mice [261, 295]. AEPS also induced the production of Th1 (IL-2 and IFN-γ) and Th2 (IL-10) cytokines and upregulated the message for IL-2, IFN-γ, IL-4, IL-10, T-bet and GATA-3 in splenocytes from the immunised mice [261]. Indeed, the authors concluded that these polysaccharides could be safe and efficacious adjuvants for use in vaccines against both pathogens and cancer cells. Popov and co-workers [158] reported that the level of serum antibodies against OVA was four-fold higher in mice immunised with the mixture of OVA and lemnan, an apiogalacturonanic pectin of duckweed *L. minor*, as compared to the response in mice immunised with OVA only. Lemnan was shown to increase the delayed-type hypersensitivity (DTH) reaction and the levels of OVA-specific serum IgG1 and IgG2a [158]. Likewise, citrus pectin was found to inhibit oral tolerance to OVA, resulting in an increased DTH and anti-OVA IgG in mice fed citrus pectin [296].

In contrast to the adjuvant properties described previously botanical polysaccharides have, in some cases, been reported to attenuate immune responses and have been proposed as promising agents for the treatment of IgE-dependent diseases (e.g., atopic dermatitis, asthma, atopic rhinitis, urticaria and food allergies). For example, Danilets and co-workers [297] studied the effects of water-soluble polysaccharides isolated from six different plants on anaphylactic shock and production of IgE and IgG1 by lymphocytes from mice immunised with OVA. They found that treatment with polysaccharides from coltsfoot, sweet flag, clover, Artemisia, marigold and elecampane reduced animal mortality after induction of anaphylactic shock. In addition, injection of these polysaccharides reduced serum concentrations of IgE and IgG1 [297]. Likewise, pectins have been reported to exhibit antiallergic activity, resulting in a suppressed allergic asthmatic reaction in animals treated with doses of 5–12 mg/kg [298]. In contrast, higher doses of pectins (e.g., 50 mg/kg) have been reported to stimulate immune responses [158, 256]. Thus, it is clear that doses of polysaccharides do have profound effects on the nature of the host response. Further studies using different mouse strains and antigen doses could help to clarify the ambiguous effects of pectin on immune response.

Because of their adjuvant effects and additional bioactivities described above, as well as their low toxicity, botanical polysaccharides have been proposed to be ideal adjuvants for use in modern cancer therapy [299]. Indeed, Shin and co-workers [300] demonstrated a synergistic antitumour effect of polysaccharides isolated from Korean red ginseng (*P. ginseng* C.A. Meyer) and paclitaxel in mice transplanted with Sarcoma-180 tumour cells, resulting in increased life span of treated animals. The augmented antitumour effect of paclitaxel was proposed to be the result of the immunomodulating antitumour effect of the polysaccharide, which was shown to exhibit B-cell-specific mitogenic activity and also induced secretion of IL-6 by spleen cells [300].

6.10 Antitumour Effects of Botanical Polysaccharides

Botanical polysaccharides have been shown to exhibit a range of antitumour properties utilising various mechanisms (**Figure 6.5**). Most antitumour activities of botanical polysaccharides are thought to be mediated via activation of host defence mechanisms, rather than direct cytotoxicity to tumour cells [301]. Note, however, that this paradigm is not absolute, and several mushroom polysaccharides have also been shown to exhibit direct inhibitory effects on cancer cell growth by modulating cell-cycle progression

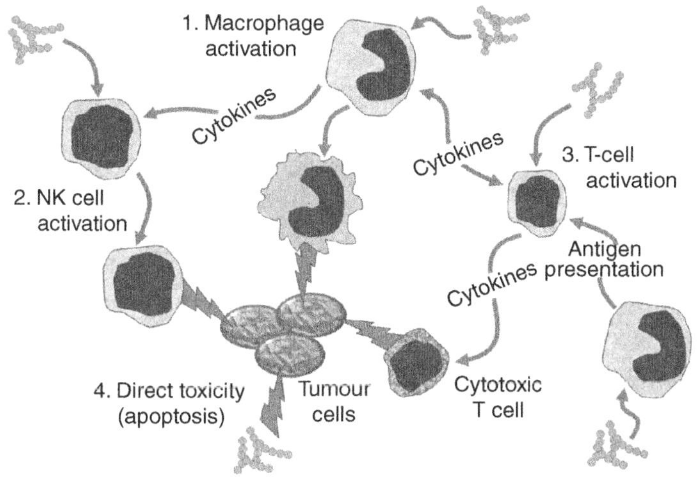

Figure 6.5 Antitumour mechanisms of botanical polysaccharides. Botanical polysaccharides utilise a variety of mechanisms to exert their antitumour effects. (1) Botanical polysaccharides activate macrophages to directly attack tumour cells, as well as generate cytokines (e.g., interleukin (IL)-2) that activate T lymphocytes and natural killer (NK) cells. Likewise, polysaccharides stimulate dendritic cells to produce immunomodulatory cytokines, such as IL-12 (not shown); (2) Botanical polysaccharides can directly stimulate or synergise with cytokines to activate NK cells to attack and destroy tumour cells; (3) Botanical polysaccharides can stimulate T helper cells to produce cytokines (e.g., IL-2, IL-4, IL-10 and interferon (IFN)-γ) that activate macrophages and induce the formation of cytotoxic T cells that kill tumour cells; and (4) Botanical polysaccharides can be directly cytotoxic to tumour cells by inhibiting cell-cycle progression and inducing apoptosis

and inducing apoptosis [302] (**Figure 6.5**). Polysaccharides with antitumour effects are mainly present as glucans with various types of glycosidic linkages, such as β-(1,3;1,6)-glucans and α-(1,3)-glucans, and as true heteroglycans, while others consist of polysaccharide-protein complexes. Three antitumour mushroom polysaccharides, that is, lentinan, schizophyllan and protein-bound polysaccharide (PSK, Krestin), isolated from *L. edodes*, *S. commune* and *C. versicolor*, respectively, have been extensively marketed in Japan [301]. Data from multiple epidemiological and clinical studies on immune effects of conventional cancer treatment and the clinical benefits of polysaccharide immune therapy suggest that immune therapy utilising polysaccharide constituents isolated from mushrooms, particularly *Trametes versicolor*, may be useful as part of a comprehensive cancer treatment and prevention strategy [303].

Oral administration of exopolysaccharides from rice bran fermented with *L. edodes* induced the activation of NK cells in a dose-dependent manner and prolonged the life span of mice transplanted with Sarcoma-180 cells or B16/Bl6 melanoma cells [304]. Lentinan increased the cytotoxic activity of macrophages for metastatic lesions [31], and this activity was suppressed by pretreatment of experimental animals with carrageenan, an anti-macrophage agent [305] (**Figure 6.5**). Similarly, schizophyllan and its soluble β-(1,6)-branched derivatives also activated the antitumour activity of macrophages [306]. Polysaccharides isolated from the mushroom *G. frondosa* increased IL-2 production in patients with lung and liver cancer, presumably as a result of macrophage activation [22]. These polysaccharides have been reported to activate macrophages, dendritic cells and T cells and to activate the cytotoxic activity of NK cells against YAC-1 cells through the production of IL-12 by macrophages [307]. Polysaccharides isolated from the blue-green algae *S. platensis* have also been shown to exhibit antimetastatic activity [308, 309]. Similarly, oral treatment of mice with polysaccharides from the algae *Enteromorpha intestinalis* resulted in up to 70% inhibition of tumour growth (Sarcoma-180 cells), which was proposed to be due to the potent immunostimulating effect of these polysaccharides [60].

Oral administration of arabinogalactan isolated from *Ribes nigrum* (designated cassis polysaccharide or CAPS) to Ehrlich carcinoma-bearing mice inhibited the solid tumour growth by 51%, and induced production of IL-2, IL-4, IL-10 and IFN-γ from splenocytes [187]. Likewise, treatment of mice with polysaccharides from *Solanum nigrum* Linne decreased the number of U14 cervical tumour cells and prolonged the survival time of U14 cervical cancer-bearing animals [310]. Acemannan has also been shown to be effective in the treatment of spontaneous canine and feline fibrosarcomas [311, 312]. When administered intraperitoneally to tumour-implanted mice, acemannan was shown to cure completely or significantly reduce the tumour burden [313]. Intravenous treatment of mice with polysaccharides isolated from *E. purpurea* significantly increased the survival rate of mice injected with lethal doses of *C. albicans* or *Listeria monocytogenes* [147, 314]. Finally, polysaccharides from the roots of *A. eriantha* significantly inhibited the growth of transplantable tumours in mice and also promoted splenocyte proliferation, NK cell and cytotoxic T-lymphocyte activity, IL-2 and IFN-γ production from splenocytes and serum antigen-specific antibody levels in tumour-bearing mice [315] (**Figure 6.5**).

Recently, a phase I/II trial testing the effects of polysaccharide extracts from *G. frondosa* (Maitake mushroom) in breast cancer patients demonstrated that orally administered polysaccharides had both stimulatory and inhibitory effects on peripheral blood immune cells [316]. The authors' conclusion was that patients taking such polysaccharides should be aware that immunological responses to these substances are more complex than previously appreciated [316].

6.11 Conclusion

Polysaccharides extracted from mushrooms, algae and higher plants have significant therapeutic potential and represent a rich source for future discovery and development of novel compounds of medicinal value (**Figure 6.1**). The nonspecific nature of the immunomodulatory activity of polysaccharides makes them attractive because they can be used to treat a broad spectrum of infections and are not susceptible to antibiotic resistance [317]. In general, immunomodulators mimic the natural mechanisms used by pathogens to stimulate innate immunity and thus are potentially beneficial in preventing infection [317]. Additionally, many plant polysaccharides exhibit antioxidant activity, adding to the beneficial effects for the host. Thus, the balance between therapeutic and proinflammatory properties is important when evaluating polysaccharide immunomodulators, and the goal should be to enhance local host defence without inducing excessive or systemic inflammation. This balance is dependent on pharmacodynamic and pharmacokinetic properties of the polysaccharide and will be important to evaluate as further research focuses on the development of such compounds as novel therapeutics to treat disease.

References

1. B.S. Paulsen, *Current Organic Chemistry*, 2001, 5, 939.
2. M.E. Heitzman, C.C. Neto, E. Winiarz, A.J. Vaisberg and G.B. Hammond, *Phytochemistry*, 2005, **66**, 1, 5.
3. I.A. Schepetkin and M.T. Quinn, *International Immunopharmacology*, 2006, **6**, 3, 317.
4. A.O. Tzianabos, *Clinical Microbiology Reviews*, 2000, **13**, 523.
5. S.P. Wasser, *Applied Microbiology and Biotechnology*, 2002, **60**, 3, 258.
6. D.S. McLellan and K.M. Jurd, *Blood Coagulation and Fibrinolysis*, 1992, **3**, 1, 69.
7. S.L. Kosaraju, *Critical Reviews in Food Science and Nutrition*, 2005, **45**, 4, 251.
8. M.Y. Leung, C. Liu, J.C. Koon and K.P. Fung, *Immunology Letters*, 2006, **105**, 2, 101.
9. S.H. Hu, J.C. Wang, J.L. Lien, E.T. Liaw and M.Y. Lee, *Applied Microbiology and Biotechnology*, 2006, **70**, 1, 107.

10. V.V. Glinsky and A. Raz, *Carbohydrate Research*, 2009, **344**, 14, 1788.
11. G. Chihara, J. Hamuro, Y.Y. Maeda, T. Shiio, T. Suga, N. Takasuka and T. Sasaki, *Cancer Detection and Prevention*, 1987, **1**, 423–443.
12. S.Y. Wang, M.L. Hsu, H.C. Hsu, C.H. Tzeng, S.S. Lee, M.S. Shiao and C.K. Ho, *International Journal of Cancer*, 1997, **70**, 6, 699.
13. B. Beutler, *Molecular Immunology*, 2004, **40**, 12, 845.
14. T.C. Sorrell and S.C. Chen, *Advances in Experimental Medicine and Biology*, 2009, **666**, 108.
15. H.X. Wang, T.B. Ng, W.K. Liu, V.E. Ooi and S.T. Chang, *International Journal of Biochemistry and Cell Biology*, 1996, **28**, 5, 601.
16. W.K. Liu, T.B. Ng, S.F. Sze and K.W. Tsui, *Immunopharmacology*, 1993, **26**, 2, 139.
17. Z.J. Pang, M. Zhou, Y. Chen and and J. Wan, *American Journal of Chinese Medicine*, 1998, **26**, 2, 133.
18. Y.J. Chen, M.S. Shiao, S.S. Lee and S.Y. Wang, *Life Sciences*, 1997, **60**, 25, 2349.
19. J.K.H. Cheung, J. Li, A.W.H. Cheung, Y. Zhu, K.Y.Z. Zheng, C.W.C. Bi, R. Duan, R.C.Y. Choi, D.T.W. Lau, T.T.X. Dong, B.W.C. Lau and K.W.K. Tsim, *Journal of Ethnopharmacology*, 2009, **124**, 1, 61.
20. H.L. Chang, L.S. Lei, C.L. Yu, Z.G. Zhu, N.N. Chen and S.G. Wu, *Journal of Chinese Medicinal Materials*, 2009, **32**, 4, 561.
21. H.Y. Hsu, K.F. Hua, C.C. Lin, C.H. Lin, J. Hsu and C.H. Wong, *Journal of Immunology*, 2004, **173**, 10, 5989.
22. N. Kodama, K. Komuta and H. Nanba, *Alternative Medicine Review*, 2002, **7**, 3, 236.
23. I. Sanzen, N. Imanishi, N. Takamatsu, S. Konosu, N. Mantani, K. Terasawa, K. Tazawa, Y. Odaira, M. Watanabe, M. Takeyama and H. Ochiai, *Journal of Experimental and Clinical Cancer Research*, 2001, **20**, 4, 591.
24. M. Okazaki, Y. Adachi, N. Ohno and T. Yadomae, *Biological and Pharmaceutical Bulletin*, 1995, **18**, 10, 1320.

25. Y. Adachi, N. Ohno and T. Yadomae, *Biological and Pharmaceutical Bulletin*, 1998, **21**, 3, 278.

26. Y. Adachi, M. Okazaki, N. Ohno and T. Yadomae, *Biological and Pharmaceutical Bulletin*, 1994, **17**, 12, 1554.

27. N. Ohno, T. Hashimoto, Y. Adachi and T. Yadomae, *Immunology Letters*, 1996, **52**, 1, 1.

28. T. Hashimoto, N. Ohno, Y. Adachi and T. Yadomae, *Biological and Pharmaceutical Bulletin*, 1997, **20**, 9, 1006.

29. K. Ishibashi, N.N. Miura, Y. Adachi, N. Ohno and T. Yadomae, *Bioscience, Biotechnology and Biochemistry*, 2001, **65**, 9, 1993.

30. J.S. Lee, K.M. Min, J.Y. Cho and E.K. Hong, *Journal of Microbiology and Biotechnology*, 2009, **19**, 9, 951.

31. A. Ladanyi, J. Timar and K. Lapis, *Cancer Immunology Immunotherapy*, 1993, **36**, 2, 123.

32. J.P. Fruehauf, G.D. Bonnard and R.B. Herberman, *Immunopharmacology*, 1982, **5**, 1, 65.

33. S. Sipka, G. Abel, J. Csongor, G. Chihara and J. Fachet, *International Journal of Immunopharmacology*, 1985, **7**, 5, 747.

34. G. Abel, J. Szollosi, G. Chihara and J. Fachet, *International Journal of Immunopharmacology*, 1989, **11**, 6, 615.

35. M. Liu, J. Li, F. Kong, J. Lin and Y. Gao, *Immunopharmacology*, 1998, **40**, 3, 187.

36. M. Mizuno, Y. Shiomi, K. Minato, S. Kawakami, H. Ashida and H. Tsuchida, *Immunopharmacology*, 2000, **46**, 2, 113.

37. Y. Ohta, J.B. Lee, K. Hayashi, A. Fujita, D.K. Park and T. Hayashi, *Journal of Agricultural and Food Chemistry*, 2007, **55**, 25, 10194.

38. M. Jin, H.J. Jung, J.J. Choi, H. Jeon, J.H. Oh, B. Kim, S.S. Shin, J.K. Lee, K. Yoon and S. Kim, *Experimental Biology and Medicine*, 2003, **228**, 6, 749.

39. C.J. Duncan, N. Pugh, D.S. Pasco and S.A. Ross, *Journal of Agricultural and Food Chemistry*, 2002, **50**, 20, 5683.

40. G.Y. Kim, Y.H. Oh and Y.M. Park, *Biochemical and Biophysical Research Communications*, 2003, **309**, 2, 399.

41. S.K. Roy, D. Das, S. Mondal, D. Maiti, B. Bhunia, T.K. Maiti and S.S. Islam, *Carbohydrate Research*, 2009, **344**, 18, 2596.

42. K.Y. Lee, H.J. You, H.G. Jeong, J.S. Kang, H.M. Kim, S.D. Rhee and Y.J. Jeon, *International Immunopharmacology*, 2004, **4**, 8, 1029.

43. K.Y. Lee and Y.J. Jeon, *International Immunopharmacology*, 2003, **3**, 10–11, 1353.

44. I. Sugawara, K.C. Lee and M. Wong, *Cancer Immunology Immunotherapy*, 1984, **16**, 3, 137.

45. N. Hirata, A. Tsuzuki, N. Ohno, M. Saita, Y. Adachi and T. Yadomae, *Zentralblatt für Bakteriologie*, 1998, **288**, 3, 403.

46. H.A. Pretus, H.E. Ensley, R.B. McNamee, E.L. Jones, I.W. Browder and D.L. Williams, *Journal of Pharmacology and Experimental Therapeutics*, 1991, **257**, 1, 500.

47. G.R. Noleto, A.L. Merce, M. Iacomini, P.A. Gorin, V.T. Soccol and M.B. Oliveira, *Molecular and Cellular Biochemistry*, 2002, **233**, 1–2, 73.

48. P.M. Stuelp-Campelo, M.B. de Oliveira, A.M. Leao, E.R. Carbonero, P.A. Gorin and M. Iacomini, *International Immunopharmacology*, 2002, **2**, 5, 691.

49. S. Omarsdottir, J. Freysdottir and E.S. Olafsdottir, *Phytomedicine*, 2007, **14**, 2–3, 179.

50. Q. Gao, M.K. Killie, H. Chen, R. Jiang and R. Seljelid, *Planta Medica*, 1997, **63**, 5, 457.

51. A.S. Daba and O.U. Ezeronye, *African Journal of Biotechnology*, 2003, **2**, 12, 672.

52. H. Nakano, K. Namatame, H. Nemoto, H. Motohashi, K. Nishiyama and K. Kumada, *Hepatogastroenterology*, 1999, **46**, 28, 2662.

53. K. Okamura, M. Suzuki, T. Chihara, A. Fujiwara, T. Fukuda, S. Goto, K. Ichinohe, S. Jimi, T. Kasamatsu, N. Kawai, K. Mizuguchi, S. Mori, H. Nakano, K. Noda, K. Sekiba, K. Suzuki, T. Suzuki, K. Takahashi, K. Takeuchi, S. Takeuchi, A. Yajima and N. Ogawa, *Cancer*, 1986, **58**, 4, 865.

54. A.T. Borchers, J.S. Stern, R.M. Hackman, C.L. Keen and M.E. Gershwin, *Proceedings of the Society for Experimental Biology and Medicine*, 1999, **221**, 4, 281.

55. K. Ingolfsdottir, K. Jurcic, B. Fischer and H. Wagner, *Planta Medica*, 1994, **60**, 6, 527.

56. E.S. Olafsdottir, S. Omarsdottir, B.S. Paulsen, K. Jurcic and H. Wagner, *Phytomedicine*, 1999, **6**, 4, 273.

57. N. Pugh and D.S. Pasco, *Phytomedicine*, 2001, **8**, 6, 445.

58. N. Pugh, S.A. Ross, H.N. ElSohly, M.A. ElSohly and D.S. Pasco, *Planta Medica*, 2001, **67**, 8, 737.

59. H. Shibata, I. Kimura-Takagi, M. Nagaoka, S. Hashimoto, R. Aiyama, M. Iha, S. Ueyama and T. Yokokura, *Biofactors*, 2000, **11**, 4, 235.

60. L. Jiao, X. Li, T. Li, P. Jiang, L. Zhang, M. Wu and L. Zhang, *International Immunopharmacology*, 2009, **9**, 3, 324.

61. Y. Kondo, A. Kato, H. Hojo, S. Nozoe, M. Takeuchi and K. Ochi, *Journal of Pharmacobio-Dynamics*, 1992, **15**, 11, 617.

62. Y. Yoshizawa, J. Tsunehiro, K. Nomura, M. Itoh, F. Fukui, A. Ametani and S. Kaminogawa, *Bioscience, Biotechnology and Biochemistry*, 1996, **60**, 10, 1667.

63. Y. Okai, K. Higashi-Okai, S. Ishizaka and U. Yamashita, *Nutrition and Cancer*, 1997, **27**, 1, 74.

64. V. Vetvicka, B. Dvorak, J. Vetvickova, J. Richter, J. Krizan, P. Sima and J.C. Yvin, *International Journal of Biological Macromolecules*, 2007, **40**, 4, 291.

65. Y. Yoshizawa, A. Enomoto, H. Todoh, A. Ametani and S. Kaminogawa, *Bioscience, Biotechnology and Biochemistry*, 1993, **57**, 11, 1862.

66. Y. Yoshizawa, A. Ametani, J. Tsunehiro, K. Nomura, M. Itoh, F. Fukui and S. Kaminogawa, *Bioscience, Biotechnology and Biochemistry*, 1995, **59**, 10, 1933.

67. H. Itoh, H. Noda, H. Amano, C. Zhuaug, T. Mizuno and H. Ito, *Anticancer Research*, 1993, **13**, 6A, 2045.

68. Reyes Suárez E., S.M. Bugden, F.B. Kai, J.A. Kralovec, M.D. Noseda, C.J. Barrow and T.B. Grindley, *Carbohydrate Research*, 2008, **343**, 15, 2623.

69. E. Lowe, P. Rice, T. Ha, C. Li, J. Kelley, H. Ensley, J. Lopez-Perez, J. Kalbfleisch, D. Lowman, P. Margl, W. Browder and D. Williams, *Microbes and Infection*, 2001, **3**, 10, 789.

70. H.Y. Hsu, N. Jeyashoke, C.H. Yeh, Y.J. Song, K.F. Hua and L.K. Chao, *Journal of Agricultural and Food Chemistry*, 2010, **58**, 2, 927.

71. S. Soltanian, E. Stuyven, E. Cox, P. Sorgeloos and P. Bossier, *Critical Reviews in Microbiology*, 2009, **35**, 2, 109.

72. H.S. Goodridge, A.J. Wolf and D.M. Underhill, *Immunological Reviews*, 2009, **230**, 1, 38.

73. L.T. Rasmussen and R. Seljelid, *Journal of Cellular Biochemistry*, 1991, **46**, 1, 60.

74. M. Doita, L.T. Rasmussen, R. Seljelid and P.E. Lipsky, *Journal of Leukocyte Biology*, 1991, **49**, 4, 342.

75. G. Abel and J.K. Czop, *International Journal of Immunopharmacology*, 1992, **14**, 8, 1363.

76. T. Sakurai, I. Suzuki, A. Kinoshita, S. Oikawa, A. Masuda, M. Ohsawa and T. Yadomae, *Chemical and Pharmaceutical Bulletin*, 1991, **39**, 1, 214.

77. E.K. Gallin, S.W. Green and M.L. Patchen, *International Journal of Immunopharmacology*, 1992, **14**, 2, 173.

78. T. Sakurai, T. Kaise, T. Yadomae and C. Matsubara, *European Journal of Pharmacology*, 1997, **334**, 2–3, 255.

79. J.K. Czop and K.F. Austen, *Proceedings of the National Academy of Sciences of the United States of America*, 1985, **82**, 9, 2751.

80. J.K. Czop, A.V. Puglisi, D.Z. Miorandi and K.F. Austen, *Journal of Immunology*, 1988, **141**, 9, 3170.

81. H.A. Pretus, I.W. Browder, P. Lucore, R.B. McNamee, E.L. Jones and D.L. Williams, *Journal of Trauma*, 1989, **29**, 8, 1152.

82. D.D. Poutsiaka, M. Mengozzi, E. Vannier, B. Sinha and C.A. Dinarello, *Blood*, 1993, **82**, 12, 3695.

83. T.J. Babineau, A. Hackford, A. Kenler, B. Bistrian, R.A. Forse, P.G. Fairchild, S. Heard, M. Keroack, P. Caushaj and P. Benotti, *Archives of Surgery*, 1994, **129**, 11, 1204.

84. T.J. Babineau, P. Marcello, W. Swails, A. Kenler, B. Bistrian and R.A. Forse, *Annals of Surgery*, 1994, **220**, 5, 601.

85. J.K. Czop and K.F. Austen, *Journal of Immunology*, 1985, **134**, 4, 2588.

86. V. Vetvicka, B.P. Thornton and G.D. Ross, *Journal of Clinical Investigation*, 1996, **98**, 1, 50.

87. E.I. Vereschagin, A.A. van Lambalgen, M.I. Dushkin, Y.S. Schwartz, L. Polyakov, A. Heemskerk, E. Huisman, L.G. Thijs and G.C. van den Bos, *Shock*, 1998, **9**, 3, 193.

88. G.D. Brown and S. Gordon, *Nature*, 2001, **413**, 6851, 36.

89. E.P. Lowe, D. Wei, P.J. Rice, C. Li, J. Kalbfleisch, I.W. Browder and D.L. Williams, *American Surgeon*, 2002, **68**, 6, 508.

90. P.R. Taylor, G.D. Brown, D.M. Reid, J.A. Willment, L. Martinez-Pomares, S. Gordon and S.Y. Wong, *Journal of Immunology*, 2002, **169**, 7, 3876.

91. A.D. Kennedy, J.A. Willment, D.W. Dorward, D.L. Williams, G.D. Brown and F.R. DeLeo, *European Journal of Immunology*, 2007, **37**, 467.

92. J. Herre, A.S. Marshall, E. Caron, A.D. Edwards, D.L. Williams, E. Schweighoffer, V. Tybulewicz, C. Reis e Sousa, S. Gordon and G.D. Brown, *Blood*, 2004, **104**, 13, 4038.

93. D.M. Underhill, E. Rossnagle, C.A. Lowell and R.M. Simmons, *Blood*, 2005, **106**, 7, 2543.

94. N.C. Rogers, E.C. Slack, A.D. Edwards, M.A. Nolte, O. Schulz, E. Schweighoffer, D.L. Williams, S. Gordon, V.L. Tybulewicz, G.D. Brown and C. Reis e Sousa, *Immunity*, 2005, **22**, 4, 507.

95. T. Daum and M.S. Rohrbach, *FEBS Letters*, 1992, **309**, 2, 119.

96. A. Mueller, J. Raptis, P.J. Rice, J.H. Kalbfleisch, R.D. Stout, H.E. Ensley, W. Browder and D.L. Williams, *Glycobiology*, 2000, **10**, 4, 339.

97. S. Xu, J. Huo, M. Gunawan, I.H. Su and K.P. Lam, *Journal of Biological Chemistry*, 2009, **284**, 33, 22005.

98. D.L. Williams and N.R. Di Luzio, *Science*, 1980, **208**, 4439, 67.

99. A. Estrada, C.H. Yun, K.A. Van, B. Li, S. Hauta and B. Laarveld, *Microbiology and Immunology*, 1997, **41**, 12, 991.

100. I.W. Browder, D.L. Williams, A. Kitahama and N.R. Di Luzio, *International Journal of Immunopharmacology*, 1984, **6**, 1, 19.

101. P.J. Rice, E.L. Adams, T. Ozment-Skelton, A.J. Gonzalez, M.P. Goldman, B.E. Lockhart, L.A. Barker, K.F. Breuel, W.K. Deponti, J.H. Kalbfleisch, H.E. Ensley, G.D. Brown, S. Gordon and D.L. Williams, *Journal of Pharmacology and Experimental Therapeutics*, 2005, **314**, 3, 1079.

102. T. Sakurai, K. Hashimoto, I. Suzuki, N. Ohno, S. Oikawa, A. Masuda and T. Yadomae, *International Journal of Immunopharmacology*, 1992, **14**, 5, 821.

103. I. Suzuki, K. Hashimoto, N. Ohno, H. Tanaka and T. Yadomae, *International Journal of Immunopharmacology*, 1989, **11**, 7, 761.

104. I. Suzuki, H. Tanaka, A. Kinoshita, S. Oikawa, M. Osawa and T. Yadomae, *International Journal of Immunopharmacology*, 1990, **12**, 6, 675.

105. D. Bimczok, J. Wrenger, T. Schirrmann, H.J. Rothkotter, V. Wray and U. Rau, *Applied Microbiology and Biotechnology*, 2009, **82**, 2, 321.

106. M.R. Ehlers, *Microbes and Infection*, 2000, **2**, 3, 289.

107. B.P. Thornton, V. Vetvicka, M. Pitman, R.C. Goldman and G.D. Ross, *Journal of Immunology*, 1996, **156**, 3, 1235.

108. G.D. Ross, V. Vetvicka, J. Yan, Y. Xia and J. Vetvicková, *Immunopharmacology*, 1999, **42**, 1–3, 61.

109. B.R. van Bruggen, A. Drewniak, M. Jansen, H.M. van, D. Roos, H. Chapel, A.J. Verhoeven and T.W. Kuijpers, *Molecular Immunology*, 2009, **47**, 2–3, 575.

110. B.M. Shao, W. Xu, H. Dai, P. Tu, Z. Li and X.M. Gao, *Biochemical and Biophysical Research Communications*, 2004, **320**, 4, 1103.

111. B.N. Gantner, R.M. Simmons, S.J. Canavera, S. Akira and D.M. Underhill, *Journal of Experimental Medicine*, 2003, **197**, 9, 1107.

112. A. Muller, P.J. Rice, H.E. Ensley, P.S. Coogan, J.H. Kalbfleish, J.L. Kelley, E.J. Love, C.A. Portera, T. Ha, I.W. Browder and D.L. Williams, *Journal of Immunology*, 1996, **156**, 9, 3418.

113. K. Sakurai, M. Mizu and S. Shinkai, *Biomacromolecules*, 2001, **2**, 3, 641.

114. T. Matsuo, Y. Kurahashi, S. Nishida, K. Kumada, T. Hayami and T. Takagi, *Gan To Kagaku Ryoho*, 1987, **14**, 5, Part 1, 310.

115. A. Tsuzuki, T. Tateishi, N. Ohno, Y. Adachi and T. Yadomae, *Bioscience, Biotechnology and Biochemistry*, 1999, **63**, 1, 104.

116. K. Ito, Y. Masuda, Y. Yamasaki, Y. Yokota and H. Nanba, *International Immunopharmacology*, 2009, **9**, 10, 1189.

117. M. Jin, H. Jeon, H.J. Jung, B. Kim, S.S. Shin, J.J. Choi, J.K. Lee, C.Y. Kang and S. Kim, *Experimental Biology and Medicine*, 2003, **228**, 6, 759.

118. S.B. Han, Y.D. Yoon, H.J. Ahn, H.S. Lee, C.W. Lee, W.K. Yoon, S.K. Park and H.M. Kim, *International Immunopharmacology*, 2003, **3**, 9, 1301.

119. L. Ramamoorthy, M.C. Kemp and I.R. Tizard, *Molecular Pharmacology*, 1996, **50**, 4, 878.

120. L. Zhang and I.R. Tizard, *Immunopharmacology*, 1996, **35**, 2, 119.

121. N. Pugh, S.A. Ross, M.A. ElSohly and D.S. Pasco, *Journal of Agricultural and Food Chemistry*, 2001, **49**, 2, 1030.

122. C. Liu, M.Y. Leung, J.C. Koon, L.F. Zhu, Y.Z. Hui, B. Yu and K.P. Fung, *International Immunopharmacology*, 2006, **6**, 11, 1634.

123. M.P. Moretao, D.F. Buchi, P.A. Gorin, M. Iacomini and M.B. Oliveira, *Immunology Letters*, 2003, **89**, 2–3, 175.

124. S.B. Han, Y.H. Kim, C.W. Lee, S.M. Park, H.Y. Lee, K.S. Ahn, I.H. Kim and H.M. Kim, *Immunopharmacology*, 1998, **40**, 1, 39.

125. Y.J. Jeon, S.B. Han, K.S. Ahn and H.M. Kim, *Immunopharmacology*, 1999, **43**, 1, 1.

126. Y.J. Jeon, S.B. Han, K.S. Ahn and H.M. Kim, *Immunopharmacology*, 2000, **49**, 3, 275.

127. X. Yang, Y. Zhao, H. Wang and Q. Mei, *Journal of Biochemistry and Molecular Biology*, 2007, **40**, 5, 636.

128. J. Puhlmann, M.H. Zenk and H. Wagner, *Phytochemistry*, 1991, **30**, 4, 1141.

129. G. Xie, I.A. Schepetkin, D.W. Siemsen, L.N. Kirpotina, J.A. Wiley and M.T. Quinn, *Phytochemistry*, 2008, **69**, 6, 1359.

130. K.Y. Lee and Y.J. Jeon, *International Immunopharmacology*, 2005, **5**, 7–8, 1225.

131. S.V. Popov, G.Y. Popova, S.Y. Nikolaeva, V.V. Golovchenko and R.G. Ovodova, *Phytotherapy Research*, 2005, **19**, 12, 1052.

132. M. Inngjerdingen, K.T. Inngjerdingen, T.R. Patel, S. Allen, X. Chen, B. Rolstad, G.A. Morris, S.E. Harding, T.E. Michaelsen, D. Diallo and B.S. Paulsen, *Glycobiology*, 2008, **18**, 12, 1074.

133. H. Diao, X. Li, J. Chen, Y. Luo, X. Chen, L. Dong, C. Wang, C. Zhang and J. Zhang, *Journal of Bioscience and Bioengineering*, 2008, **105**, 2, 85.

134. T. Matsumoto and H. Yamada, *Journal of Pharmacy and Pharmacology*, 1995, **47**, 2, 152.

135. I. Ando, Y. Tsukumo, T. Wakabayashi, S. Akashi, K. Miyake, T. Kataoka and K. Nagai, *International Immunopharmacology*, 2002, **2**, 8, 1155.

136. K. Hase, P. Basnet, S. Kadota and T. Namba, *Planta Medica*, 1997, **63**, 3, 216.

137. B.P. da Silva, J.B. Tostes and J.P. Parente, *Fitoterapia*, 2000, **71**, 5, 516.

138. J.Y. Song, H.O. Yang, S.N. Pyo, I.S. Jung, S.Y. Yi and Y.S. Yun, *Archives of Pharmacal Research*, 2002, **25**, 2, 158.

139. J. Escribano, M.J.M. Diaz-Guerra, H.H. Riese, J. Ontanon, D. Garcia-Olmo, D.C. Garcia-Olmo, A. Rubio and J.A. Fernandez, *Cancer Letters*, 1999, **144**, 1, 107.

140. X. Yang, Y. Zhao and Y. Lv, *Journal of Agricultural and Food Chemistry*, 2007, **55**, 12, 4684.

141. K.I. Kim, K.S. Shin, W.J. Jun, B.S. Hong, D.H. Shin, H.Y. Cho, H.I. Chang, S.M. Yoo and H.C. Yang, *Bioscience, Biotechnology and Biochemistry*, 2001, **65**, 11, 2369.

142. A.J. Kim, Y.O. Kim, J.S. Shim and J.K. Hwang, *Bioscience, Biotechnology and Biochemistry*, 2007, **71**, 6, 1428.

143. E.M. Choi and J.K. Hwang, *Fitoterapia*, 2002, **73**, 7–8, 629.

144. J.Y. Liu, F.L. Yang, C.P. Lu, Y.L. Yang, C.L. Wen, K.F. Hua and S.H. Wu, *Journal of Agricultural and Food Chemistry*, 2008, **56**, 21, 9892.

145. Y. Zhang, H. Kiyohara, T. Matsumoto and H. Yamada, *Planta Medica*, 1997, **63**, 5, 393.

146. J. Roesler, A. Emmendörffer, C. Steinmuller, B. Luettig, H. Wagner and M.L. Lohmann-Matthes, *International Journal of Immunopharmacology*, 1991, **13**, 7, 931.

147. C. Steinmüller, J. Roesler, E. Grottrup, G. Franke, H. Wagner and M.L. Lohmann-Matthes, *International Journal of Immunopharmacology*, 1993, **15**, 5, 605.

148. M. Stimpel, A. Proksch, H. Wagner and M.L. Lohmann-Matthes, *Infection and Immunity*, 1984, **46**, 3, 845.

149. B. Luettig, C. Steinmuller, G.E. Gifford, H. Wagner and M.L. Lohmann-Matthes, *Journal of the National Cancer Institute*, 1989, **81**, 9, 669.

150. V. Goel, C. Chang, J. Slama, R. Barton, R. Bauer, R. Gahler and T. Basu, *Journal of Nutritional Biochemistry*, 2002, **13**, 8, 487.

151. T. Koga and M. Kikuchi, *Bioscience, Biotechnology and Biochemistry*, 1993, **57**, 3, 367.

152. G. Yang and Y. Yu, *Proceedings of the Chinese Academy of Medical Sciences and the Peking Union Medical College*, 1990, **5**, 4, 188.

153. M. Nose, K. Terawaki, K. Oguri, Y. Ogihara, K. Yoshimatsu and K. Shimomura, *Biological and Pharmaceutical Bulletin*, 1998, **21**, 10, 1110.

154. A. Cheng, F. Wan, Z. Jin, J. Wang and X. Xu, *Journal of Ethnopharmacology*, 2008, **118**, 1, 59.

155. A. Cheng, F. Wan, J. Wang, Z. Jin and X. Xu, *International Immunopharmacology*, 2008, **8**, 1, 43.

156. X.B. Yang, Y. Zhao, Y. Yang and Y. Ruan, *Journal of Agricultural and Food Chemistry*, 2008, **56**, 16, 6905.

157. I.A. Schepetkin, C.L. Faulkner, L.K. Nelson-Overton, J.A. Wiley and M.T. Quinn, *International Immunopharmacology*, 2005, **5**, 13–14, 1783.

158. S.V. Popov, V.V. Golovchenko, R.G. Ovodova, V.V. Smirnov, D.S. Khramova, G.Y. Popova and Y.S. Ovodov, *Vaccine*, 2006, **24**, 26, 5413.

159. J. Hauer and F.A. Anderer, *Cancer Immunology Immunotherapy*, 1993, **36**, 4, 237.

160. D. Kostalova, A. Kardosova and V. Hajnicka, *Fitoterapia*, 2001, **72**, 7, 802.

161. A. Hirazumi and E. Furusawa, *Phytotherapy Research*, 1999, **13**, 5, 380.

162. A. Togola, M. Inngjerdingen, D. Diallo, H. Barsett, B. Rolstad, T.E. Michaelsen and B.S. Paulsen, *Journal of Ethnopharmacology*, 2008, **115**, 3, 423.

163. I.A. Schepetkin, G. Xie, L.N. Kirpotina, R.A. Klein, M.A. Jutila and M.T. Quinn, *International Immunopharmacology*, 2008, **8**, 10, 1455.

164. T.P. Smolina, T.F. Solov'eva and N.N. Besednova, *Antibiotika I Khimioterapiya*, 2001, **46**, 7, 19.

165. K.M. Park, Y.S. Kim, T.C. Jeong, C.O. Joe, H.J. Shin, Y.H. Lee, K.Y. Nam and J.D. Park, *Planta Medica*, 2001, **67**, 2, 122.

166. J.Y. Shin, J.Y. Song, Y.S. Yun, H.O. Yang, D.K. Rhee and S. Pyo, *Immunopharmacology and Immunotoxicology*, 2002, **24**, 3, 469.

167. Y.S. Lee, I.S. Chung, I.R. Lee, K.H. Kim, W.S. Hong and Y.S. Yun, *Anticancer Research*, 1997, **17**, 1A, 323.

168. D.S. Lim, K.G. Bae, I.S. Jung, C.H. Kim, Y.S. Yun and J.Y. Song, *Journal of Infection*, 2002, **45**, 1, 32.

169. J.Y. Song, S.K. Han, E.H. Son, S.N. Pyo, Y.S. Yun and S.Y. Yi, *International Immunopharmacology*, 2002, **2**, 7, 857.

170. Y. Sonoda, T. Kasahara, N. Mukaida, N. Shimizu, M. Tomoda and T. Takeda, *Immunopharmacology*, 1998, **38**, 3, 287.

171. H. Gao, F. Wang, E.J. Lien and M.D. Trousdale, *Pharmaceutical Research*, 1996, **13**, 8, 1196.

172. V.A. Assinewe, J.T. Amason, A. Aubry, J. Mullin and I. Lemaire, *Phytomedicine*, 2002, **9**, 5, 398.

173. K.H. Kwon, K.I. Kim, W.J. Jun, D.H. Shin, H.Y. Cho and B.S. Hong, *Biological and Pharmaceutical Bulletin*, 2002, **25**, 3, 367.

174. H. Sakagami, M. Ikeda, S. Unten, K. Takeda, J. Murayama, A. Hamada, K. Kimura, N. Komatsu and K. Konno, *Anticancer Research*, 1987, **7**, 6, 1153.

175. W. Westerhof, P.K. Das, E. Middelkoop, J. Verschoor, L. Storey and C. Regnier, *Drugs under Experimental and Clinical Research*, 2001, **27**, 5–6, 165.

176. S.B. Han, S.H. Park, K.H. Lee, C.W. Lee, S.H. Lee, H.C. Kim, Y.S. Kim, H.S. Lee and H.M. Kim, *International Immunopharmacology*, 2001, **1**, 11, 1969.

177. Y.D. Yoon, S.B. Han, J.S. Kang, C.W. Lee, S.K. Park, H.S. Lee, J.S. Kang and H.M. Kim, *International Immunopharmacology*, 2003, **3**, 13–14, 1873.

178. X. Fang, M.M. Yu, W.H. Yuen, S.Y. Zee and R.C. Chang, *International Journal of Molecular Medicine*, 2005, **16**, 6, 1109.

179. S.A. Im, K.J. Kim and C.K. Lee, *International Immunopharmacology*, 2006, **6**, 9, 1451.

180. S.V. Popov, G.Y. Popova, R.G. Ovodova, O.A. Bushneva and Y.S. Ovodov, *International Journal of Immunopharmacology*, 1999, **21**, 9, 617.

181. G. Xie, I.A. Schepetkin and M.T. Quinn, *International Immunophar macology*, 2007, **7**, 13, 1639.

182. V.R. Desai, R. Ramkrishnan, G.J. Chintalwar and K.B. Sainis, *International Immunopharmacology*, 2007, **7**, 10, 1375.

183. R. Raghu, D. Sharma, R. Ramakrishnan, S. Khanam, G.J. Chintalwar and K.B. Sainis, *Immunology Letters*, 2009, **123**, 1, 60.

184. P.K.R. Nair, S.J. Melnick, R. Ramachandran, E. Escalon and C. Ramachandran, *International Immunopharmacology*, 2006, **6**, 12, 1815.

185. H.P. Ramesh, K. Yamaki and T. Tsushida, *Carbohydrate Polymers*, 2002, **50**, 1, 79.

186. J.M. Luk, W. Lai, P. Tam and M.W. Koo, *Life Sciences*, 2000, **67**, 2, 155.

187. R. Takata, R. Yamamoto, T. Yanai, T. Konno and T. Okubo, *Bioscience, Biotechnology and Biochemistry*, 2005, **69**, 11, 2042.

188. W. Ni, X. Zhang, H. Bi, J. Iteku, L. Ji, C. Sun, J. Fang, G. Tai, Y. Zhou and J. Zhao, *Carbohydrate Research*, 2009, **344**, 18, 2512.

189. L. Liu, Z. Guo, Z. Lv, Y. Sun, W. Cao, R. Zhang, Z. Liu, C. Li, S. Cao and Q. Mei, *International Immunopharmacology*, 2008, **8**, 11, 1481.

190. D. Diallo, B.S. Paulsen, T.H. Liljeback and T.E. Michaelsen, *Journal of Ethnopharmacology*, 2001, **74**, 2, 159.

191. D. Diallo, B.S. Paulsen, T.H. Liljeback and T.E. Michaelsen, *Journal of Ethnopharmacology*, 2003, **84**, 2–3, 279.

192. L. Dong, S. Xia, Y. Luo, H. Diao, J. Zhang, J. Chen and J. Zhang, *Journal of Controlled Release*, 2009, **134**, 3, 214.

193. M.F. Tosi, *Journal of Allergy and Clinical Immunology*, 2005, **116**, 2, 241.

194. V. Witko-Sarsat, P. Rieu, B. Descamps-Latscha, P. Lesavre and L. Halbwachs-Mecarelli, *Laboratory Investigation*, 2000, **80**, 5, 617.

195. A. El Abbouyi, M. Toumi, Y. El Hachimi and A. Jossang, *Journal of Ethnopharmacology*, 2004, **91**, 1, 159.

196. Y. Zhu, F. Pettolino, S.L. Mau, Y.C. Shen, C.F. Chen, Y.C. Kuo and A. Bacic, *Planta Medica*, 2006, **72**, 13, 1193.

197. E. Wakshull, D. Brunke-Reese, J. Lindermuth, L. Fisette, R.S. Nathans, J.J. Crowley, J.C. Tufts, J. Zimmerman, W. Mackin and D.S. Adams, *Immunopharmacology*, 1999, **41**, 89.

198. E. Sonck, E. Stuyven, B. Goddeeris and E. Cox, *Veterinary Immunology and Immunopathology*, 2009, **135**, 3–4, 199.

199. M. Faurschou and N. Borregaard, *Microbes and Infection*, 2003, **5**, 14, 1317.

200. C.C. Winterbourn, M.B. Hampton, J.H. Livesey and A.J. Kettle, *Journal of Biological Chemistry*, 2006, **281**, 52, 39860.

201. T.A. Kuznetsova, T.S. Zaporozhets, N.N. Besednova, N.M. Shevchenko, T.N. Zviagintseva, A.N. Mamaev and A.P. Momot, *Antibiotika I Khimioterapiya*, 2003, **48**, 4, 11.

202. R.G. Ovodova, V.V. Golovchenko, A.S. Shashkov, S.V. Popov and I. Ovodov, *Bioorganicheskaya Khimiya*, 2000, **26**, 10, 743.

203. M.J. Hsu, S.S. Lee, S.T. Lee and W.W. Lin, *British Journal of Pharmacology*, 2003, **139**, 2, 289.

204. B. Petri, M. Phillipson and P. Kubes, *Journal of Immunology*, 2008, **180**, 10, 6439.

205. Y. Hao, Q.Y. Qiu and J. Wu, *Chinese Journal of Integrated Traditional and Western Medicine*, 2004, **24**, 5, 427.

206. R. Fei, Y. Fei, S. Zheng, Y.G. Gao, H.X. Sun and X.L. Zeng, *Acta Pharmacologica Sinica*, 2008, **29**, 4, 499.

207. S.V. Popov, R.G. Ovodova, G.I. Popova, I.R. Nikitina and I. Ovodov, *Bioorganicheskaya Khimiya*, 2007, **33**, 1, 187.

208. J.C. Graff, E.M. Kimmel, B. Freedman, I.A. Schepetkin, J. Holderness, M.T. Quinn, M.A. Jutila and J.F. Hedges, *International Immunopharmacology*, 2009, **9**, 11, 1313.

209. B. Halliwell, *FASEB Journal*, 1987, **1**, 358.

210. K.B. Beckman and B.N. Ames, *Physiological Reviews*, 1998, **78**, 2, 547.

211. P. Karihtala and Y. Soini, *APMIS*, 2007, **115**, 2, 81.

212. K.J.A. Davies, *Biochemical Society Symposium*, 1995, **61**, 1.

213. J.P. Kehrer, *Critical Reviews in Toxicology*, 1993, **23**, 21.

214. B. Halliwell, *Lancet*, 1994, **344**, 8924, 721.

215. C.C. Winterbourn, *Clinical and Experimental Pharmacology and Physiology*, 1995, **22**, 11, 877.

216. O.I. Aruoma, *Free Radicals in Biology and Medicine*, 1996, **20**, 5, 675.

217. G. Fernandes, *Immunologic Research*, 2008, **40**, 3, 244.

218. J.H. Wu, C. Xu, C.Y. Shan and R.X. Tan, *Life Sciences*, 2006, **78**, 6, 622.

219. X. Yang, Y. Zhao, Y. Zhou, Y. Lv, J. Mao and P. Zhao, *Biological and Pharmaceutical Bulletin*, 2007, **30**, 10, 1884.

220. X. Huang, Q. Li, H. Li and L. Guo, *Cellular and Molecular Neurobiology*, 2009, **29**, 8, 1211.

221. X. Yang, Y. Zhao, Y. Lv, Y. Yang and Y. Ruan, *Journal of Biochemistry and Molecular Biology*, 2007, **40**, 6, 928.

222. D. Slamenova, J. Labaj, L. Krizkova, G. Kogan, J. Sandula, N. Bresgen and P. Eckl, *Cancer Letters*, 2003, **198**, 2, 153.

223. L.N. Liu, Q.B. Mei, L. Liu, F. Zhang, Z.G. Liu, Z.P. Wang and R.T. Wang, *World Journal of Gastroenterology*, 2005, **11**, 10, 1503.

224. R. Kohen, V. Shadmi, A. Kakunda and A. Rubinstein, *British Journal of Nutrition*, 1993, **69**, 3, 789.

225. A. Kardošová and E. Machová, *Fitoterapia*, 2006, **77**, 5, 367.

226. Z.Q. Zhang, Y.J. Tian and J. Zhang, *Journal of Chinese Medicinal Materials*, 2008, **31**, 2, 268.

227. X.M. Yang, W. Yu, Z.P. Ou, H.L. Ma, W.M. Liu and X.L. Ji, *Plant Foods for Human Nutrition*, 2009, **64**, 2, 167.

228. Q. Zhang, P. Yu, Z. Li, H. Zhang, Z. Xu and P. Li, *Journal of Applied Phycology*, 2003, **15**, 4, 305.

229. A. Luo, X. He, S. Zhou, Y. Fan, T. He and Z. Chun, *International Journal of Biological Macromolecules*, 2009, **45**, 4, 359.

230. C.L. Lin, C.C. Wang, S.C. Chang, B.S. Inbaraj and B.H. Chen, *International Journal of Biological Macromolecules*, 2009, **45**, 2, 146.

231. Y. Zhao, Y.O. Son, S.S. Kim, Y.S. Jang and J.C. Lee, *Journal of Biochemistry and Molecular Biology*, 2007, **40**, 5, 670.

232. Y. Fan, X. He, S. Zhou, A. Luo, T. He and Z. Chun, *International Journal of Biological Macromolecules*, 2009, **45**, 2, 169.

233. J. Zhu and M. Wu, *Journal of Agricultural and Food Chemistry*, 2009, **57**, 3, 812.

234. W. Xu, F. Zhang, Y. Luo, L. Ma, X. Kou and K. Huang, *Carbohydrate Research*, 2009, **344**, 2, 217.

235. B.C. Gilbert, D.M. King and C.B. Thomas, *Carbohydrate Research*, 1984, **125**, 2, 217.

236. Y. Wu and D. Wang, *Journal of Natural Products*, 2008, **71**, 2, 241.

237. M.D. Rees, E.C. Kennett, J.M. Whitelock and M.J. Davies, *Free Radicals in Biology and Medicine*, 2008, **44**, 12, 1973.

238. M.M.S. Asker, M.G. Mahmoud and G.S. Ibrahim, *Journal of Applied Sciences Research*, 2007, **3**, 10, 1170.

239. C.K. Veena, A. Josephine, S.P. Preetha and P. Varalakshmi, *Human and Experimental Toxicology*, 2007, **26**, 12, 923.

240. H. Qi, Q. Zhang, T. Zhao, R. Hu, K. Zhang and Z. Li, *Bioorganic and Medicinal Chemistry Letters*, 2006, **16**, 9, 2441.

241. F. Liu, V.E. Ooi and S.T. Chang, *Life Sciences*, 1997, **60**, 10, 763.

242. E. Tsiapali, S. Whaley, J. Kalbfleisch, H.E. Ensley, I.W. Browder and D.L. Williams, *Free Radicals in Biology and Medicine*, 2001, **30**, 4, 393.

243. H. Yuan, W. Zhang, X. Li, X. Lu, N. Li, X. Gao and J. Song, *Carbohydrate Research*, 2005, **340**, 4, 685.

244. J. Wang, Q. Zhang, Z. Zhang, J. Zhang and P. Li, *International Journal of Biological Macromolecules*, 2009, **44**, 2, 170.

245. T.J. Hu, X.H. Shuai, J.R. Chen, Y.Y. Wei and R.L. Zheng, *International Journal of Biological Macromolecules*, 2009, **45**, 3, 279.

246. Y.K. Hong, H.T. Wu, T. Ma, W.J. Liu and X.J. He, *International Journal of Biological Macromolecules*, 2009, **45**, 1, 61.

247. F. Virgili and M. Marino, *Free Radicals in Biology and Medicine*, 2008, **45**, 1205.

248. I.A. Schepetkin, L.N. Kirpotina, L. Jakiw, A.I. Khlebnikov, C.L. Blaskovich, M.A. Jutila and M.T. Quinn, *Journal of Immunology*, 2009, **183**, 10, 6754.

249. P.C. Nowell, *Cancer Research*, 1960, **20**, 462.

250. P. Capek and V. Hribalová, *Phytochemistry*, 2004, **65**, 13, 1983.

251. G. Chintalwar, A. Jain, A. Sipahimalani, A. Banerji, P. Sumariwalla, R. Ramakrishnan and K. Sainis, *Phytochemistry*, 1999, **52**, 6, 1089.

252. S.H. Gohla, H.D. Haubeck and R.D. Neth, *Leukemia*, 1988, **2**, 8, 528.

253. B.A. Cobb, Q. Wang, A.O. Tzianabos and D.L. Kasper, *Cell*, 2004, **117**, 5, 677.

254. J.F. Zhao, H. Kiyohara, H. Yamada, N. Takemoto and H. Kawamura, *Carbohydrate Research*, 1991, **219**, 149.

255. H. Kiyohara, N. Takemoto, J.F. Zhao, H. Kawamura and H. Yamada, *Planta Medica*, 1996, **62**, 1, 14.

256. M.H. Sakurai, T. Matsumoto, H. Kiyohara and H. Yamada, *Immunology*, 1999, **97**, 3, 540.

257. S. Omarsdottir, J. Freysdottir, H. Barsett, B.S. Paulsen and E.S. Olafsdottir, *Phytomedicine*, 2005, **12**, 6–7, 461.

258. H. Yamada, K.S. Ra, H. Kiyohara, J.C. Cyong and Y. Otsuka, *Carbohydrate Research*, 1989, **189**, 209.

259. C.S. Nergard, H. Kiyohara, J.C. Reynolds, J.E. Thomas-Oates, T. Matsumoto, H. Yamada, T. Patel, D. Petersen, T.E. Michaelsen, D. Diallo and B.S. Paulsen, *Biomacromolecules*, 2006, **7**, 1, 71.

260. G. Wang, W. Lin, R. Zhao and N. Lin, *Journal of Hygiene Research*, 2008, **37**, 5, 577.

261. H.X. Sun, H. Wang, H.S. Xu and Y. Ni, *Vaccine*, 2009, **27**, 30, 3984.

262. H. Kiyono and S. Fukuyama, *Nature Reviews Immunology*, 2004, **4**, 9, 699.

263. J.R. McGhee, S.M. Michalek, H. Kiyono, J.H. Eldridge, D.E. Colwell, S.I. Williamson, M.J. Wannemuehler, E. Jirillo, L.M. Mosteller, D.M. Spalding, S. Hamada, K.A. Gollahon, I. Morisaki, R.L. Gregory and W.J. Koopman, *Microbiology and Immunology*, 1984, **28**, 3, 261.

264. J. Kunisawa and H. Kiyono, *Cellular and Molecular Life Sciences*, 2005, **62**, 12, 1308.

265. H. Kiyohara, K. Nonaka, M. Sekiya, T. Matsumoto, T. Nagai, Y. Tabuchi and H. Yamada, *Evidence-based Complementary and Alternative Medicine*, 2009, **6**, 2, 195.

266. J. Sakushima, M. Nose and Y. Ogihara, *Biological and Pharmaceutical Bulletin*, 1997, **20**, 11, 1175.

267. T. Hong, T. Matsumoto, H. Kiyohara and H. Yamada, *Phytomedicine*, 1998, **5**, 5, 353.

268. H. Kiyohara, T. Matsumoto and H. Yamada, *Phytomedicine*, 2002, **9**, 7, 614.

269. K.W. Yu, H. Kiyohara, T. Matsumoto, H.C. Yang and H. Yamada, *Carbohydrate Polymers*, 2001, **46**, 2, 147.

270. H. Zhao, Y. Luo, C. Lu, N. Lin, C. Xiao, S. Guan, D.A. Guo, Z. Liu, D. Ju, X. He and A. Lu, *Planta Medica*, 2010, **76**, 3, 223.

271. P. Balachandran, N.D. Pugh, G. Ma and D.S. Pasco, *International Immunopharmacology*, 2006, **6**, 12, 1808.

272. M. Wang, L.J. Guilbert, L. Ling, J. Li, Y. Wu, S. Xu, P. Pang and J.J. Shan, *Journal of Pharmacy and Pharmacology*, 2001, **53**, 11, 1515.

273. P.D. Biondo, S.J. Robbins, J.D. Walsh, L.J. McCargar, V.J. Harber and C.J. Field, *Applied Physiology Nutrition and Metabolism*, 2008, **33**, 5, 966.

274. N.J. Cox and K. Subbarao, *Lancet*, 1999, **354**, 9186, 1277.

275. M. Schiller, D. Metze, T.A. Luger, S. Grabbe and M. Gunzer, *Experimental Dermatology*, 2006, **15**, 5, 331.

276. M. Witvrouw and E. DeClercq, *General Pharmacology*, 1997, **29**, 4, 497.

277. D.J. Schaeffer and V.S. Krylov, *Ecotoxicology and Environmental Safety*, 2000, **45**, 3, 208.

278. P. Gerber, J.D. Dutcher, E.V. Adams and J.H. Sherman, *Proceedings of the Society for Experimental Biology and Medicine*, 1958, **99**, 3, 590.

279. J. Wang, Y. Hu, D. Wang, F. Zhang, X. Zhao, S. Abula, Y. Fan and L. Guo, *International Journal of Biological Macromolecules*, 2010, **46**, 2, 212.

280. C.R. Parish, L. Low, H.S. Warren and A.L. Cunningham, *Journal of Immunology*, 1990, **145**, 4, 1188.

281. J.M. Davis, E.A. Murphy, A.S. Brown, M.D. Carmichael, A. Ghaffar and E.P. Mayer, *Medicine and Science in Sports and Exercise*, 2004, **36**, 8, 1321.

282. Y. Ohta, J.B. Lee, K. Hayashi, A. Fujita, D.K. Park and T. Hayashi, *Journal of Agricultural and Food Chemistry*, 2007, **55**, 25, 10194.

283. S.H. Seo and R.G. Webster, *Journal of Virology*, 2002, **76**, 3, 1071.

284. Z.G. Peng, H.S. Chen, Z.M. Guo, B. Dong, G.Y. Tian and G.Q. Wang, *Acta Pharmaceutica Sinica*, 2008, **43**, 7, 702.

285. W. Zheng, C. Chen, Q. Cheng, Y. Wang and C. Chu, *International Immunopharmacology*, 2006, **6**, 7, 1093.

286. Y. Qiu, Y.L. Hu, B.A. Cui, H.Y. Zhang, X.F. Kong, D.Y. Wang and Y.G. Wang, *Poultry Science*, 2007, **86**, 12, 2530.

287. Y. Chen, D. Wang, Y. Hu, Z. Guo, J. Wang, X. Zhao, Y. Fan, L. Guo, S. Yang, F. Sai and Y. Xing, *International Journal of Biological Macromolecules*, 2010, **46**, 4, 425.

288. Y. Qiu, Y.L. Hu, B.A. Cui, H.Y. Zhang, X.F. Kong, D.Y. Wang and Y.G. Wang, *Poultry Science*, 2007, **86**, 12, 2530.

289. M. Selegean, M.V. Putz and T. Rugea, *International Journal of Molecular Sciences*, 2009, **10**, 8, 3616.

290. D. Womble and J.H. Helderman, *International Journal of Immunophar macology*, 1988, **10**, 8, 967.

291. A.D. Chinnah, M.A. Baig, I.R. Tizard and M.C. Kemp, *Vaccine*, 1992, **10**, 8, 551.

292. W.R. Usinger, *Vaccine*, 1997, **15**, 17–18, 1902.

293. M.A. Sheets, B.A. Unger, G.F. Giggleman, Jr. and I.R. Tizard, *Molecular Biotherapy*, 1991, **3**, 1, 41.

294. K.M. Yates, L.J. Rosenberg, C.K. Harris, D.C. Bronstad, G.K. King, G.A. Biehle, B. Walker, C.R. Ford, J.E. Hall and I.R. Tizard, *Veterinary Immunology and Immunopathology*, 1992, **35**, 1–2, 177.

295. Y.X. Sun, *Chemistry and Biodiversity*, 2009, **6**, 6, 890.

296. D.S. Khramova, S.V. Popov, V.V. Golovchenko, F.V. Vityazev, N.M. Paderin and Y.S. Ovodov, *Nutrition*, 2009, **25**, 2, 226.

297. M.G. Danilets, N.V. Belska, Y.P. Bel'sky, E.G. Uchasova, E.S. Trophimova, A.A. Ligatcheva, A.M. Guriev, M.V. Belousov, R.R. Ahmedganov, M.S. Usubov, and V.I. Agaphonov, *Bulletin of Experimental Biology and Medicine*, 2008, **146**, 5, 585.

298. J.C. Lee, S.C. Pak, S.H. Lee, C.S. Na, S.C. Lim, C.H. Song, Y.H. Bai and C.H. Jang, *Journal of Alternative and Complementary Medicine*, 2004, **10**, 3, 527.

299. R. Chang, *Journal of Alternative and Complementary Medicine*, 2002, **8**, 5, 559.

300. H.J. Shin, Y.S. Kim, Y.S. Kwak, Y.B. Song, Y.S. Kim and J.D. Park, *Planta Medica*, 2004, **70**, 11, 1033.

301. V.E. Ooi and F. Liu, *Current Medicinal Chemistry*, 2000, **7**, 7, 715.

302. I. Lavi, D. Friesem, S. Geresh, Y. Hadar and B. Schwartz, *Cancer Letters*, 2006, **244**, 1, 61.

303. L.J. Standish, C.A. Wenner, E.S. Sweet, C. Bridge, A. Nelson, M. Martzen, J. Novack and C. Torkelson, *Journal of the Society for Integrative Oncology*, 2008, **6**, 3, 122.

304. H.Y. Kim, J.H. Kim, S.B. Yang, S.G. Hong, S.A. Lee, S.J. Hwang, K.S. Shin, H.J. Suh and M.H. Park, *Journal of Medicinal Food*, 2007, **10**, 1, 25.

305. Y.Y. Maeda, S.T. Watanabe, G. Chihara and M. Rokutanda, *International Journal of Immunopharmacology*, 1984, **6**, 5, 493.

306. S. Usui, Y. Tomono, M. Sakai, T. Kiho and S. Ukai, *Biological and Pharmaceutical Bulletin*, 1995, **18**, 12, 1630.

307. N. Kodama, A. Asakawa, A. Inui, Y. Masuda and H. Nanba, *Oncology Reports*, 2005, **13**, 3, 497.

308. D.L. Ren, J.Z. Wang, H. Noda, H. Amano and S. Ogawa, *Planta Medica*, 1995, **61**, 2, 120.

309. T. Mishima, J. Murata, M. Toyoshima, H. Fujii, M. Nakajima, T. Hayashi, T. Kato and I. Saiki, *Clinical and Experimental Metastasis*, 1998, **16**, 6, 541.

310. J. Li, Q.W. Li, D.W. Gao, Z.S. Han and W.Z. Lu, *Phytotherapy Research*, 2009, **23**, 11, 1524.

311. C. Harris, K. Pierce, G. King, K.M. Yates, J. Hall and I. Tizard, *Molecular Biotherapy*, 1991, **3**, 4, 207.

312. G.K. King, K.M. Yates, P.G. Greenlee, K.R. Pierce, C.R. Ford, B.H. McAnalley and I.R. Tizard, *Journal of the American Animal Hospital Association*, 1995, **31**, 5, 439.

313. S.Y. Peng, J. Norman, G. Curtin, D. Corrier, H.R. McDaniel and D. Busbee, *Molecular Biotherapy*, 1991, **3**, 2, 79.

314. J. Roesler, C. Steinmuller, A. Kiderlen, A. Emmendörffer, H. Wagner and M.L. Lohmann-Matthes, *International Journal of Immunopharmacology*, 1991, **13**, 1, 27.

315. H.S. Xu, Y.W. Wu, S.F. Xu, H.X. Sun, F.Y. Chen and L. Yao, *Journal of Ethnopharmacology*, 2009, **125**, 2, 310.

316. G. Deng, H. Lin, A. Seidman, M. Fornier, G. D'Andrea, K. Wesa, S. Yeung, S. Cunningham-Rundles, A.J. Vickers and B. Cassileth, *Journal of Cancer Research and Clinical Oncology*, 2009, **135**, 9, 1215.

317. B.B. Finlay and R.E.W. Hancock, *Nature Reviews Microbiology*, 2004, **2**, 497.

Acknowledgements

This work was supported in part by National Institutes of Health grant P20 RR-020185 and the Montana State University Agricultural Experimental Station.

7 Pharmaceutical Applications of Cyclodextrins

Dominique Duchêne and Amélie Bochot

7.1 Introduction

Cyclodextrins (CD) were discovered by Villiers at the end of the 19th century in a product of potato starch fermentation [1]. Later they were isolated by Schardinger at the beginning of the 20th century [2]. Surprisingly, they constituted no more than a scientific curiosity for many years and it was only by the end of the 1970s and beginning of the 1980s that they became attractive for their industrial uses. Then, they were the subject of many patents and research works, in particular in food industries [3, 4] and cosmetology [4–6], but the articles on their potential pharmaceutical applications were those that contributed the most to their development [7, 8]. Presently, every year some 1500–2000 articles, book chapters or lectures are devoted to CD.

7.2 Main Cyclodextrins

7.2.1 Nature and Characteristics of Natural Cyclodextrins

Chemically, CD are cyclic oligosaccharides produced by the enzymatic degradation of starch. The cycloglycosyl transferase can be isolated from the culture broth of various bacilli. Depending on the exact reaction conditions, three main CD can be obtained in various proportions [9].

The main natural CD (α, β and γ) consist of 6, 7 and 8 glucopyranose units, respectively. They are cone shaped (**Figure 7.1**) with hydroxyl groups on each side. On the other hand, their cavity is constituted by the glucosidal moieties. Such a three-dimensional ring structure results in high external hydrophilicity and internal hydrophobicity [9, 10].

Due to their hydrophilicity, CD are easily water-soluble, but surprisingly β-CD is less soluble (18.5 g/l at 25 °C; 145 g/l for α-CD and 233 g/l for γ-CD). However, the aqueous solubility of CD is much lower than that of similar acyclic saccharides. This is the consequence of strong binding of CD molecules inside the crystal lattice.

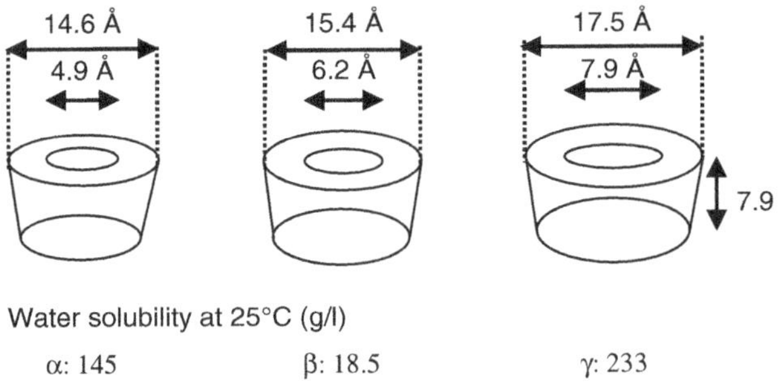

Figure 7.1 Natural CD: sizes and water solubility

Furthermore, for β-CD, with an odd number of glucopyranose units, intramolecular hydrogen bonds appear between hydroxyl groups, preventing hydrogen bond formation with the surrounding water molecules and resulting in poor water solubility [11].

7.2.2 Hydrophilic Derivatives

The low solubility of β-CD in water restricts its applications. For this reason, a number of derivatives have been synthesised [12, 13]. The most commonly used hydrophilic derivatives are the methyl- [14] and the hydroxypropyl-CD [15, 16].

Methylation of CD can be either partial or complete (trimethyl-CD). Presently, the randomly methylated β-CD is very often used. Surprisingly, these derivatives are much more soluble than the parent CD, which presents a number of drawbacks for their utilisation: their solubility decreases with an increase in temperature, and, for pharmaceutical purpose, they are rather haemolytic.

From this standpoint, hydroxypropyl-CD are more interesting. Hydroxypropylation occurs in a random manner, and the products obtained are not pure chemical entities, but amorphous complicate mixtures, with high water solubility. More recently, sulfobutyl ethers of β-CD have been synthesised [17]. These anionic CD are highly water-soluble, and have a very light haemolytic effect. These products can be considered as competitors of the hydroxypropyl-CD for parenteral or ophthalmic use [18].

7.2.3 Amphiphilic Cyclodextrins Derivatives

The high external hydrophilicity of CD or inclusions does not allow them to have a close contact with biological lipid membranes *in vivo*. This is one of the reasons why many scientists synthesised amphiphilic CD [19–21] in which a number of hydroxyl groups are substituted with a lipophilic hydrocarbon chain. Depending on their structure, these CD (**Figure 7.2**) have been named: lollipop [22], cup-and-ball [23], medusa [24], bouquet [25] or skirt-CD [26]. Finally, lollipop type CD have been equipped with a 'head' capable of molecular recognition [27].

For the moment, these CD are just exiting research objects for drug targeting as nanoparticles [20] (see **Section 7.5.4.2**), micelles or liposomes [21].

Figure 7.2 Examples of amphiphilic CD

7.2.4 Inclusion Formation

The hydrophobic cavity in CD confers on them the ability to include apolar molecules or part of molecules, depending on the reaction conditions. When compared with the initial invited molecule, the resulting inclusion complex exhibits new physicochemical characteristics (especially stability and apparent solubility) often very interesting for various industrial applications. No covalent bonds are formed or broken during drug-CD complex formation, and in aqueous solutions, the complexes are readily dissociated. Free drug molecules are in equilibrium with the molecules bound within the CD cavity [11].

Formation of inclusion results from a series of equilibrium reactions [28] (**Figure 7.3**) in which the affinity constant, or stability constant, K_C, governs the affinity of the guest molecule for the host CD [29–37]. Higher is the K_C, more stable is the inclusion and less dissociation occurs. The value of K depends, among others, on the size of the CD cavity and that of the guest molecule (or part of molecule). It also depends on the more or less good fitting of the guest molecule inside the CD cavity. As a general rule, the complex is strong when there is size complementarity between the guest and the CD cavity [38]. Depending on their respective size, the guest molecule enters the CD cavity by the narrow side (primary hydroxyl groups) or by the wide one (secondary hydroxyl groups).

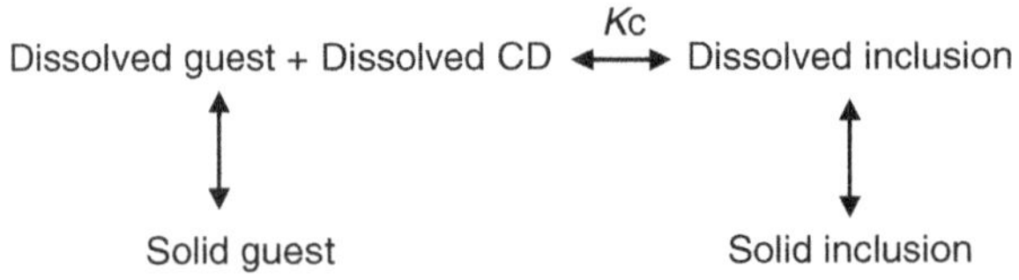

Figure 7.3 Series of equilibria leading to the formation of guest molecule inclusion in CD in water solution or in the presence of humidity

For many years, β-CD and its water-soluble derivatives have been the most often employed ones for inclusion complex preparation. There are two reasons for their usage: the cavity of α-CD is generally too small to include interesting therapeutic molecules, and, for a number of years, because of the difficulty in its preparation, the cost of γ-CD was too high to allow its use in either pharmaceutical or food industry. Presently, there is no significant difference in the cost of the various CD and hence the cost is no more a problem. Besides, sulfobutylether-β-CD, because of its polyanionic nature, interacts particularly well with cationic drugs.

The method used to prepare an inclusion complex between a CD and a guest compound has a significant impact on the characteristics of the final product, such as yield, solubility and stability of the complex. Most often, the choice of the CD type depends on the future role of the inclusion. In the pharmaceutical filed, the CD chosen depends on the drug administration route, the fact that the CD is registered in one (or more) of the main pharmacopoeia (**Table 7.1**), and the price of the CD.

Table 7.1 Cyclodextrins registered in the main pharmacopoeias

Cyclodextrin	Pharmacopoeia		
	European Pharmacopoeia	**USP/NF**	**JPC**
α-CD	Yes	No	Yes
β-CD	Yes	Yes	Yes
HP-β-CD	Yes	Yes	No
RM-β-CD	No	No	No
SEB-β-CD	No	No	No
γ-CD	In progress	Yes	Yes
HP-γ-CD	No	No	No

HP: Hydroxypropyl
JPC: Japan Pharmacopeia
RM: Randomly methylated
SEB: Sulfobutyl ether
USP-NF: United States Pharmacopeia - National Formulary

The preparation method has to be adapted to the production level (industry or laboratory scale) and the objective (increase in solubility, stability and so on). Finally, the necessity to add a third or a fourth component for a better solubility of the product obtained has to be considered. Many methods have been described for the preparation of inclusion complex. Anyhow, it should be kept in mind that, except in the case where the inclusion precipitates spontaneously from the preparation medium, the product obtained is a mixture of three compounds: inclusion complex, empty CD and free guest [39].

7.2.5 Toxicity

The toxicity of natural CD and their water-soluble derivatives has been much studied. There are excellent reviews that provide detailed information on this topic, and the interested reader is referred to these [40–42].

When used by the oral route, CD can be hydrolysed by amylases from the digestive tract, in a manner similar to starch [43]. The hydrolysis of α-CD, with a small diameter, is more difficult than that of β-CD, and much more difficult than that of γ-CD. For this reason, there is not enough time for α-CD to be hydrolysed in the intestine; on the other hand, γ-CD is almost completely hydrolysed and β-CD is only partially hydrolysed. Whatever the extent of hydrolysis, natural CD do not present any direct toxicity at the concentrations used either in pharmacy or in food industry. Furthermore, because of their external hydrophilicity and their molecular weight ranging from 1000 to 2000 Da, CD are not absorbed, under their original form, through the intestinal barrier [44]. However, in the digestive tract, CD are capable of including, in a reversible manner, various molecules that are already present, among which are the biliary salts.

When administered by the parenteral route [45, 46] and especially intravenously, β-CD can destabilise the red blood cell membrane by inclusion of some of their constituents, leading to haemolysis. It is also capable of including the blood cholesterol; both the inclusion and β-CD itself, being poorly soluble, crystallise in kidneys leading to necrosis. These drawbacks are not encountered with α- or γ-CD, which are much more soluble and present either a too narrow or a too wide cavity to include the cholesterol in a state stable enough to result in pernicious effects. The haemolytic effect depends on the type (for the parent CD, it is in the order β-CD > α-CD > γ-CD) and concentration (strongly) of CD [47]; the methylated CD shows a higher haemolytic effect than other CD [47]. Besides, hydroxypropylated derivatives [48] and the sulfobutyl ether of β-CD have undergone extensive safety studies and are currently used in products approved by the Food and Drug Administration for the parenteral route [42] (**Table 7.2**).

Table 7.2 Drug dosage forms containing CD (nonexhaustive table)

CD/active ingredient	Trade name	Form	Company
α-CD			
Alprostadil (prostaglandin E_1)	Prostavasin Rigidur Prostandin Prostandin 500	Parenteral Parenteral Parenteral Parenteral	Schwarz Pharma Rigips Ono Ono
Limaprost (prostaglandin E_1 analogue)	Opalmon	Tablet	Ono
Cefotiam, HCl	Pansporin Taketiam Texodil	Tablet Tablet Tablet	Takeda Takeda Grunenthal
β-CD			
Benexate, HCl	Ulgut Lonmiel	Capsule Capsule	Teikoku Shionogi
Calcium carbonate	Calprimum	Tablet	Debregeas et Associés Pharma
Cetirizine chlorhydrate	Alairgix	Tablet	Coopérative Pharmaceutique Française
Cefditorem, pivoxil	Meiact	Tablet	Meiji Seika
Chlordiazepoxide	Transillium	Tablet	Gador
Dexamethasone	Glymesason	Ointment	Fujinaga
Dinoprostone (prostaglandin E_2)	Prostarmon E	Sublingual tablet	Ono
Diphenhydramin, HCl	Stada-Travel	Chewable tablets	Stada
Iodine	Mena-Gargle	Solution (gargle)	Kyushin
Meloxicam	Mobitil	Tablet suppository	Medical Union Pharmaceuticals Medical Union Pharmaceuticals
Nicotine	Nicorette microtab Nicogum	Sublingual tablet Chewing-gum	Pfizer (Pharmacia) Pierre Fabre
Nimesulide	Nimedex	Tablet	Novartis
Nitroglycerine	Nitropen	Sublingual tablet	Nippon Kayaku

Table 7.2 Continued

CD/active ingredient	Trade name	Form	Company
Omeprazol	Omebeta	Tablet	Betafarm
Piroxicam	Bréxin Brexidol Cycladol Flogene Brexidol 20	Tablet Tablet Tablet Liquid Suppository	Pierre Fabre (Chiesi) Nycomed Pharma (Chiesi) Chiesi Ache Biovail Pharmaceuticals
Tiaprofenic acid	Surgamyl	Tablet	Roussel-Maestrelli
Minoxidil	Alopexy	Solution	Pierre Fabre dermatologie
2-Hydroxypropyl-β-CD Cisapride	Propulsid	Suppository	Janssen
Indomethacine and gentamicin	Indobiotic	Eye drop	Chauvin
Indomethacine	Indocollyre	Eye drop	Chauvin
Itraconazole	Sporanox	Oral solution Parenteral	Janssen Janssen
Mitomycin	Mitozytrex	Parenteral	Janssen
Methyl-β-CD Cloramphenicol	Clorocil	Eye drop	Oftalder
Sulfobutylether-β-CD Aripiprazole	Abilify	Parenteral	Pfizer
Maropitant	Cerenia	Parenteral	Pfizer Animal
Voriconazole	Vfend	Parenteral	Pfizer
Ziprasidone mezylate	Geodon, Zeldox	Parenteral	Pfizer
2-Hydroxypropyl-γ-CD Diclofenac sodium	Voltaren	Eye drop	Novartis
^{99}Tc, teoboroxime	Cardiotec	Parenteral	Bracco
Octakis-(6-deoxy-6-S-mercaptopropionyl)-γ-CD sodium salt) Sugammadex	Bridion	Parenteral	Schering-Plough

^{99}Tc Technetium-99

*Adapted from T. Loftsson and D. Duchene, International Journal of Pharmaceutics, 2007, **329**, 1. ©2007, Elsevier [10]*

Natural CD are regarded as non-irritant to skin [49], eyes and mucosa upon inhalation [50]. However, α-CD may cause some noncorrosive eye irritation. Among the CD derivatives, hydroxypropyl-β-CD and sulfobutylether-β-CD are safe for use in aqueous eye drop solutions, while methylated CD can cause serious irritations and even corrosion to the eye [51]. However, methylated CD in low concentrations can be safely applied on skin and mucosa. Application of CD solution directly onto nasal, buccal or vaginal mucosa is regarded as safe for hydrophilic and natural CD in a wide range of concentrations, while for methylated CD derivatives, the concentration and the application time should be controlled [50].

7.3 Interest of Cyclodextrins in Pharmaceutical Uses

The industrial use of CD results from their ability to form inclusion compounds with various substances, as long as the steric hindrance or charge of the guest molecule is not an obstacle. In the early 1980s, it was mostly in cosmetology, food and pharmacy industries that CD were considered as raw materials [52]. For pharmaceutical purposes their use is obviously easier if they are registered at one of the three main world pharmacopoeias (Europe, the United States, Japan). However, this is not a prerequisite. CD status for these three pharmacopoeias is summarised in **Table 7.1**.

7.3.1 Reduction or Elimination of Bitter Taste of Drugs

The majority of the orally administered and at least moderately water-soluble drug substances have a very to extremely bitter taste [53]. Only dissolved substances elicit the taste sensation. This mainly causes a problem for paediatric medicines administered by the oral route to children in the form of liquid formulations. Interestingly, the bitter, astringent taste of drugs, which get in the mouth as they dissolve in aqueous medium or in the saliva, can be strongly reduced or fully eliminated if the bitter component forms an inclusion complex with an appropriate CD [53]. On rinsing the mouth cavity with concentrated β-CD solution and thereafter using, for example, cetirizine hydrochloride, one can immediately feel the bitter taste. If, however, the cetirizine is administered in saturated β-CD solution, no bitter taste is observed [53]. The use of CD in excess in the formulation shifts the dissociation equilibrium towards the association, allowing a further significant reduction of the bitterness. Szejtli and Szente reported in their review [53] numerous examples for elimination/reduction of bitter/irritating or unpleasant odours of orally administered drugs in the presence of CD. Among them, we can mention the oral administration of nicotine in the form of sublingual tablet (Nicorette Microtab®) or chewing gum (Nicogum®) formulated with β-CD (**Table 7.2**).

7.3.2 Change of State

Molecules (or parts of molecules) from any type of product can enter the CD cavity to lead to an inclusion. Thus, gases or liquids can be included. This, generally results in a solidification of the product, because of the solid state of the inclusion obtained. By this method, many volatile or aromatic substances can be stabilised with respect to their volatilisation, and these are much easier to handle and incorporate in a final product [54]. By this means, liquids and oils can be converted to free-flowing powders.

In pharmacy, the polymorphism of drugs can affect their solubility, dissolution rate, stability but also bioavailability. For this reason, it is important to control habit and polymorphic transition. Amorphous CD such as hydroxypropyl-β-CD are able to control the polymorphism transition and the crystallisation of poorly soluble drugs [55] such as spironolactone [56], nifedipine [57], chloramphenicol acetate [58] or tolbutamide [59]. Thus, these CD can be introduced in the formulation to maintain the higher dissolution characteristics and oral bioavailability of drugs.

7.3.3 Improvement in Stability

Stabilisation of labile compounds into the formulation is essential to maintain the efficacy of drugs. Inclusion constitutes a true molecular encapsulation, protecting the included molecule (or part of molecule) [11, 50, 60, 61] against ambient attacks such as dehydration, hydrolysis (chemical or enzymatic), oxidation and photodecomposition. Such attacks can occur either on solid or on dissolved products. In the case of solid products, the resulting degradation can be the consequence of air oxidation or the result of photoinstability. In the absence of water or humidity, the inclusion is physically stable if the weak part of the invited molecule is inside the inclusion and there is no dissociation releasing the unstable part of the molecule. Such a protection is efficient and can be used for pharmaceutical or cosmetic products [6, 62–65]. When the inclusion is dissolved or in the presence of humidity, the stabilisation obtained is more disputable due to the presence of water, which leads to its dissociation (or at least to an equilibrium between inclusion and free products). A potential hydrolysis could be significantly slowed down only if the association constant is high enough [66]. Nevertheless, such a limited result can be of true value in the case of molecules with high therapeutic importance [67, 68]. Improvement in solution stability was demonstrated for several active therapeutic molecules, for example, antibiotics (ampicillin and methicillin [69], amphotericin B [70]), prostaglandins [71], anticancer products (chlorambucil [72], melphalan [66, 72], doxorubicin [73], daunorubicin [73, 74]) or corticoids [75].

In any case, it is preferable to present the product in the form of a powder for dissolution. Some prostaglandins (PGE_1 and PGE_2) are marketed as a CD complex with α- and β-CD, respectively, to increase their solid-state stability [76] (**Table 7.2**).

However, it is important to mention that in some cases, inclusion of drugs into CD may also lead to a faster degradation considering the characteristics of the inclusion complex [77, 78].

7.3.4 Increase in Solubility

Numerous active ingredients are poorly water-soluble, preventing their use in liquid form (oral, parenteral or ocular), or decreasing their bioavailability in molecular form. Several strategies are developed by the pharmaceutical industry to tackle the problems associated with the delivery of hydrophobic drugs (e.g., use of salts, co-solvents, surfactants, lipid-based formulations). The use of CD for the solubilisation of hydrophobic drugs is another approach, which ended in the marketing of several medicines (**Table 7.2**). This is the domain in which CD are most often used [11, 42, 61, 79–82]. Valentino Stella in two excellent papers reported how one can determine, with minimum experimentation, whether CD might be the right choice as solubilisation enhancers for a given poorly water-soluble drug [42, 83]. The reader is directed to these publications for practical examples.

For basic amine drugs, increased solubility can be achieved by decreasing the pH of the solution to below the pK_a of the amine group. However, because of toxicity considerations, it is important to minimise the pH decrease by adding CD to achieve the same solubility at a pH value closer to physiological pH [42]. This strategy was carried out to formulate voriconazole, ziprasidone, aripiprazole, maropitant or itraconazole (**Table 7.2**) as parenteral or oral solutions in which pH adjustment and CD (sulfobutylether-β- or hydroxypropyl-β-CD) are used to achieve the desired formulation solubility [42].

One has to keep in mind that the intrinsic solubility of the guest product is not improved, and the desired formulation solubility achieved is due to the inclusion supermolecule, which presents a higher solubility than the parent product. When in solution, either in the drug dosage form itself or in the biological fluids, the formation/dissociation equilibrium occurs. In this case again, extemporaneous preparation of solutions is recommended. Interestingly, it can be noted that association to a polymer can prolong the inclusion state in solution [83, 84].

Finally, the solubilisation of drugs with CD allows administration of these compounds as a solution instead of a suspension, facilitating the administration of reproducible doses whereas suspensions must be shaken well before use to ensure reproducible doses.

7.3.5 Improvement in Bioavailability

More than 40% of drug failures in development can be traced to poor biopharmaceutical properties, specifically poor dissolution or poor permeability. As indicated by the large number of research papers and reviews on this topic, a number of works have demonstrated the potential of CD to improve the oral bioavailability of active compounds. The mechanism by which inclusion of an active ingredient in a CD can lead to an increase in bioavailability is summarised in **Figure 7.4**. The hydrophilic solid inclusion from the solid dosage form can be easily dissolved in biological fluids and then dissociate more or less, according to the affinity constant between the drug and the CD, releasing the (lipophilic) drug in a molecular form capable of permeating biological membranes. According to this mechanism, if the stability constant is high, few active ingredient molecules will be released and available for absorption. On the other hand, if the affinity constant is low, many molecules will be released, but if their absorption through biological membranes is slow, they can re-precipitate. However, inclusions undergoing a permanent and very rapid dissociation/formation phenomenon [79, 85] can release active molecules as rapidly as these molecules permeate biological membranes. Thus, the presence of drug/CD inclusion at the hydrated biological membrane surface most often results in an increase in bioavailability [86, 87]. It has been shown that CD increase the oral bioavailability of FDA class II active ingredients (low aqueous solubility, high absorption through biological membranes), but they can decrease the bioavailability of class I (high solubility, high absorption) and III (high solubility, low absorption) active ingredients [88].

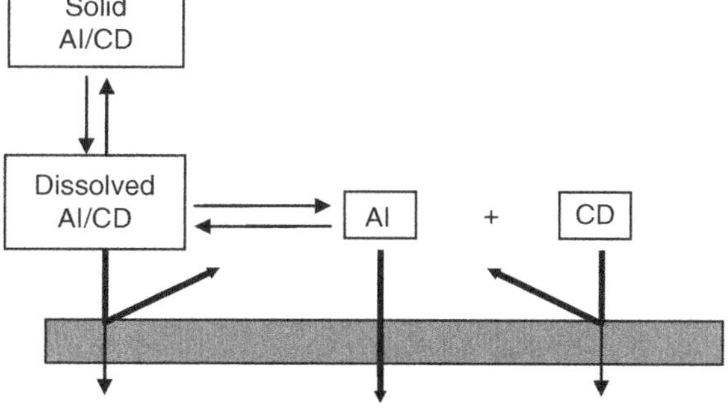

Figure 7.4 Schematic representation of the mechanism of an increase in the bioavailability of a water-insoluble (or poorly soluble) active ingredient by inclusion in CD (active ingredient (AI) and cyclodextrin (CD))

Some CD are well-known absorption enhancers and they increase the permeability of drugs through the skin [89] and the nasal mucosa [90–93] not only by their ability to solubilise active ingredients and release them in a molecular form easily absorbable through the biological membranes, but also by a direct absorption enhancer activity. The most effective absorption-enhancing CD for peptides and proteins are the methylated β-CD derivatives, dimethyl-β-CD and randomly methylated β-CD [90]. They are active at relatively low concentrations ranging between 2% and 5% (*w/v*). Also α-CD can substantially increase the bioavailability of a variety of peptides and proteins, but it is less potent [90]. The mechanism of action of methylated β-CD as absorption enhancers occurs probably by transiently changing permeability [94] of the mucosa due to the displacement by the CD of cell constituents such as cholesterol and phospholipids [95] and opening of tight junctions [96, 97].

7.3.6 Decrease in Side Effects

CD have also demonstrated their potential to reduce the irritation caused by drugs. In the same way that inclusion of a guest inside the cavity of a CD can decrease its attack by the surrounding elements, it can also protect the surrounding medium for undesirable side effects of the host molecule. An interesting application is the decrease in the ulcerous effect of nonsteroidal anti-inflammatory drugs administered by the oral route. This effect has a double origin: local and systemic. The protection that can be conferred by inclusion in a CD can only be local and, furthermore, is incomplete, because of the equilibrium existing between the dissolved complex and the free active ingredients and empty CD. However, inclusion of phenylbutazone [98] or naproxen [99] in β-CD or of indomethacin in β- or hydroxylpropyl-β-CD [100] results at least in a decrease in the ulcerous effects on the gastrointestinal tract. The reduction of corrosive activity of drug molecules on ocular tissues [50, 101] and skin [50] has also been demonstrated. Indeed, following treatment with a commercial preparation of retinoic acid, a drug successfully employed in the treatment of acne vulgaris and in some disorders of keratinisation, all patients suffered from erythema and burning sensation. In groups treated with hydrogel or moisturising base containing retinoic acid/β-CD complexes, none of the patients reported marked local side effects [102].

7.4 Pharmaceutical Industry

For reasons mentioned above, CD seem to be major tools when formulating a number of therapeutic substances. However, their appearance in pharmaceutical forms can be considered as relatively slow.

New drugs have to undergo chemical, pharmacological, technological and clinical advanced studies, before the marketing approval can be obtained. One can be faced with various problems when using a CD in formulation, especially when the active ingredient is included. In fact, what can be then considered as the therapeutic molecule? This question is of importance for products already on the market, for which a simple 'rejuvenation' is looked at by the use of a CD. If it is the inclusion complex that is considered as the therapeutic substance, it has to undergo all the studies as a new molecule. If it is the guest molecule itself, then it is necessary to demonstrate that the bioavailability is not modified by the CD. Anyhow, a number of patent medicines (sometimes under different dosage forms) contain CD (**Table 7.2**) [10]. Most of these medicines contain β-CD or some of its water-soluble derivatives: hydroxypropylated or its sulfobutyl ether (**Table 7.2**). Only few medicines contain α-CD or hydroxypropyl-γ-CD (**Table 7.2**). If CD are mainly used in solid dosage forms for oral administration, they can also be used for other administration routes, and they are the subject of a number of research works [103]. It is the case for parenteral [104–106], ocular [50, 101, 107, 108], nasal [67, 68], rectal [60, 69] or dermal [50, 60, 109] routes.

When the use of CD is introduced in drug formulation, it is important to consider specific aspects. Addition of CD in pharmaceutical formulations increases the formulation bulk of solid dosage forms. The relatively high molecular weight of CD molecules and mass size limitations of oral dosing restrict the use of CD in solid oral forms to potent drugs having good complexing properties [87, 88].

Formulations of medicines consist of highly complex mixtures. One must be very careful with regard to the complexity of the formula, because CD are able to modify the characteristics of the formulation and also to interact with formulation ingredients and cause physicochemical stability problems. It is also important to consider a potential competition and/or interaction between excipient (polymer [110–113], co-solvent [114, 115], emulsifying agent [50], preservatives such as parabens [116–119]) and CD. In any case, it appears to be of greatest importance to carry out preliminary and ageing tests to assess the stability of the formulations.

7.5. Recent Aspects of Pharmaceutical Applications of Cyclodextrins

There are a number of possibilities for novel applications of CD including the formulation improvements or drug delivery with protein, peptide and oligonucleotide dosage forms, and the development of new derivatives but also new uses for existing derivatives to formulate dispersed systems.

7.5.1 Use of Cyclodextrins to Formulate Peptides or Proteins

Aachmann and co-workers reported different effects of CD on different proteins: aggregation suppression (if residues responsible for aggregation are highly solvent accessible; e.g., insulin [120], growth hormone [121], interleukin -2 [122]), protection against degradation (if point of attack of a protease is sterically 'masked' by CD) and alteration of function (if residues involved in the function are 'masked' by CD) [123]. Peptide and protein-CD interactions determine a change in the chemical and biological properties of the peptides and proteins as they are able to alter their three-dimensional structure [124].

7.5.2 New Cyclodextrins Derivatives

In 2009, Sugammadex, a modified γ-CD (octakis-(6-deoxy-6-*S*-mercaptopropionyl-γ-CD sodium salt) has been approved in Europe (Bridion®) (**Table 7.2**) as a new compound in anaesthesia [125, 126]. This new CD is able to encapsulate specifically and only nondepolarising steroidal muscle relaxants, rocuronium or vecuronium. The association constants were determined to be very high: 10^7 M^{-1} for vecuronium and 25×10^6 M^{-1} for rocuronium [126]. The clinical trials that have been performed for Sugammadex approval have demonstrated promising results. Indeed, this CD is able to rapidly reverse (2–5 min) different levels of neuromuscular blockade. Sugammadex's onset time is about 10 times that of neostigmine without the need of concomitant atropine administration [125].

7.5.3 Cyclodextrins and Lipid-based Systems

7.5.3.1 Emulsions

Emulsions are thermodynamically unstable liquid/liquid dispersed systems that cause phase separation. Although natural CD do not possess any surface-active properties, they are able to stabilise simple oil-water [127–130] and multiple oil-water-oil emulsions [131–133]. At the oil/water interface, a partial inclusion is formed between the main components of the vegetable oils (triglycerides) and CD. Only one fatty acid chain of the triglyceride can be entrapped with two or three CD molecules depending on the length of the hydrocarbon chain. The two other fatty acid chains are not included in the CD. Due to its amphiphilic property (a hydrophilic head and a hydrophobic tail), the partial inclusion complex could play the role of a surface-active agent [134]. The best emulsifying effect is observed with α- and β-CD whereas the γ-CD is too wide to lead to an optimal interaction with the fatty acid chains [127, 133]. Active ingredients may be added to emulsions formulated with CD and vegetable oils, but they must not interact with the CD cavity, otherwise they displace the fatty acid chains and destabilise the emulsion. From this standpoint, high-molecular-weight active ingredients probably may not destabilise emulsions prepared with α-CD [128, 132].

7.5.3.2 Beads

Beads made of α-CD and soybean oil are promising lipid carriers and have been described in the literature [135–137]. The continuous external shaking of a mixture composed of an aqueous solution of α-CD and soybean oil results for a few days in calibrated particles (1.6 ± 0.2 mm) with a yield of more than 80% [135]. The final bead structure is a partial crystalline matrix of CD molecules surrounding the microdomains of oil [135]. These new lipid-based particles are efficient in encapsulating and deliverig lipophilic drugs such as isotretinoin or adapelene by the oral and topical routes, respectively [136, 137]. Furthermore, they are well tolerated by the skin and the oil content of the beads is sufficient to provide an occlusive effect [136].

7.5.3.3 Liposomes

Liposomes are vesicles with size lower than 1 μm, and are composed of one or more phospholipid bilayers enclosing an aqueous phase. Liposomes can entrap hydrophilic molecules inside the vesicles, and hydrophobic molecules within the lipid bilayer. However, retention of lipophilic compounds in the lipid bilayer can be problematic because some molecules destabilise its structure, thus limiting the diversity and quantity of molecules that can be carried by liposomes [138]. CD improve the drug stability [139] (against light or hydrolysis), enhance the loading efficiency of hydrophobic drugs and modify their localisation within liposomes [140]. As a result, the included drug is predominantly incorporated within the aqueous core rather than in the lipid bilayer, changing its release profile [138]. However, it is also known that CD extract lipid components from the bilayer of liposomes. This could undermine the potential benefits of liposomes as drug carriers [141].

Amphiphilic CD derivatives were designed to form a variety of supramolecular assemblies, such as vesicles [21].

7.5.4 Cyclodextrins in Nanoparticulate Drug Delivery Systems

Recently, nanoparticles, after liposomes and microparticles, have become the necessary carriers of fragile drug molecules such as anticancer products, proteins, genes and so on. The main reasons are that nanoparticles are more stable than liposomes, and, because of their small size, they have a larger surface area than microparticles, leading to higher bioavailability due to better contact with biological membranes. In this new type of formulation, CD revealed their ability to be more than simple additives, and were rather intelligent parts of the systems themselves [142, 143].

Different types of CD-containing nanoparticles can be considered: those in which natural CD or their hydrophilic derivatives are used in association to a polymer, and those in which amphiphilic CD derivatives are used as the major nanoparticle matrix constituents.

7.5.4.1 Cyclodextrins in Polymer Nanoparticles

There are many reasons to use CD in the formulation of polymer nanoparticles, such as the polymer nanoparticles may have a poor loading capacity, especially for weakly water-soluble drugs, or they do not protect enough the fragile drug molecule. In this domain, two types of nanoparticles have to be considered: that in which the CD is either physically associated to the (co)polymer or chemically linked to the polymer to form a new copolymer, and that in which the CD is used in polyplex systems for gene delivery [144].

7.5.4.1.1 Cyclodextrins and Polymers in Nanoparticles

CD can be added to different polymers in order to increase the nanoparticle loading capacity. This is the case of poly(alkylcyanoacrylate) and poly(anhydride) nanoparticles.

Poly(alkylcyanoacrylates) are particularly interesting not only because they are biodegradable but also because they lead to nanoparticles by a simple nanoprecipitation method occurring in aqueous medium; however, their loading capacity can be rather low, especially in the case of poorly water-soluble drugs. The feasibility of nanoparticles in the presence of α-, β- and γ-CD, their hydroxylpropyl derivatives, and the sulfobutyl ether of β-CD has been investigated [145, 146]. Very small particles were obtained in the presence of hydroxylpropyl-CD, which are most probably located at the nanoparticle surface. The increase in the nanoparticle loading capacity depends on the CD content and could be as high as 28- to 50-fold for progesterone or even 129-fold in the case of prednisolone [147]. The *in vitro* drug release occurs in two steps: after a fast release, a plateau is observed, with the complete release being obtained in the presence of esterases [148].

A general mechanism for drug loading has been proposed. In the polymerisation medium, the drug/CD inclusion reversibly dissociates according to its stability constant (D/CD ↔ D + CD), and this leads to free entities. The free hydrophobic drug can be displaced from the solution to the polymer network, driven by its partition coefficient in favour of the polymer. The level of drug incorporation into the polymer network depends on the affinity between the drug and the polymer, compared with the affinity of the drug for the CD cavity. The remaining drug/CD complex, as well as empty CD, can be adsorbed at the nanoparticle surface.

Saquinavir, a potent HIV-1 and HIV-2 protease inhibitor, has low oral bioavailability, because of hepatic first-pass effect, limited absorption due to low water solubility and the effect of P-glycoprotein responsible for an efflux mechanism. Association of hydroxylpropyl- or methyl-β-CD to poly(alkylcyanoacrylate) nanoparticles has been attempted for the preparation of nanoparticles intended for oral administration [149, 150]. Unfortunately, the simple inclusion of saquinavir in one of these CD did not improve its absorption assessed on Caco-2 cell monolayers. On the other hand, the addition of 2.5% free methyl-β-CD increased the absorption, probably by a decrease of the basolateral to apical transport of the drug [151, 152].

Poly(anhydride) nanoparticles such as poly(methyl vinyl ether-*co*-maleic anhydride) (Gantrez®AN) nanoparticles are interesting because of their bioadhesive properties, but their loading capacity is limited. An attempt has been made to increase the incorporation of lipophilic drugs by combining the poly(anhydride) to CD. The best results were obtained with β-CD, the less water-soluble CD of the investigated series [153].

Chitosan is a polysaccharide well known to improve drug bioavailability not only because of its mucoadhesive properties, which can prolong the residence time at the absorption site, but also because of its ability to open the tight junctions. A series of works were carried out on chitosan nanoparticles to investigate the value of inclusion of chitosan. The main objective was to combine the advantages of β-chitosan with that of CD, which can both carry hydrophobic drugs and thus increase their bioavailability by increasing their apparent water solubility. The association of poorly water-soluble drugs such as triclosan or furosemide to hydroxylpropyl-β-CD facilitated their entrapment into the nanoparticles. The *in vitro* release profile was very similar to that observed with poly(alkylcyanoacrylate) nanoparticles: an initial rapid release followed by a plateau, with the remaining drug being released only in the presence of the enzyme chitosanase [154]. Work carried out on insulin in chitosan nanoparticles intended for nasal administration led to the conclusion that in the presence of sulfobutyl ether of β-CD, it was possible to decrease the amount of chitosan in order to obtain the same insulin bioavailability [155].

A very interesting work is that in which CD are used both for the nanoparticle formation and for the nanoparticle loading. The nanoparticles or nanogels (because of the large amount of water retained in the structure) are formed spontaneously in aqueous medium upon the association of two water-soluble polymers: a hydrophobically modified dextran (obtained by grafting alkyl side chains) and a β-CD polymer. Part of the alkyl chains formed inclusion complexes with some CD cavities, leaving most of the CD available to include hydrophobic molecules of interest such as therapeutic agents; under stirring the system rolls up on itself and forms nanoaggregates [156, 157]. It is possible to load the nanoassemblies with hydrophobic drugs such as benzophenone, by adding the drug to each of the polymer solutions (modified dextran and β-CD polymer) before they are mixed [158].

CD can be chemically linked to polymers in nanoparticles. In this case, a copolymer is firstly formed between the CD and the polymer, and then it is used to prepare the nanoparticles.

Because of their biocompatibility and biodegradability, poly(lactide), poly(glycolide) and their copolymers poly(lactide-*co*-glycolide) are among the most widely used polymers for the preparation of nanoparticles. Their association to CD was studied with the aim of encapsulating proteins [159–161]. Work was carried out with β-CD derivatives: mono(6-(2-aminoethyl)amino-6-deoxy)-β-CD, ethylenediamino-bridged bis-β-CD and diethylenetriamino-bridged bis-β-CD. The presence of a CD derivative increased the entrapment efficiency of bovine serum albumin (model protein) and its stability was preserved [159–161].

A conjugate was prepared between β-CD and the amphiphilic biocompatible poly(4-acryloylmorpholine) (PACM) capable of forming nanoparticles for the delivery of acyclovir [162, 163]. The antiviral activity assessed on cell cultures showed that the antiviral potency of the acyclovir nanoparticles was higher than that of free acyclovir. This result was due to the intracellular accumulation of acyclovir in the case of nanoparticles. The authors speculated that the higher antiviral activity of acyclovir/β-CD-PACM nanoparticles is due to the internalisation and perinuclear accumulation of the nanoparticles, delivering the drug in the vicinity of the nucleus, which is the compartment where acyclovir exerts its antiviral activity.

Other β-CD based copolymers have been prepared with the aim of creating either a material for nanobiotechnology, that is, ethylene glycol diglycidyl ether-*co*-aminoethylcarbamoyl-β-CD [164, 165], or 'ideal' nanoparticles composed polymer families constituted of the same backbone and bearing adapted chemical groups conferring on them desired properties, that is, poly(γ-benzyl-L-glutamate)-β-CD [166, 167].

7.5.4.1.2 Cyclodextrins and Polyplexes

Polyplexes are nonviral transfection systems with low immunogenicity and the capacity to handle larger size of DNA than viruses. They are constituted by the association of cationic polymer and nucleic acid. Cationic polymers can deliver plasmid DNA into cells by self-assembling with the anionic DNA via electrostatic interactions and condensation into nanoparticles, which can be easily endocytosed. Whatever their nature, cationic polymers have similar mechanisms of delivery, but their transfection efficiency differs, which depends not only on their chemical nature, but also on their exact structure and their molecular weight. Considering the ability of CD to form inclusion complexes exploited in drug formulation, and also their low toxicity and lack of immunogenicity, works were undertaken on CD-containing copolymer polyplexes for gene delivery. In these systems, CD have been employed either in linear copolymer-containing polyplex systems or in more elaborated structures such as star, dendron or polyrotaxane-containing polyplexes.

Linear cationic β-CD-containing copolymer polyplexes were prepared by the polymerisation of a difunctionalised β-CD monomer (A) with a difunctionalised monomer (B) to give an ABAB copolymer. The difunctionalised β-CD were either 6A,6D-dideoxy-6A,6D-diamino-β-CD or 6A,6D-dideoxy-6A,6D-di (2-aminoethanethio)-β-CD. The difunctionalised monomers were either a series of diimidates differing by the number of methylene units or dithiobis(succinimidyl propionate). The resulting copolymers form polyplexes with different plasmids by rapid self-assembly in aqueous solutions [168, 169].

Poly(ethylene imine) (PEI) has also been used for polyplex preparation but its use is limited by its toxicity and difficulties in formulation. For this reason, different authors tried to combine the qualities of CD-based polymers (among which low toxicity) with that of linear or branched PEI [170–173].

Whatever their efficiency *in vitro*, polyplexes present a number of drawbacks when used *in vivo*: they interact with serum proteins and are rapidly eliminated by phagocytosis; furthermore, they aggregate at physiological ionic strength. Finally, they should not be taken by Kupffer cells, but should target specific tumour cells, which can be achieved due to the presence of specific ligands. For all these reasons, modified polyplexes have been developed by grafting on the polyplexes moieties displaying specific biological activities such as polyethylene glycol to obtain stealth particles [174], or ligands having affinity for defined molecules or cells, like galactose [174] or transferrin [175–177].

When CD are employed in elaborated structures of copolymer-containing polyplexes, these 'elaborated' structures can be star-shaped CD, dendrimers or polyrotaxanes.

Star-shaped CD derivatives are supramolecular assemblies in which CD are used as cores for the synthesis of medium-sized conjugates obtained by per-substitution of the primary hydroxyls [178]. Most often such assemblies have been synthesised with the objective of coating the supramolecular structure with saccharide ligands towards biological receptors such as lectins [179–184] or Gd^{3+} paramagnetic chelates for enhancement of magnetic resonance imaging resolution [185]. Such type of star-shaped polymer has also been investigated in the preparation of polyplexes for gene delivery [186–188].

Dendrimers are a class of highly branched, monodisperse, globular macromolecules. They have a molecular architecture characterised by regular dendritic branching with radial symmetry. Dendrimers have been prepared with different types of subunits: polyamidoamine [189–192] and polypropylenimine [193–195]. They are characterised by high surface charge density and water solubility, enabling electrostatic interactions with nucleic acids and allowing the formation of dendrimer-DNA complexes which can be used for efficient transfection of cells [189]. However, dendrimers with low generations (generation 1–3) have only low gene transfer activity [196, 197], whereas dendrimers with higher generations exhibit cytotoxicity [198]. Different authors demonstrated the advantage of grafting β-CD to dendrimers in order to increase the transfection efficiency and decrease the cytotoxicity [189, 193].

CD-based polyrotaxanes and pseudopolyrotaxanes are constituted by multiple CD rings threaded on a polymer chain with or without bulky end-caps [199–201]. When a cationic CD is used, the corresponding polyrotaxanes or pseudopolyrotaxanes can be used for gene delivery [201].

7.5.4.2 Amphiphilic Cyclodextrins Nanoparticles

As mentioned above, different nonionic amphiphilic CD (**Figure 7.2**) have been prepared and among them skirt-shaped and medusa-like β-CD have been used in the preparation of nanoparticles (**Figure 7.5**). They differ not only by the grafted hydrocarbon chain length (C_6 or C_{14}) and the fact that it is linear or branched, but also by the substitution position (primary or secondary face: I or II) and the nature of the bond (ester, O, or amide, N). These CD revealed real surfactant properties [202–204]. It seems that CD substituted with C_6 linear chains are the most interesting ones.

The type of CD used can lead to either nanospheres or nanocapsules [205, 206], and nanospheres (matrix systems) have been investigated to a larger extent.

For nanosphere preparation, different methods can be employed. The most simple one is the so-called nanoprecipitation method [207, 208]: a water-miscible organic phase containing the amphiphilic CD is added with stirring to an aqueous solution

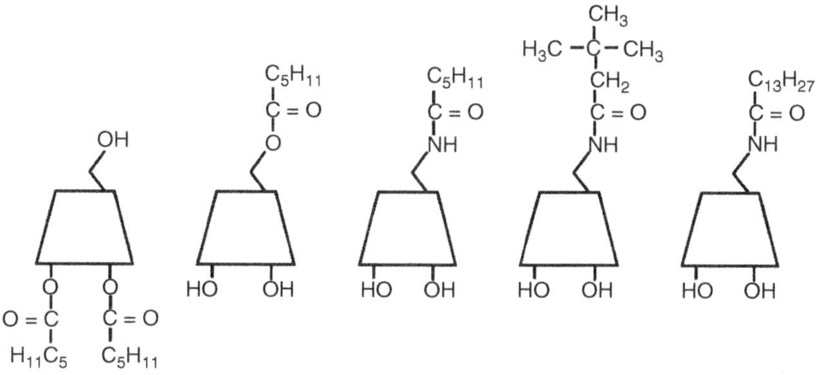

Figure 7.5 Different types of nonionic amphiphilic CD used in nanoparticle preparation

in the presence or not of surfactant, resulting in nanospheres being formed within one hour at room temperature. In this process, several loading methods can be employed [209, 210]: either the drug is added to the organic phase during the preparation, or the drug is in a preformed inclusion, which is also added to the organic phase, or these two methods can be mixed together leading to highly loaded nanospheres. Another method, that is, emulsion/solvent evaporation process, can be employed: amphiphilic CD is dissolved in a water-immiscible organic phase with the active ingredient poured into water under stirring and finally the organic solvent is evaporated [211].

The *in vitro* release profile of the drug, studied on progesterone [208, 209, 211], clotrimazole and bifonazole [210], depends on the loading technique, the affinity constant between the drug and the mother CD (β- or γ-CD) and the substitution side (the primary face being narrower than the secondary face).

Nonionic amphiphilic CD nanospheres have been loaded with different anticancer drugs such as tamoxifen [212, 213], paclitaxel [214, 215] and camptothecin [216]. Among the advantages presented by the nanospheres, it was demonstrated on MCF-7 cells that the anticancer efficacy of tamoxifen and paclitaxel was equivalent to that of drug solutions in organic solvents (ethanol or Cremophor), without displaying the toxic effects of the solvent, and camptothecin stability was significantly improved.

In conclusion, it seems that amphiphilic CD are promising derivatives to obtain nanospheres with high loading capacity, stabilising the properties of labile drugs. In particular, the possibility to use different derivatives allows modulation of drug loading and release [20].

7.6 Conclusion

CD are of great value for preparing pharmaceutical forms. Due to their entrapping capacity, they have been increasingly considered for application in various fields and are more and more acceptable to both their potential users and the regulatory authorities. The consequence is an acceleration in their industrial uses.

CD are also very useful tools in modern drug delivery systems as carriers capable of specific targeting in the organism. Further research in this domain will contribute to the increase in CD uses.

References

1. A. Villiers, *Comptes Rendus de l' Académie des Sciences*, 1891, **112**, 536.
2. F. Schardinger, *Zeitschrift für Lebensmittel-Untersuchung und - Forschung A*, 1903, **6**, 865.
3. L. Szente and J. Szejtli, *Trends in Food Science & Technology*, 2004, **15**, 137.
4. J. Szejtli, *Starch*, 1982, **34**, 379 .
5. H.J. Buschmann and E. Schollmeyer, *Journal of Cosmetic Science*, 2002, **53**, 185.
6. A. Bochot and D. Duchêne in *Actifs et Additifs en Cosmétologie*, Eds., C. Martini and M. Seiller, Tec et Doc Lavoisier, Paris, France, 2006, p.993.
7. D. Duchêne and D. Wouessidjewe, *PharmTech International*, 1990, **2**, 21.
8. D. Duchêne and D. Wouessidjewe in *Polysaccharides in Medicinal Applications*, Ed., S. Dumitriu, Marcel Dekker, New York, NY, USA, 1996, p.575.
9. D. Duchêne, B. Debruères and A. Brétillon, *Labo Pharma, Prolems et Techniques*, 1984, **32**, 842.
10. T. Loftsson and D. Duchêne, *International Journal of Pharmaceutics*, 2007, **329**, 1.
11. T. Loftsson and M.E. Brewster, *Journal of Pharmaceutical Sciences*, 1996, **85**, 1017.

12. D. Duchêne and D. Wouessidjewe, *PharmTech International*, 1990, **2**, 21.

13. D. Duchêne and D. Wouessidjewe, *Journal of Coordination Chemistry*, 1992, **27**, 223.

14. K. Uekama and T. Irie in *Cyclodextrins and Their Industrial Uses*, Ed., D. Duchêne, Editions de Santé, Paris, France, 1987, p.393.

15. L. Szente and C.E. Strattan in *New Trends in Cyclodextrins and Derivatives*, Ed., D. Duchêne, Editions de Santé, Paris, France, 1991, p.55.

16. J.L. Mesens, P. Putteman and P. Verheyen in *New Trends in Cyclodextrins and Derivatives*, Ed., D. Duchêne, Editions de Santé, Paris, France, 1991, p.369.

17. E.A. Luna, E.R. Bornancini, D.O. Thompson, R.A. Rajewski and V.J. Stella, *Carbohydrate Research*, 1997, **299**, 103.

18. K. Jarvinen, T. Jarvinen, D.O. Thompson and V.J. Stella, *Current Eye Research*, 1994, **13**, 897.

19. E. Memişoğlu-Bilensoy, A.A. Hincal, A. Bochot, L. Trichard and D. Duchêne in *Microencapsulation*, Ed., S. Benita, Taylor & Francis, New York, NY, USA, 2006, p.269.

20. E. Bilensoy and A.A. Hincal, *Expert Opinion on Drug Delivery*, 2009, **6**, 1161.

21. M. Roux, B. Perly and F. Djedaini-Pilard, *European Biophysics Journal with Biophysics Letters*, 2007, **36**, 861.

22. N. Bellanger and B. Perly, *Journal of Molecular Structure*, 1992, **273**, 215.

23. J. Lin, *Synthèse des Cyclodextrines Amphiphiles et Étude de Leur Incorporation dans des Phases Phospholipidiques*, Université Paris VI, Paris, France, 1995. [Ph.D Thesis]

24. F. Djédaïni, A.W. Coleman and B. Perly *in Proceedings of the International Symposium on Cyclodextrins*, Paris, France, 1990, p.328.

25. J. Canceill, L. Jullien, L. Lacombe and J.M. Lehn, *Helvetica Chimica Acta*, 1992, **75**, 791.

26. P. Zhang, C.C. Ling, A.W. Coleman, H. Parrotlopez and H. Galons, *Tetrahedron Letters*, 1991, **32**, 2769.

27. R. Auzely-Velty, B. Perly, O. Tache, T. Zemb, P. Jehan, P. Guenot, J.P. Dalbiez and F. Djedaini-Pilard, *Carbohydrate Research*, 1999, **318**, 82.

28. F. Hirayama and K. Uekama in *Cyclodextrins and Their Industrial Uses*, Ed., H. Duchêne, Editions de Santé, Paris, France, 1987, p.131.

29. T. Higuchi and K. Connors, *Advances in Analytical Chemistry and Instrumentation*, 1965, **4**, 117.

30. G. Zingone and F. Rubessa, *International Journal of Pharmaceutics*, 2005, **291**, 3.

31. N. Ono, F. Hirayama, H. Arima and K. Uekama, *Chemical & Pharmaceutical Bulletin*, 2001, **49**, 78.

32. M. Cirri, F. Maestrelli, S. Orlandini, S. Furlanetto, S. Pinzauti and P. Mura, *Journal of Pharmaceutical and Biomedical Analysis*, 2005, **37**, 995.

33. V. Lemesle-Lamache, M. Taverna, D. Wouessidjewe, D. Duchêne and D. Ferrier, *Journal of Chromatography A*, 1996, **735**, 321.

34. Y. Saito, K. Hashizaki, H. Taguchi, K. Tomono, H. Goto and N. Ogawa, *Drug Development and Industrial Pharmacy*, 2000, **26**, 1111.

35. J. Chen, C.M. Ohnmacht and D.S. Hage, *Journal of Chromatography A*, 2004, **1033**, 115.

36. C.M. Moraes, P. Abrami, E. de Paula, A.F.A. Braga and L.F. Fraceto, *International Journal of Pharmaceutics*, 2007, **331**, 99.

37. N. Ono, F. Hirayama, H. Arima and K. Uekama, *European Journal of Pharmaceutical Sciences*, 1999, **8**, 133.

38. V. Gabelica and N. Galic, *Journal of the American Society of Mass Spectrometry*, 2002, **13**, 946.

39. D. Duchêne in *Cyclodextrins in Pharmaceutics, Cosmetics and Biomedicine*, Ed., E. Memişoğlu-Bilensoy, Wiley, Hoboken, NJ, USA, 2011, p.3.

40. D.O. Thompson, *Critical Reviews in Therapeutic Drug Carrier Systems*, 1997, **14**, 1.

41. T. Irie and K. Uekama, *Journal of Pharmaceutical Sciences*, 1997, **86**, 147.

42. V.J. Stella and Q.R. He, *Toxicologic Pathology*, 2008, **36**, 30.

43. A. Gerloczy, A. Fonagy, P. Keresztes, L. Perlaky and J. Szejtli, *Arzneimittel Forschung/Drug Research*, 1985, **35**, 2, 1042.

44. J. Szejtli, A. Gerlóczy and A. Fónagy, *Arzneimittel Forschung*, 1980, **30**, 808.

45. T. Irie, M. Otagiri, M. Sunada, K. Uekama, Y. Ohtani, Y. Yamada and Y. Sugiyama, *Journal of Pharmacobio-Dynamics*, 1982, **5**, 741.

46. Y. Kubota, M. Fukuda, M. Muroguchi and K. Koizumi, *Biological & Pharmaceutical Bulletin*, 1996, **19**, 1068.

47. J. Szejtli, *Journal of Inclusion Phenomena and Macrocyclic Chemistry*, 2005, **52**, 1.

48. F. Leroy-Lechat, D. Wouessidjewe, J.P. Andreux, F. Puisieux and D. Duchêne, *International Journal of Pharmaceutics*, 1994, **101**, 97.

49. G. Piel, S. Moutard, E. Uhoda, F. Pilard, G.E. Pierard, B. Perly, L. Delattre and B. Evrard, *European Journal of Pharmaceutics and Biopharmaceutics*, 2004, **57**, 479.

50. K. Cal and K. Centkowska, *European Journal of Pharmaceutics and Biopharmaceutics*, 2008, **68**, 467.

51. I.P. Kaur, S. Chhabra and D. Aggarwal, *Current Drug Delivery*, 2004, **1**, 351.

52. H. Hashimoto, *Journal of Inclusion Phenomena and Macrocyclic Chemistry*, 2002, **44**, 57.

53. J. Szejtli and L. Szente, *European Journal of Pharmaceutics and Biopharmaceutics*, 2005, **61**, 115.

54. M. Gal-Fuzy, L. Szente, J. Szejtli and J. Harangi, *Pharmazie*, 1984, **39**, 558.

55. K. Uekama, *Journal of Inclusion Phenomena and Macrocyclic Chemistry*, 2002, **44**, 3.

56. O.A.E. Soliman, K. Kimura, F. Hirayama, K. Uekama, H.M. ElSabbagh, A.E-G.H.A. El-Gawad and F.M. Hashim, *International Journal of Pharmaceutics*, 1997, **149**, 73.

57. F. Hirayama, Z. Wang and K. Uekama, *Pharmaceutical Research*, 1994, **11**, 1766.

58. F. Hirayama, M. Usami, K. Kimura and K. Uekama, *European Journal of Pharmaceutical Sciences*, 1997, **5**, 23.

59. K. Kimura, F. Hirayama and K. Uekama, *Journal of Pharmaceutical Sciences*, 1999, **88**, 385.

60. H. Matsuda and H. Arima, *Advanced Drug Delivery Reviews*, 1999, **36**, 81.

61. R. Challa, A. Ahuja, J. Ali and R.K. Khar, *AAPS PharmSciTech*, 2005, **6**, E329.

62. M. Pourmokhtar and G.A. Jacobson, *Pharmazie*, 2005, **60**, 837.

63. M.A. Bayomi, K.A. Abanumay and A.A. Al-Angary, *International Journal of Pharmaceutics*, 2002, **243**, 107.

64. A. Iaconinoto, M. Chicca, S. Pinamonti, A. Casolari, A. Bianchi and S. Scalia, *Pharmazie*, 2004, **59**, 30.

65. S. Scalia, A. Molinari, A. Casolari and A. Maldotti, *European Journal of Pharmaceutical Sciences*, 2004, **22**, 241.

66. P. Lejoyeux, D Wouessidjewe and D. Duchêne in *Proceedings of the International Symposium on Cyclodextrins*, Paris, France, 1990, p.348.

67. J.C. Kang, V. Kumar, D. Yang, P.R. Chowdhury and R.J. Hohl, *European Journal of Pharmaceutical Sciences*, 2002, **15**, 163.

68. M. Jumaa, L. Chimilio, S. Chinnaswamy, S. Silchenko and V.J. Stella, *Journal of Pharmaceutical Sciences*, 2004, **93**, 532.

69. P. Hsyu, R.P. Hegde, B.K. Birmingham and C.T. Rhodes, *Drug Development and Industrial Pharmacy*, 1984, **10**, 601.

70. N. Rajagopalan, S.C. Chen and W.S. Chow, *International Journal of Pharmaceutics*, 1986, **29**, 161.

71. K. Uekama, H. Adachi, T. Irie, T. Yano, M. Saita and K. Noda, *Journal of Pharmacy and Pharmacology*, 1992, **44**, 119.

72. T. Loftsson, S. Bjornsdottir, G. Palsdottir and N. Bodor, *International Journal of Pharmaceutics*, 1989, **57**, 63.

73. M.E. Brewster, T. Loftsson, K.S. Estes, J.L. Lin, H. Fridriksdottir and N. Bodor, *International Journal of Pharmaceutics*, 1992, **79**, 289.

74. O. Bekers, J.H. Beijnen, B.J. Vis, A. Suenaga, M. Otagiri, A. Bult and W.J.M. Underberg, *International Journal of Pharmaceutics*, 1991, **72**, 123.

75. R.F.V. Lopez, J.H. Collett and M.V.L.B. Bentley, *International Journal of Pharmaceutics*, 2000, **200**, 127.

76. M.E. Davis and M.E. Brewster, *Nature Reviews Drug Discovery*, 2004, **3**, 1023.

77. H. Adachi, T. Irie, F. Hirayama and K. Uekama, *Chemical & Pharmaceutical Bulletin*, 1992, **40**, 1586.

78. S. Weisse, B. Perly, C. Creminon, F. Ouvrard-Baraton and F. Djedaini-Pilard, *STP Pharma Sciences*, 2004, **14**, 77.

79. M.E. Brewster and T. Loftsson, *Advanced Drug Delivery Reviews*, 2007, **59**, 645.

80. N. Ozdemir and S. Ordu, *Drug Development and Industrial Pharmacy*, 1998, **24**, 19.

81. P. Bertholet, M. Gueders, G. Dive, A. Albert, V. Barillaro, B. Perly, D. Cataldo, G. Piel, L. Delattre and B. Evrard, *Journal of Pharmaceutical Sciences*, 2005, **8**, 163.

82. T. Loftsson, D. Hreinsdottir and M. Masson, *International Journal of Pharmaceutics*, 2005, **302**, 18.

83. V.M. Rao and V.J. Stella, *Journal of Pharmaceutical Sciences*, 2003, **92**, 927.

84. M. Cirri, F. Maestrelli, G. Corti, S. Furlanetto and P. Mura, *Journal of Pharmaceutical and Biomedical Analysis*, 2006, **42**, 126.

85. V.J. Stella and R.A. Rajewski, *Pharmaceutical Research*, 1997, **14**, 556.

86. T. Loftsson, F. Konradsdottir and M. Masson, *Pharmazie*, 2006, **61**, 83.

87. R.L. Carrier, L.A. Miller and I. Ahmed, *Journal of Controlled Release*, 2007, **123**, 78.

88. T. Loftsson, *Journal of Inclusion Phenomena and Macrocyclic Chemistry*, 2002, **44**, 63.

89. U. Vollmer, B.W. Muller, J. Peeters, J. Mesens, B. Wilffert and T. Peters, *Journal of Pharmacy and Pharmacology*, 1994, **46**, 19.

90. F.W.H.M. Merkus, J.C. Verhoef, E. Marttin, S.G. Romeijn, P.H.M. van der Kuy, W.A.J.J. Hermens and N.G.M. Schipper, *Advanced Drug Delivery Reviews*, 1999, **36**, 41.

91. W.A.J.J. Hermens, M.J.M. Deurloo, S.G. Romeyn, J.C. Verhoef and F.W.H.M. Merkus, *Pharmaceutical Research*, 1990, **7**, 500.

92. T. Irie, K. Wakamatsu, H. Arima, H. Aritomi and K. Uekama, *International Journal of Pharmaceutics*, 1992, **84**, 129.

93. E. Marttin, J.C. Verhoef and F.W.H.M. Merkus, *Journal of Drug Targeting*, 1998, **6**, 17.

94. L. Hovgaard and H. Brondsted, *Pharmaceutical Research*, 1995, **12**, 1328.

95. C.A. Ventura, M. Fresta, D. Paolino, S. Pedotti, A. Corsaro and G. Puglisi, *Journal of Drug Targeting*, 2001, **9**, 379.

96. E. Marttin, J.C. Verhoef, F. Spies, J. van der Meulen, J.F. Nagelkerke, H.K. Koerten and F.W.H.M. Merkus, *Journal of Controlled Release*, 1999, **57**, 205.

97. F. Ahsan, J.J. Arnold, T.Z. Yang, E. Meezan, E.M. Schwiebert and D.J. Pillion, *European Journal of Pharmaceutical Sciences*, 2003, **20**, 27.

98. N. Nambu, K. Kikuchi, T. Kikuchi, Y. Takahashi, H. Ueda and T. Nagai, *Chemical & Pharmaceutical Bulletin*, 1978, **26**, 3609.

99. F.J.O. Espinar, S.A. Igea, J.B. Mendez and J.L.V. Jato, *International Journal of Pharmaceutics*, 1991, **70**, 35.

100. S.Z. Lin, D. Wouessidjewe, M.C. Poelman and D. Duchêne, *International Journal of Pharmaceutics*, 1994, **106**, 63.

101. T. Loftsson and T. Jarvinen, *Advanced Drug Delivery Reviews*, 1999, **36**, 59.

102. R.Y. Anadolu, T. Sen, N. Tarimci, A. Birol and C. Erdem, *Journal of the European Academy of Dermatology and Venereology*, 2004, **18**, 416.

103. R.A. Rajewski and V.J. Stella, *Journal of Pharmaceutical Sciences*, 1996, **85**, 1142.

104. T. Irie, K. Fukunaga and J. Pitha, *Journal of Pharmaceutical Sciences*, 1992, **81**, 521.

105. T. Irie, K. Fukunaga, M.K. Garwood, T.O. Carpenter, J. Pitha and J. Pitha, *Journal of Pharmaceutical Sciences*, 1992, **81**, 524.

106. E. Albers and B.W. Müller, *Critical Reviews in Therapeutic Drug Carrier Systems*, 1995, **12**, 311.

107. E. Stefansson and T. Loftsson, *Journal of Inclusion Phenomena and Macrocyclic Chemistry*, 2002, **44**, 23.

108. T. Loftsson and M. Masson, *International Journal of Pharmaceutics*, 2001, **225**, 15.

109. T. Loftsson and J.H. Olafsson, *International Journal of Dermatology*, 1998, **37**, 241.

110. E. Bilensoy, M.A. Rouf, I. Vural, M. şen and A.A. Hincal, *AAPS PharmSciTech*, 2006, **7**, 1.

111. H. Blanco-Fuente, B. Esteban-Fernández, J. Blanco-Méndez and F.J. Otero-Espinar, *Chemical & Pharmaceutical Bulletin*, 2002, **40**, 40.

112. L. Boulmedarat, J.L. Grossiord, E. Fattal and A. Bochot, *International Journal of Pharmaceutics*, 2003, **254**, 59.

113. E. Kim, Z. Gao, J. Park, H. Li and K. Han, *International Journal of Pharmaceutics*, 2002, **233**, 159.

114. A. Doliwa, S. Santoyo and P. Ygartua, *Skin Pharmacology and Applied Skin Physiology*, 2001, **14**, 97.

115. A. Harada, *Coordination Chemistry Reviews*, 1996, **148**, 115.

116. K. Uekama, Y. Ikeda, F. Hirayama, M. Otagiri and M. Shibata, *Yakugaku Zasshi*, 1980, **100**, 994.

117. L.W. Chan, T.R.R. Kurup, A. Muthaiah and J.C. Thenmozhiyal, *International Journal of Pharmaceutics*, 2000, **195**, 71.

118. S.J. Lehner, B.W. Muller and J.K. Seydel, *International Journal of Pharmaceutics*, 1993, **93**, 201.

119. S.J. Lehner, B.W. Muller and J.K. Seydel, *Journal of Pharmacy and Pharmacology*, 1994, **46**, 186.

120. A.K. Banga and R. Mitra, *Journal of Drug Targeting*, 1993, **1**, 341.

121. H.H. Kwak, W.S. Shim, M.K. Choi, M.K. Son, Y.J. Kim, H.C. Yang, T.H. Kim, G.I. Lee, B.M. Kim, S.H. Kang and C.K. Shim, *Journal of Controlled Release*, 2009, **137**, 160.

122. M.E. Brewster, M.S. Hora, J.W. Simpkins and N. Bodor, *Pharmaceutical Research*, 1991, **8**, 792.

123. F.L. Aachmann, D.E. Otzen, K.L. Larsen and R. Wimmer, *Protein Engineering*, 2003, **16**, 905.

124. A.F. Soares, R. de Albuquerque, R.A. Carvalho and F. Veiga, *Nanomedicine*, 2007, **2**, 183.

125. B. Plaud, *Annales Françaises d'Anesthésie et de Réanimation*, 2009, **28**, S64.

126. L.H.D.J. Booij, *Anaesthesia*, 2009, **64**, 31.

127. K. Shimada, Y. Ohe, T. Ohguni, K. Kawano, J. Ishii and T. Nakamura, *Journal of the Japanese Society for Food Science and Technology-Nippon Shokuhin Kagaku Kogaku Kaishi*, 1991, **38**, 16.

128. S.C. Yu, A. Bochot, G. Le Bas, M. Cheron, J.L. Grossiord, M. Seiller and D. Duchêne, *STP Pharma Sciences*, 2001, **11**, 385.

129. K. Hashizaki, T. Kageyama, M. Inoue, H. Taguchi, H. Ueda and Y. Saito, *Chemical & Pharmaceutical Bulletin*, 2007, **55**, 1620.

130. M. Inoue, K. Hashizaki, H. Taguchi and Y. Saito, *Chemical & Pharmaceutical Bulletin*, 2008, **56**, 1335.

131. S.C. Yu, A. Bochot, M. Cheron, M. Seiller, J.L. Grossiord, G. Le Bas and D. Duchêne, *STP Pharma Sciences*, 1999, **9**, 273.

132. S.C. Yu, A. Bochot, G. Le Bas, M. Cheron, J. Mahuteau, J.L. Grossiord, M. Seiller and D. Duchêne, *International Journal of Pharmaceutics*, 2003, **261**, 1.

133. D. Duchêne, A. Bochot, S.C. Yu, C. Pepin and M. Seiller, *International Journal of Pharmaceutics*, 2003, **266**, 85.

134. K. Shimada, K. Kawano, J. Ishii and T. Nakamura, *Journal of Food Science*, 1992, **57**, 655.

135. A. Bochot, L. Trichard, G. Le Bas, H. Alphandary, J.L. Grossiord, D. Duchêne and E. Fattal, *International Journal of Pharmaceutics*, 2007, **339**, 121.

136. L. Trichard, M.B. Delgado-Charro, R.H. Guy, E. Fattal and A. Bochot, *Pharmaceutical Research*, 2008, **25**, 435.

137. L. Trichard, E. Fattal, M. Besnard and A. Bochot, *Journal of Controlled Release*, 2007, **122**, 47.

138. L. Trichard, A. Bochot and D. Duchêne in *Cyclodextrins and Their Complexes*, Ed., H. Dodziuk, Wiley-VCH, Warsaw, Poland, 2006, p.423.

139. Y.L. Loukas, V. Vraka and G. Gregoriadis, *International Journal of Pharmaceutics*, 1998, **162**, 137.

140. B. McCormack and G. Gregoriadis, *International Journal of Pharmaceutics*, 1994, **112**, 249.

141. V. Joguparthi and B.D. Anderson, *Pharmaceutical Research*, 2008, **25**, 2505.

142. D. Duchêne, G. Ponchel and D. Wouessidjewe, *Advanced Drug Delivery Reviews*, 1999, **36**, 29.

143. D. Duchêne, D. Wouessidjewe and G. Ponchel, *Journal of Controlled Release*, 1999, **62**, 263.

144. D. Duchêne and R. Gref in *Cyclodextrins in Pharmaceutics, Cosmetics and Biomedicine*, Ed., E. Memişoğlu-Bilensoy, Wiley, Hoboken, NJ, USA, 2011, p.371.

145. A. Monza da Silveira, G. Ponchel, F. Puisieux and D. Duchêne, *Pharmaceutical Research*, 1998, **15**, 1051.

146. D. Duchêne, G. Ponchel, H. Boudad and A.M. da Silveira, *Annales Pharmaceutiques Francaises*, 2001, **59**, 384.

147. A. Monza da Silveira, D. Duchêne and G. Ponchel, *STP Pharma Sciences*, 2000, **10**, 309.

148. A.M. da Silveira, D. Duchêne and G. Ponchel, *Polymer Science Series A*, 2004, **46**, 1192.

149. H. Boudad, P. Legrand, G. Lebas, M. Cheron, D. Duchêne and G. Ponchel, *International Journal of Pharmaceutics*, 2001, **218**, 113.

150. H. Boudad, P. Legrand, M. Appel, M.H. Coconnier and G. Ponchel, *STP Pharma Sciences*, 2001, **11**, 369.

151. H. Boudad, P. Legrand, M. Besnard, D. Duchêne and G. Ponchel in *Proceedings of the 4th World Meeting on Pharmaceutics, Biopharmaceutics, Pharmaceutical Technology*, Florence, Italy, 2002, p.807.

152. H. Boudad, P. Legrand, M-H. Coconnier, M. Appel, D. Duchêne and G. Ponchel, in *Proceedings of the 4th World Meeting on Pharmaceutics, Biopharmaceutics, Pharmaceutical Technology*, Florence, Italy, 2002, p.1385.

153. M. Agueros, L. Ruiz-Gaton, C. Vauthier, K. Bouchemal, S. Espuelas, G. Ponchel and J.M. Irache, *European Journal of Pharmaceutical Sciences*, 2009, **38**, 405.

154. F. Maestrelli, M. Garcia-Fuentes, P. Mura and M.J. Alonso, *European Journal of Pharmaceutics and Biopharmaceutics*, 2006, **63**, 79.

155. D. Teijeiro-Osorio, C. Remunan-Lopez and M.J. Alonso, *Biomacromolecules*, 2009, **10**, 243.

156. R. Gref, C. Amiel, K. Molinard, S. Daoud-Mahammed, B. Sebille, B. Gillet, J.C. Beloeil, C. Ringard, V. Rosilio, J. Poupaert and P. Couvreur, *Journal of Controlled Release*, 2006, **111**, 316.

157. S. Daoud-Mahammed, P. Couvreur and R. Gref, *International Journal of Pharmaceutics*, 2007, **332**, 185.

158. S. Daoud-Mahammed, P. Couvreur, K. Bouchemal, M. Cheron, G. Lebas, C. Amiel and R. Gref, *Biomacromolecules*, 2009, **10**, 547.

159. H. Gao, Y.W. Yang, Y.G. Fan and J.B. Ma, *Journal of Controlled Release*, 2006, **112**, 301.

160. H. Gao, Y.N. Wang, Y.G. Fan and J.B. Ma, *Journal of Controlled Release*, 2005, **107**, 158.

161. H. Gao, Y. Wang, Y. Fan and J. Ma, *Journal of Biomedical Materials Research A*, 2007, **80**, 111.

162. M. Bencini, E. Ranucci, P. Ferruti, A. Manfredi, F. Trotta and R. Cavalli, *Journal of Polymer Science, Part A: Polymer Chemistry Edition*, 2008, **46**, 1607.

163. R. Cavalli, M. Donalisio, A. Civra, P. Ferruti, E. Ranucci, F. Trotta and D. Lembo, *Journal of Controlled Release*, 2009, **137**, 116.

164. M. Eguchi, Y-Z. Du, S. Taira and M. Kodaka, *NanoBiotechnology*, 2005, **1**, 165.

165. M. Eguchi, Y.Z. Du, Y. Ogawa, T. Okada, N. Yumoto and M. Kodaka, *International Journal of Pharmaceutics*, 2006, **311**, 215.

166. M.E.M. Barbosa, V. Montembault, S. Cammas-Marion, G. Ponchel and L. Fontaine, *Polymer International*, 2007, **56**, 317.

167. M.E. Barbosa, L. Bouteiller, S. Cammas-Marion, V. Montembault, L. Fontaine and G. Ponchel, *Journal of Molecular Recognition*, 2008, **21**, 169.

168. H. Gonzalez, S.J. Hwang and M.E. Davis, *Bioconjugate Chemistry*, 1999, **10**, 1068.

169. S.J. Hwang, N.C. Bellocq and M.E. Davis, *Bioconjugate Chemistry*, 2001, **12**, 280.

170. S.H. Pun, N.C. Bellocq, A.J. Liu, G. Jensen, T. Machemer, E. Quijano, T. Schluep, S.F. Wen, H. Engler, J. Heidel and M.E. Davis, *Bioconjugate Chemistry*, 2004, **15**, 831.

171. M.E. Davis, S.H. Pun, N.C. Bellocq, T.M. Reineke, S.R. Popielarski, S. Mishra and J.D. Heidel, *Current Medicinal Chemistry*, 2004, **11**, 179.

172. I.K. Park, H.A. Von Recum, S.Y. Jiang and S.H. Pun, *Langmuir*, 2006, **22**, 8478.

173. M.L. Forrest, N. Gabrielson and D.W. Pack, *Biotechnology and Bioengineering*, 2004, **89**, 416.

174. S.H. Pun and M.E. Davis, *Bioconjugate Chemistry*, 2002, **13**, 630.

175. N.C. Bellocq, S.H. Pun, G.S. Jensen and M.E. Davis, *Bioconjugate Chemistry*, 2003, **14**, 1122.

176. D.W. Bartlett and M.E. Davis, *Bioconjugate Chemistry*, 2007, **18**, 456.

177. S.H. Pun, N.C. Bellocq, J. Cheng, B.H. Grubbs, G.S. Jensen, M.E. Davis, F. Tack, M. Brewster, M. Janicot, B. Janssens, W. Floren and A. Bakker, *Cancer Biology and Therapy*, 2004, **3**, 641.

178. R. Roy, F. Hernandez-Mateo and F. Santoyo-Gonzalez, *Journal of Organic Chemistry*, 2000, **65**, 8743.

179. A. Vargas-Berenguel, F. Ortega-Caballero, F. Santoyo-Gonzalez, J.J. Garcia-Lopez, J.J. Gimenez-Martinez, L. Garcia-Fuentes and E. Ortiz-Salmeron, *Chemistry - A European Journal*, 2002, **8**, 812.

180. J.J. Garcia-Lopez, F. Hernandez-Mateo, J. Isac-Garcia, J.M. Kim, R. Roy, F. Santoyo-Gonzalez and A. Vargas-Berenguel, *Journal of Organic Chemistry*, 1999, **64**, 522.

181. A. Garcia-Barrientos, J.J. Garcia-Lopez, J. Isac-Gaarcia, F. Ortega-Caballero, C. Uriel, A. Vargas-Berenguel and F. Santoyo-Gonzalez, *Synthesis*, 2001, 7, 1057.

182. F. Ortega-Caballero, J.J. Gimenez-Martinez, L. Garcia-Fuentes, E. Ortiz-Salmeron, F. Santoyo-Gonzalez and A. Vargas-Berenguel, *Journal of Organic Chemistry*, 2001, **66**, 7786.

183. C.O. Mellet, J. Defaye and J.M.G. Fernandez, *Chemistry - A European Journal*, 2002, **8**, 1982.

184. M. Adeli, Z. Zarnegar and R. Kabiri, *European Polymer Journal*, 2008, **44**, 1921.

185. J.M. Bryson, W.J. Chu, J.H. Lee and T.M. Reineke, *Bioconjugate Chemistry*, 2008, **19**, 1505.

186. C.A. Yang, H.Z. Li, S.H. Goh and J. Li, *Biomaterials*, 2007, **28**, 3245.

187. S. Srinivasachari, K.M. Fichter and T.M. Reineke, *Journal of the American Chemical Society*, 2008, **130**, 4618.

188. F.J. Xu, Z.X. Zhang, Y. Ping, J. Li, E.T. Kang and K.G. Neoh, *Biomacromolecules*, 2009, **10**, 285.

189. B.J. Roessler, A.U. Bielinska, K. Janczak, I. Lee and J.R. Baker, *Biochemical and Biophysical Research Communications*, 2001, **283**, 124.

190. H. Arima, F. Kihara, F. Hirayama and K. Uekama, *Bioconjugate Chemistry*, 2001, **12**, 476.

191. F. Kihara, H. Arima, T. Tsutsumi, F. Hirayama and K. Uekama, *Bioconjugate Chemistry*, 2002, **13**, 1211.

192. F. Kihara, H. Arima, T. Tsutsumi, F. Hirayama and K. Uekama, *Bioconjugate Chemistry*, 2003, **14**, 342.

193. W. Zhang, Z. Chen, X.X. Song, J.M. Si and G.P. Tang, *Technology in Cancer Research & Treatment*, 2008, 7, 103.

194. C.A. Nijhuis, J.K. Sinha, G. Wittstock, J. Huskens, B.J. Ravoo and D.N. Reinhoudt, *Langmuir*, 2006, **22**, 9770.

195. C.A. Nijhuis, K.A. Dolatowska, B.J. Ravoo, J. Huskens and D.N. Reinhoudt, *Chemistry - A European Journal*, 2007, **13**, 69.

196. A. Bielinska, J.F. Kukowska-Latallo, J. Johnson, D.A. Tomalia and J.R. Baker, Jr., *Nucleic Acids Research*, 1996, **24**, 2176.

197. L.H. Qin, D.R. Pahud, Y.Z. Ding, A.U. Bielinska, J.F. Kukowska-Latallo, J.R. Baker and J.S. Bromberg, *Human Gene Therapy*, 1998, **9**, 553.

198. J.F. Kukowska-Latallo, A.U. Bielinska, J. Johnson, R. Spindler, D.A. Tomalia and J.R. Baker, Jr., *Proceedings of the National Academy of Sciences of the United States of America*, 1996, **93**, 4897.

199. G. Wenz, *Angewandte Chemie: International Edition*, 1994, **33**, 803.

200. G. Wenz, B.H. Han and A. Muller, *Chemical Reviews*, 2006, **106**, 782.

201. J. Li and X.J. Loh, *Advanced Drug Delivery Reviews*, 2008, **60**, 1000.

202. P.C. Tchoreloff, M.M. Boissonnade, A.W. Coleman and A. Baszkin, *Langmuir*, 1995, **11**, 191.

203. M. Munoz, R. Deschenaux and A.W. Coleman, *Journal of Physical Organic Chemistry*, 1999, **12**, 364.

204. C. Ringard-Lefebvre, A. Bochot, E. Memisoglu, D. Charon, D. Duchêne and A. Baszkin, *Colloids and Surfaces B: Biointerfaces*, 2002, **25**, 109.

205. M. Skiba, D. Wouessidjewe, H. Fessi, J.P. Devissaguet, D. Duchêne and F. Puisieux, inventors; CNRS/University Paris-Sud 11, assignee; FR 9207287, 1992.

206. M. Skiba, D. Wouessidjewe, H. Fessi, J.P. Devissaguet, D. Duchêne and F. Puisieux, inventors; CNRS/University Paris-Sud 11, assignee; FR 9207285, 1992.

207. D. Wouessidjewe, M. Skiba, F. LeroyLechat, E. LemosSenna, F. Puisieux and D. Duchêne, *STP Pharma Sciences*, 1996, **6**, 21.

208. E. Lemos-Senna, D. Wouessidjewe, S. Lesieur, F. Puisieux, G. Couarrazze and D. Duchêne, *Pharmaceutical Development and Technology*, 1998, **3**, 1.

209. E. Memisoglu, A. Bochot, M. Sen, D. Duchêne and A.A. Hincal, *International Journal of Pharmaceutics*, 2003, **251**, 143.

210. E. Memisoglu, A. Bochot, M. Ozalp, M. Sen, D. Duchêne and A.A. Hincal, *Pharmaceutical Research*, 2003, **20**, 117.

211. E. Lemos-Senna, D. Wouessidjewe, S. Lesieur and D. Duchêne, *International Journal of Pharmaceutics*, 1998, **170**, 119.

212. E. Memisoglu-Bilensoy, L. Vural, A. Bochot, J.M. Renoir, D. Duchêne and A.A. Hincal, *Journal of Controlled Release*, 2005, **104**, 489.

213. I. Vural, E. Memisoglu-Bilensoy, J.M. Renoir, A. Bochot, D. Duchêne and A.A. Hincal, *Journal of Drug Delivery Science and Technology*, 2005, **15**, 339.

214. E. Bilensoy, O. Gurkaynak, A.L. Dogan and A.A. Hincal, *International Journal of Pharmaceutics*, 2008, **347**, 163.

215. E. Bilensoy, O. Gurkaynak, M. Ertan, M. Sen and A.A. Hincal, *Journal of Pharmaceutical Sciences*, 2008, **97**, 1519.

216. Y. Cirpanli, E. Bilensoy, A.L. Dogan and S. Calis, *European Journal of Pharmaceutics and Biopharmaceutics*, 2009, **73**, 82.

8 Bioactivity of Chondroitin Sulfate

Iossif A. Strehin and Jennifer H. Elisseeff

8.1 Introduction

Chondroitin sulfate (ChS) is a constituent of tissue extracellular matrix (ECM), cell surface receptors and intracellular organelles such as lysosomes. It is a complex polysaccharide composed of the disaccharide repeat unit glucuronic acid (GlcA) and *N*-acetyl-D-galactosamine (GalNAc). The complexity of the molecule is compounded by the presence of sulfate groups on the four hydroxyl groups of the disaccharide repeat unit. The number and location of sulfation groups per disaccharide unit can vary, which makes the encoding of information possible. This information can then be used to guide cell behaviour via changes in intracellular signalling, which could alter viability, proliferation, differentiation, ECM synthesis and matrix degradation. In this chapter, we will go into detail about the biological activity of ChS at the cellular level, and we will also discuss its use in treating arthritis and in guiding cell function in biomaterials.

ChS is a complex glycosaminoglycan (GAG) composed of the disaccharide repeat unit GlcA and GalNAc. ChS is found as part of proteoglycans in the ECM of tissues, on cell surface receptors or inside cells. Variants of the molecule exist and are classified based on the sulfation pattern of the hydroxyl groups found on the second and third carbon of GlcA and the fourth and sixth carbon of GalNAc (**Figure 8.1**). ChS has been shown to modulate numerous biological functions including inflammation, enzymatic activity, ECM synthesis, adhesion, nerve growth, apoptosis, mitogenesis, differentiation and stem cell niches. This modulation is dependent on the type of ChS used. For example, $R_4 = SO^{3-}$; $R_2, R_3, R_6 = H$ (ChS-A) prevents nerve growth while $R_4, R_6 = SO^{3-}$; $R_2, R_3 = H$ (ChS-E) has the opposite effect.

One predominant variant of ChS exists per specific tissue and species of animal (**Table 8.1**). For example, the predominant type of ChS in shark cartilage is $R_6 = SO^{3-}$; $R_2, R_3, R_4 = H$ (ChS-C) (**Figure 8.1**) while in squid cartilage it is ChS-E. The predominant type of ChS in human blood plasma is $R_2, R_3, R_4, R_6 = H$ (ChS-O) while that in human bone is ChS-C. Additionally, specific types of ChS can be found in one tissue but not another (**Table 8.1**). This high variability in sulfation pattern from tissue to tissue and the presence of ChS in many different types of tissue suggest that

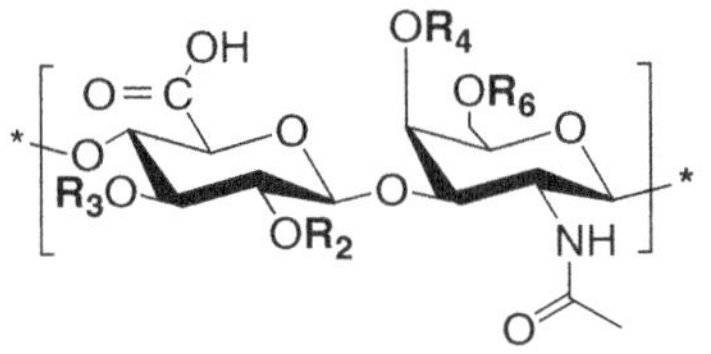

Figure 8.1 Chemical structures of ChS variants

ChS is involved in important biological functions and its specific sulfation pattern plays a role in its activity.

In this chapter, some details about some of the biological functions that ChS has been shown to modulate, will be discussed. How the GAG has been used to treat diseases such as arthritis will also be considered. Finally, biomaterials containing ChS and how the molecule improves their function will be described.

8.2 Cytoprotection

Cell death can be initiated via exposure to numerous stresses including reactive oxygen species (ROS) [13–15], serum starvation [16–19] and prolonged neuronal excitation [20]. Under such stresses, ChS has been shown to play a protective role.

ROS under controlled concentrations can be beneficial in body functions such as defence against infection. However, when too much ROS are around, this can lead to damage of organic molecules including lipids, proteins and DNA. Thus prolonged exposure to such oxidative stresses leads to cell damage and eventually death via apoptosis or necrosis. ChS-A is able to reduce apoptosis by interfering with the damaging effects of ROS via at least three mechanisms:

- Interacting directly with metal ions responsible for the production of ROS and neutralising their effects [21];
- Activating intracellular signalling pathways that synthesise antioxidant proteins [13, 14, 22]; and
- Preventing the ROS-initiated upregulation of proinflammatory molecules [13, 14, 23].

The first mechanism is simple to understand. By chelating iron and copper ions and removing them from solution, ChS-A is in effect preventing the initiation of ROS production [21, 24]. Interestingly, this response is specific for ChS-A since ChS-C

Table 8.1 Sulfation pattern, concentration, molecular weight, and extent of sulfation for ChS molecules isolated from different tissues and animal sources

Tissue	Source	Species	Δdi-0S	Δdi-4S	Δdi-6S	Δdi-(2,6)S	Δdi-(2,4)S	Δdi-(4,6)S	Δdi-(2,4,6)S	Total ChS	M_R (x 1,000)	SO_3^-/COO^-	Source
Cartilage	Adult (50–60 years) PFE	Human		≤ 2%	98 ± 2%					4.3 ± 1.1 cg/g [a]			[1]
	Infant cartilage from PFE	Human		44 ± 12%	56 ± 12%					8.5 ± 1.2 cg/g [a]			[1]
	N/A	Shark	7%		62%	31%							[2]
	N/A	Shark	2.2%	27.7%	49.0%	16.1%		3%	1.9%		>50.00	1.21	[3]
	N/A	Squid		37%	11%			52%					[2]
	Trachea	Bovine	4.4%	56%	37.9%	0.8%		0.8%			26.1	0.97	[4]
	Trachea	Bovine	9%	51%	40%						24	0.91	[5]
IVD	Nucleus pulposus	Human		28%	72%								[6]
AA	N/A	Human	3.5%	17.3%	78.4%	0.3%		0.5%		~ 0.01 g/g [ab]			[7]
Cornea	Stroma	Human	65%	29%	6%					1.08 pmol/μm^3 x 10^{-4} [c]			[8]
Bone	Femur	Human	6.4%	23.8%	67.9%	1.2%		0.7%		0.96 mg/g [a]			[9]

Table 8.1 (Continued)

Tissue	Source	Species	Δdi-0S	Δdi-4S	Δdi-6S	Δdi-(2,6)S	Δdi-(2,4)S	Δdi-(4,6)S	Δdi-(2,4,6)S	Total ChS	Mr (x 1,000)	SO_3^-/ COO^-	Source
Brain	N/A	Embryo rat (E18)		66.8%	30.3%	1.7%		1.2%		2.3 μmol/ brain [d]			[10]
Blood Plasma	Cerebellum	Postnatal mouse (P1)	9.1%	49.5%	35.4%	3.2%	0.15%	2.61%		249 ± 26 pmol/ mm^3 [e]		0.968	[11]
Unspecified	Cerebellum	Postnatal mouse (P20)	3.1%	84.3%	5.3%	3.2%	3.2%	0.82%		313 ± 80 pmol/ mm^3 [e]		1.04	[11]
	N/A	Human	59%	38%	3%						15.60	0.41	[3]
	N/A	Shark	5.1%	29%	39%	23.3%	0.5%	3%			55.72	1.22	[12]

[a] *Expressed as weight ChS per weight of dry tissue*
[b] *Assumes that uronic acid content is 30% of glycosaminoglycan content*
[c] *Expressed in pmols disaccharides per area of tissue*
[d] *Expressed in μmols disaccharides per whole brain*
[e] *Total amount of Δdimers*
PFE: Proximal femoral epiphysis
IVD: Intervertebral disc
AA: Abdominal aorta

cannot interfere with the formation of ROS [24]. Not surprising, as we age, the type of ChS in tissues such as arterial walls [25, 26] and cartilage [1] shifts from having predominantly ChS-A residues to ChS-C residues. In effect the tissues become less resistant to oxidative damage. Thus as we age, the ROS in tissues are elevated and are partially responsible for age-related diseases such as arthritis and atherosclerosis. In addition, in the case of atherosclerosis, ChS-C has been shown to bind low-density lipoproteins, the harmful type of lipoprotein, while ChS-A exhibits little or no such affinity [27]. So not only does the shift from ChS-A to ChS-C increases ROS formation, but it makes arteries more susceptible to develop atherosclerotic lesions [28].

Cells synthesise antioxidant proteins to protect themselves from elevated levels of ROS. However, in the case of disease, this mechanism can be overwhelmed and the excess of ROS can lead to apoptosis. Pre-exposure to ChS-A conditions the cells by activating intracellular signalling pathways such as the phosphatidylinositol 3-kinase (PI3K)/akt pathway [22], which lead to the further activation of downstream signalling molecules and an increase in the production of anti-ROS proteins such as haem oxygenase-1 [22], catalase [14], superoxide dismutase [23] and glutathione reductase/glutathione [13, 14, 23]. The increase in antioxidant protein synthesis in effect makes the cells more resistant to ROS-initiated apoptosis [13, 22].

Fibroblasts exposed to ROS have increased nuclear translocation of nuclear factor-κB (NF-κB) [23], which has been shown to lead to increased secretion of proinflammatory molecules [13, 14]. Thus, ROS can induce inflammation, which can then lead to further ROS production and eventually apoptosis. In an acute pancreatitis mouse model, in the presence of ChS-A, apoptosis levels decreased as shown by reduced expression and protein levels of the executioner caspases 3 and 7 [13]. ChS-A decreased nuclear translocation of NF-κB and decreased degradation of inhibitory κB-α (IκB-α) protein leading to decreased expression and protein levels of proinflammatory cytokines tumour necrosis factor-α (TNF-α) and interleukin (IL)-6 and decreased neutrophil infiltration. Furthermore, lower lipid peroxidation was observed in the presence of the polysaccharide. Similar results were observed *in vivo* with collagen-induced osteoarthritis (OA) in a mouse model [14].

Serum starvation can also initiate cell death. ChS-A can inhibit apoptosis initiated by serum starvation in rat vascular smooth muscle cells (rVSMC), human primary aortic smooth muscle cells (AOSMC) and human lung fibroblasts (hFB) [16–18]. The percentage of apoptotic rVSMC and AOSMC is significantly lower for cells incubated in serum-free medium supplemented with 60–250 μg/ml ChS-A when compared with cells incubated in serum-free medium alone. This antiapoptotic phenotype is in part attributed to increased phosphorylation of extracellular signal-regulated kinase 1/2 and increased protein levels of Bcl-x_L (B-cell lymphoma leukaemia-x_L). hFB show a similar decrease in apoptosis when exposed to serum-free medium supplemented with ChS-A [17], but the response in this case is attributed to signalling pathways downstream of PI3K as well as an increase in the protein levels of Bcl-x_L and a decrease in the protein levels of Bim-EL (Bcl-2-interacting mediator of cell death-EL). Additionally, the antiapoptotic phenotype is observed at lower concentrations of ChS-A for systemic sclerosis fibroblasts when compared to normal fibroblasts.

ChS-C also has antiapoptotic activity as demonstrated in serum-starved rat lung fibroblasts [19]. The effective concentration of ChS-C is between 0.5 and 4 µg/ml. ChS-C increases PI3K signalling and phosphorylation of Bcl-associated death protein (BAD) at both Ser112 and Ser136. The phosphorylation of BAD at these residues inactivates the protein and thus contributes to the observed decrease in apoptosis. Additionally, ChS-C prevents cleavage of executioner caspases 3 and 7. Thus, even though ChS-C does not have anti-ROS effects like ChS-A, it is able to alter intercellular signalling of apoptotic pathways such that the cells become more resistant to cell death.

Neuronal cell death can be induced via exposure to excessive levels of excitatory amino acids that bind to and activate glutamate cell surface receptors. This mode of cell death has been linked to numerous neurodegenerative disorders. Pre-exposure to 10 µg/ml ChS-E for as little as 3 hours significantly decreased the excitotoxicity of the glutamate analogues *N*-methyl-D-aspartic acid (NMDA) and α-amino-3-hydroxy-5-methyl-4-isoxazolepropionic acid [20]. ChS-A, dermatan sulfate, ChS-C and highly sulfated R_2, R_6 = SO^{3-}; R_3, R_4 = H (ChS-D) as well as other highly sulfated polysaccharides including heparin and dextran sulfate showed no such neuroprotective activity at concentrations between 10 and 100 µg/ml. This supports the idea that the specific sulfation pattern on the ChS-E molecule and not the charge density or backbone of the molecule is responsible for the observed outcomes. Additionally, ChS-E with molecular weights between 5 and 75 kDa exhibits this activity. It does not act as an NMDA blocker as currents in neocortical neurons treated with ChS-E are unchanged from control. Rather, a change in protein phosphorylation and a decrease in caspase-3 activity were observed, suggesting that changes in intracellular signalling could have led to a decrease in apoptosis and decreased cell death. Thus, through changes in intracellular signalling, ChS-E acted as a neuroprotective agent against excitotoxicity in neocortical neurons.

8.3 Mitogenesis

ChS can bind to proteins and alter their capacity to stimulate cell proliferation. In the presence of fibroblast growth factor-2 (FGF-2) but not epidermal growth factor, ChS-E stimulates proliferation of rat neural stem/progenitor cells isolated from the ventricular zone of the telencephalon while ChS-A, ChS-C, keratan sulfate (KS) and hyaluronic acid do not exhibit this effect [29]. Since ChS-E has been shown to bind FGF-2 with significantly greater affinity than ChS-A, ChS-C, ChS-D, KS and hyaluronic acid [30], the binding of ChS-E to FGF-2 could be presenting the growth factor to the cells in a more favourable form that stimulates cell proliferation to a greater degree. Also, ChS-E is detected in the ECM surrounding stem/progenitor cells in the telencephalon region of the brain [29], and proteoglycans isolated from mouse embryos contained ChS chains with as much as 32% ChS-E disaccharide repeat units [31], which suggests that ChS-E chains exist *in vivo* and should be able to modulate FGF-2-dependent proliferation of neural stem cells.

The same type of ChS molecule can either stimulate or suppress cell proliferation based on the cell type used. Exogenously added ChS-A and ChS-C increase the proliferation of chondrocytes cultured on collagen type II scaffolds [32]. On the other hand, ChS-A decreases the proliferative effects of platelet-derived growth factor-BB (PDGF-BB, with two B chains) on hFB [33]. It does so by reducing the phosphorylation of PDGF receptor beta (PDGF-Rβ). The method by which phosphorylation of the receptor is achieved is still unclear, but it is believed that similar to hyaluronic acid [34], ChS interaction with CD-44 could lead to intracellular signalling activating phosphatases, which in turn decrease the activation of PDGF-Rβ. Further work needs to be done to support this hypothesis. Nevertheless, one type of ChS molecule can have very different effects on different types of cells.

8.4 Differentiation

The superficial zone of articular cartilage contains a population of progenitor cells [35]. It is speculated that the pericellular environment of these cells is composed of ChS proteoglycans with a low degree of sulfation, which keeps the cells in their undifferentiated state [36]. When the progenitor cells divide, the daughter cells exit the region of low sulfation and enter a highly sulfated region where differentiation into mature chondrocytes occurs. The cells in the highly sulfated region are believed to begin differentiating because sulfated proteoglycans selectively bind growth factors depending on their sulfation pattern [30] and these growth factors are capable of driving the differentiation process. The stem cell niches in articular cartilage are rich in aggrecan and perlican containing ChS molecules with nonsulfated disaccharide and a low concentration of four sulfated disaccharides, while the regions of differentiation are rich in repeat units sulfated in the 4- or 6-position [36]. By this way ChS molecules may be able to guide differentiation of progenitor cells in articular cartilage.

ChS-A drives bone marrow-derived mesenchymal stem cells (MSC) towards chondrogenic differentiation [37]. When MSC encapsulated in hydrogels containing both ChS-A and poly(ethylene glycol) (PEG) are compared to MSC encapsulated in PEG-alone hydrogels, the cells in the ChS gels aggregate together, which is what happens *in vivo* during chondrogenesis in limb development. The aggregation observed *in vivo* is attributed to interactions between cadherin receptors [38–40]. Expression of cadherin-11 is increased in the ChS gels when compared to PEG-alone gels. The increased aggregation of MSC in ChS-A gels also leads to improved chondrogenesis and decreased osteogenesis as assessed by gene expression and biochemical analysis [37]. Thus, since the ratio of ChS-A to ChS-C disaccharides and the total ChS content found in native articular cartilage decrease as we age [1, 41], and the presence of ChS-A improves chondrogenesis of MSC, mimicking the ECM environment of the native tissue with the specific ChS molecule present during development could improve differentiation of stem cells. Further work with ChS rich in the ChS-C disaccharide unit needs to be done to confirm this hypothesis.

8.5 Neuronal Growth

ChS proteoglycans are generally considered as repulsive substrata for neurite outgrowth. For example, aggrecan extracted from the nucleus pulposus of human intervertebral discs inhibits neurite growth of dorsal root ganglion (DRG) neurons [42]. Following treatment with chondroitinase ABC, neurite growth is significantly improved while keratinase treatment exhibits a smaller effect. It is important to note that aggrecan of the human nucleus pulposus contains a mixture of ChS-A and ChS-C disaccharide units [6] and increasing amount of data suggests that these specific types of ChS, especially ChS-A, are neurorepulsive [43, 44].

Not all ChS types inhibit neurite growth. ChS-E has been shown to promote neuronal growth with hippocampal [45–47], dopaminergic [45] and DRG neurons [45]. This biological activity of ChS-E has been linked to its ability to interact with certain neuromodulatory factors including midkine [10, 45, 48, 49], brain-derived neurotrophic factor (BDNF) [45] and pleiotrophin [48, 49] at physiologically relevant concentrations. Other ChS types including ChS-A, ChS-C and R_3, R_4 = SO^{3-}; R_2, R_6 = H (ChS-K) do not exhibit such interactions at physiologically relevant concentrations of the growth factors [10]. The association between ChS-E and the neuromodulatory factors is dependent on the sulfation pattern and not the charge density of the molecule. For example, ChS-K that is sulfated at the third carbon of GlcA and the fourth carbon of GalNAc does not bind midkine even though it is highly sulfated like ChS-E [10]. Finally, the binding of midkine and BDNF to ChS-E does not interfere with the binding of these neuromodulatory factors to their respective cell surface receptors [45]. Thus the biological activity of the bound growth factors is preserved, which explains the improved neurite growth on ChS-E-coated surfaces.

Like ChS-E, ChS-D has been shown to have neuronal growth-promoting function for hippocampal neurons [46, 47, 50]. However, unlike ChS-E, ChS-D has very little affinity for growth factors [49]. Either the ChS-D molecule interacts directly with cells, or the growth factors involved in its neurite outgrowth-promoting activity have not been identified yet. More data are needed to elucidate the exact mechanism by which ChS-D promotes neuronal growth.

8.6 Enzymatic Activity

ChS molecules can both increase and decrease enzymatic activity. As mentioned in **Section 8.7.2**, ChS-C and ChS-A are potent inhibitors of leucocyte elastase, an enzyme responsible for the breakdown of articular cartilage in arthritic diseases. On the other hand, ChS-A strongly increases the stability and to a lesser degree the activity of cathepsin K such that over extended periods of time, cathepsin K will degrade collagen substrates up to 100 times more efficiently in the presence of the GAG than in its absence [51]. ChS-C also exhibits such activity but to a much lower degree [51]. ChS-A forms a complex with cathepsin K and the formation of the complex is important in the degradation of triple helical collagen but not gelatinised collagen [52]. Other cathepsins including cathepsins L, B, S and F do not form a complex with ChS-A and, under physiological conditions, cannot cleave collagen type I and II molecules [53]. However, all of the cathepsins are able to degrade gelatinised collagen, which implies that cathepsin K is involved in degrading the collagen molecules, which then become gelatinised and accessible to the other enzymes. The association of the polysaccharide and the enzyme occurs between the sulfate and carboxyl groups of three disaccharide units of ChS and the positively charged residues in the R domain of cathepsin K [54]. The R domain is far from the active site; thus the interaction most likely is an allosteric regulation, which in turn makes the enzyme able to recognise and degrade the triple helical collagen structure.

The importance of cathepsin K is evident in the genetic disorder pycnodysostosis. The mutation occurs at residue 212 (Y212C), and it abolishes the interaction between ChS-A and the protein, thus eliminating the enzyme's ability to degrade the collagen substrates but not gelatin substrates [52]. This leads to decreased bone resorption and the formation of denser bone, which during development causes a drastic decrease in stature. On the other hand, overproduction of the enzyme can lead to disorders such as osteoporosis. Consequently, research is under way to find mechanisms to interfere with the enzyme's activity, which include interfering with complex formation of ChS-A and cathepsin K, masking the collagen binding site and blocking attachment of ChS-A to collagen [55]. The appeal to using these mechanisms of inhibition is that they will interfere with the degradation of collagen while at the same time allowing the enzyme to execute other functions such as degradation of gelatinised collagen.

In addition to being potent collagenases and gelatinases, cathepsins have also been reported to degrade elastin and can thus contribute to the progression of atherosclerosis. Specifically, cathepsins K, S and V have been identified in atheroma tissue as well as macrophages found in that tissue and are responsible for approximately 60% of the degradation of elastin in neighbouring arterial ECM [56]. ChS-A binds to and inhibits the activity of cathepsins K and V while it has no effect on the activity of cathpsin S and does not bind to it [56]. Thus, as we age and lose GAG from arterial walls, the tissue becomes more susceptible to the formation of atherosclerotic lesions. It is interesting to note that ChS-A activates the collagenase activity of cathepsin K while at the same time deactivating its elastolytic activity. Nevertheless, ChS-A is an important modulator of cathepsin activity.

8.7 Arthritis

ChS is classified as a symptomatic slow-acting drug for OA and has also been shown to have structure-modifying effects in the knee and finger joints. Following oral administration in numerous clinical studies, the outcomes show relief in symptoms [57–62], improved joint space width [57, 58, 60, 63], decreased ECM degradation of cartilage and bone tissue [60] and improved joint mobility [59, 61, 62]. The effectiveness of the clinical trials is strongly dependent on the quality of ChS used and on patient selection criteria [64, 65]. Additionally, it takes several months following the initial dose of the drug before the manifestation of these benefits is seen in patients, hence the classification of slow-acting drug. When treatment was halted and outcomes measured over time, the beneficial effects of ChS were shown to persist [59]. ChS is very well tolerated with similar adverse effects as in the placebo group [57–60, 62, 63]. Thus, the effectiveness and safety of the drug have stimulated immense interest in determining the underlying factors responsible for the chondroprotective effects of ChS.

The ECM of articular cartilage is composed mainly of collagen type II and aggrecan [66]. The type of ChS found on aggrecan molecules in cartilage is ChS-A and ChS-C, so ChS-containing therapeutic agents for OA such as Condrosulf® are composed of mixtures of ChS-A and ChS-C. Therefore, studies related to the effects of ChS on chondrocytes, synoviocytes and inflammatory cells found in joint spaces look at these two types of ChS and their effects on inflammation, enzymatic activity and ECM synthesis.

8.7.1 Anti-inflammatory Effects of Chondroitin Sulfate

Inflammation is a key factor contributing to the progression of arthritic diseases, such as OA. The inflammatory process in the knee joint is initiated by microtrauma leading to the release of ECM fragments and fibronectin fragments, which in turn stimulate chondrocytes to produce matrix metalloproteinases (MMP) [67], a disintegrin-like and metalloproteinase with thrombospondin (ADAMTS)-5 [67, 68] and proinflammatory molecules such as inducible nitric oxide synthase (iNOS) [69], IL-1β [67, 70], IL-1α [70], IL-6 [70] and TNF-α [67, 70]. Such degradation by-products also decrease proteoglycan synthesis by chondrocytes [71]. The release of MMP and ADAMTS leads to further matrix degradation and stimulation of the chondrocytes while the proinflammatory molecules stimulate synoviocytes, chondrocytes and macrophages to secrete more proinflammatory and matrix degradation molecules. Thus, it is a positive feedback cycle in which the ultimate outcome is the degradation of the articulating surfaces and neighbouring bone structures leading to the painful symptoms observed in OA.

ChS decreases the production of both MMP and ADAMTS involved in collagen type II and aggrecan degradation in cartilage tissue. ChS-A and ChS-C have both been shown to decrease the IL-1β-stimulated secretion of ADAMTS-4 and ADAMTS-5 in human chondrocytes and synoviocytes as well as MMP-13 in human chondrocytes [72]. ADAMTS-4 and ADAMTS-5 are aggrecanases while MMP-13 is a collagenase, which cleaves collagen type II more efficiently than it cleaves collagen types I and III. ChS-A and ChS-C are also able to decrease MMP-1 and MMP-13 production in normal knee mouse chondrocytes following stimulation with a lipopolysaccharide (LPS) [73]. In a clinical trial, a mixture of ChS-A and ChS-C administered orally decreased collagenolytic activity [74]. Hence, one mechanism by which ChS decreases ECM degradation of cartilage tissue involves decreasing the production of matrix-destructive enzymes.

ChS has also been shown to decrease expression and secretion of proinflammatory molecules. For example, when chondrocytes are exposed to a mixture of IL-1β, ChS-A and ChS-C, the production of prostaglandin E_2 (PGE_2) is not significantly different from that of the control in the first 0–16 days [75]. However, when the cells are exposed to IL-1β alone, the levels of PGE_2 are more than two times that of the control. Additionally, the average PGE_2 production in all the treatment groups from 0 to 16 days remains about an order of magnitude greater per day when compared to average PGE_2 production between days 16 and 24. One possible explanation for this is that during the culture period, the cells synthesise aggrecan, which also has ChS, thus increasing local ChS concentrations in all groups and lowering the secretion of PGE_2. The decrease in PGE_2 levels is also supported by the results of a clinical study where the levels of phospholipase A_2 (PLA_2) in the synovial fluid of patients were significantly decreased from day 0 to day 10 for the treatment group receiving a mixture of ChS-A and ChS-C orally [74]. No such differences were observed in the placebo groups. PLA_2 is an enzyme catalysing the release of arachidonic acid from glycerol, and the fatty acid is then used as a substrate for making eicosanoids. Eicosanoids are inflammatory mediators, which include prostaglandins. Therefore, part of the anti-inflammatory activity of ChS could be due to its effect on prostaglandin production.

Following stimulation of mouse chondrocytes with LPS, ChS-A and ChS-C were both able to decrease expression and/or synthesis of IL-1β, IL-6, TNF-α, interferon-γ, iNOS and NO [73, 76, 77]. This response was attributed to the decrease in nuclear translocation of NF-κB and decreased degradation of IκB-α protein. IκB-α is normally bound to NF-κB in the cytoplasm of the cell preventing its nuclear translocation. Upon stimulation by proinflammatory factors, IκB-α is phosphorylated and the complex dissociates followed by the degradation of IκB-α and the nuclear translocation of NF-κB. Once in the nucleus, NF-κB binds to DNA and stimulates the transcription of numerous proinflammatory molecules. Thus, both ChS molecules somehow are able to keep IκB-α from being phosphorylated and degraded, thus decreasing the secretion of cytokines. Increased cell viability and a decrease in apoptosis were also observed in both ChS groups. Interestingly, dermatan sulfate or KS had no effect on the cells, showing that this is a specific response to ChS.

ChS-A decreased matrix proteoglycan loss and erosion in a mouse model of collagen-induced arthritis [14]. This is again attributed to a decrease in NF-κB nuclear translocation and decreased degradation of IκB-α protein, which leads to decreased expression of proinflammatory molecules. A decrease in MMP-13 expression and synthesis, lipid peroxidation and neutrophil activation is also seen in the presence of ChS. Finally, endogenous antioxidant levels increased with ChS treatment. Together, these anti-inflammatory responses in the presence of ChS contribute to the observed decrease in the degradation of cartilage tissue.

Oxidative stresses are present in arthritic knees and those stresses are in part responsible for the observed cartilage degeneration [78]. ChS increases the production of endogenous antioxidants [13, 14] and acts as an antioxidant itself [21]. The polysaccharide is able to reduce membrane damage by ROS [74]. For more information on the protective effects of ChS against ROS, please refer **Section 8.2**.

Another mechanism by which ChS may decrease the deleterious effects of inflammation in the arthritic knee joints is by decreasing the recruitment of inflammatory cells. Leucocyte concentration in the synovial fluid of gonarthrosic patients was lower in patients receiving a mixture of ChS-A and ChS-C when compared to placebo [74]. Also, chemotaxis of human leucocytes significantly decreased in the presence of ChS of varying molecular weights [74]. In a separate study, following subcutaneous injections of hydrogels composed of either ChS-A and PEG or PEG alone in a rat model, the number of inflammatory cells near the implant-tissue interface was significantly lower with the ChS-PEG gels when compared to PEG-alone gels [79]. Furthermore, in a collagen-induced arthritis mouse model, administration of ChS-A to knee joints decreased polymorphonuclear cell activation and recruitment [14]. Thus, ChS decreases the recruitment of inflammatory cells.

8.7.2 Modulation of Enzymatic Activity by Chondroitin Sulfate

In addition to decreasing the synthesis of enzymes responsible for the matrix degradation of cartilage, ChS can directly inhibit enzymatic activity. Leucocyte elastase is one such enzyme and is capable of degrading both collagen and proteoglycans and is associated with the progression of arthritis. Fractions containing 92% ChS-C and those containing approximately equal concentrations of ChS-A and ChS-C can inhibit the activity of leucocyte elastase with K_I = 1.8 and 140 μg/ml, respectively [80]. Therefore, ChS sulfated at the 6-position is more effective at inhibiting the enzyme. Additionally, ChS with molecular weights above 2 kDa inhibits enzymatic activity to the same extent, but below 2 kDa, the inhibition is drastically reduced as a function of molecular weight [81]. The inhibition mechanism is simple, intersecting, hyperbolic noncompetitive inhibition with a maximum inhibition of 60–70%, and the interactions between the enzyme and polysaccharide are electrostatic [80].

8.7.3 Effects of Chondroitin Sulfate on Extracellular Matrix Synthesis and Synovial Fluid Composition

ChS has a proanabolic effect on cells in the knee joints. It stimulates the synthesis of proteoglycans [32, 75, 82, 83] as well as collagen type II [32] and restores collagen type II synthesis in IL-1β-treated chondrocytes [75]. ChS has been shown to increase hyaluronic acid concentration and synovial fluid viscosity in patients with gonarthrosis [74]. One reason for the observed improvements of the synovial fluid properties in the presence of ChS could be that ChS stimulates human synoviocytes to synthesise higher molecular weight hyaluronic acid by modulating the expression of hyaluronic acid synthases 1, 2 and 3 [84]. Thus, ChS stimulates cells found in the knee joint to produce ECM molecules, which replace degraded ECM and improve synovial fluid properties.

8.8 Biomaterials

As we are learning more about the biological activity of ChS, biomaterials containing ChS are being developed for clinical applications such as cartilage regeneration [85–87], heart valve tissue engineering [88], growth factor or drug delivery [89–94], cell delivery [79, 89], surgical adhesives [79, 95] and sealants [79, 95–97], intervertebral disc regeneration [98], conjunctival regeneration [99] and wound healing [100]. These materials are composed of either completely biological or a combination of synthetic and biological polymers. Incorporating biological polymers like ChS into biomaterials is advantageous because it endows them with bioactivity that could control cell function such as proliferation, differentiation as well as dedifferentiation, matrix production, enzyme synthesis, apoptosis and cytokine secretion. On the other hand, the advantage in using synthetic polymers is that there is tight control over the chemical structure of the molecules and therefore tight control over the mechanical properties of the biomaterial. In any case, the molecules constituting the material need to be crosslinked such that the bioactivity is not lost, there is minimum toxicity to surrounding tissues and/or encapsulated cells, the material is still biodegradable and the mechanical properties can be controlled.

8.8.1 Heart Valve Tissue Engineering

Incorporating ChS in collagen hydrogels results in tissue-engineered heart valves that more closely resemble native tissue. Coculture of valve interstitial cells (VIC) and valve endothelial cells (VEC) in collagen scaffolds shows protein expression very similar to that of native tissue [88]. The same is true when a mixture of ChS-A and ChS-C is incorporated along with the collagen, but the VIC display enhanced production of elastin and laminin, and the VEC express endothelial nitric oxide synthase, which is absent in the collagen-alone gels [88]. Thus, protein synthesis by VEC and VIC with ChS added to the scaffolds more closely resembles that of native tissue. Additionally, the surface of the ChS-containing scaffolds has a more organised layer of VEC, in addition to the absence of any gaps, unlike collagen-alone scaffolds. In effect, the VEC covering the scaffolds will prevent blood-collagen interactions. This is important because collagen exposed to blood will stimulate platelet aggregation and activation, thus leading to thrombogenesis [101, 102]. One possible explanation for the observed benefits is that ChS could have concentrated growth factors in the tissue, which could have influenced cell signalling and protein synthesis. Additionally, ChS has been known to interfere with collagen fibrillogenesis, making the scaffolds more porous without disturbing the triple helical structure of the molecule [88, 103]. The porosity in collagen-ChS scaffolds has been shown to influence cell adhesion to and migration on the scaffolds [104]. Thus altering the porosity of the scaffold and recruitment of growth factors could have led to better cell adhesion and protein expression by VEC and VIC.

8.8.2 Wound Healing

ChS-PEG hydrogel films have been shown to produce favourable outcomes in wound healing models using pigs [105], guinea pigs [106], rabbits [107], and normal [108] and diabetic [94] mice. The injuries treated include skin wounds, tympanic membrane perforations and maxillary sinus mucosa wounds. Skin wounds in a pig model [105] and a mouse model [108] have been shown to re-epithelialise significantly faster when treated with ChS-PEG films plus Tegaderm™ than when treated with Tegaderm™ alone. Also, the treatments containing ChS-PEG have more collagen deposition and a more organised dermal tissue. Growth factors can also be added to the films to accelerate the healing process. When basic fibroblast growth factor (bFGF) is doped into the hydrogels, the fraction of wounds that completely close within 2 weeks is significantly greater, and the effect has a dose response [94]. In fact, monitoring the release of bFGF from ChS-PEG hydrogels for up to a month shows that more than 50% of the bioactivity of bFGF is preserved, and the released growth factor is able to stimulate neovascularisation in a subcutaneous mouse model when compared to injecting bFGF alone or films lacking bFGF [93]. Thus, the materials can be used to deliver growth factors to the wounds while preserving the activity of the growth factors for prolonged periods of time. When treated with ChS-PEG films as opposed to no film at all, wounds generated in the maxillary sinus mucosa in a rabbit model exhibit accelerated healing and the epithelium formed is respiratory epithelium [107]. In the no-film controls, the epithelium that forms is squamous. ChS-PEG hydrogels also promote wound healing in traumatic perforations generated in the tympanic membrane of guinea pigs [106]. When compared to hydrogels composed of mixtures of hyaluronic acid, gelatin and/or PEG, the ChS-PEG hydrogels significantly accelerate healing of tympanic membrane perforations. The wounds are completely closed within 1 day when ChS-PEG hydrogels are applied and take 4–6 days with the other treatments. Thus, ChS itself somehow helps accelerate the healing process.

Most likely the mechanism through which wound healing occurs in the presence of the hydrogels has to do with the material acting as a repository for growth factors, which stimulate the epithelial cells and dermal cells to regenerate the damaged tissue. When ChS-A hydrogels are incubated in distilled water supplemented with positively charged proteins such as lysozyme and aprotinin, the proteins are completely removed from solution and are stored inside the hydrogels [90]. One possible explanation is that this likely occurs through electrostatic interactions since negatively charged proteins such as bovine serum albumin do not exhibit that effect [90]. ChS is negatively charged, so it should bind positively charged proteins and repel negatively charged ones. Additionally, some proteins are more active [31] after binding to ChS polymers while others lose their activity. Thus, ChS could modulate the local environments by switching on and off various signalling molecules, which in turn drives the cell response down a certain path that is best suited for the regeneration of the tissue of interest.

An alternative scaffold composed of collagen and ChS-C has been successfully used on burn victims to regenerate wounds covering up to 95% of their body surface area [109]. Cosmetically, the regenerated tissue formed with the scaffolds more closely resembles skin tissue, and the neodermis is vascularised and contains collagen fibres in the form usually seen in control skin samples as opposed to the avascular scar tissue observed in controls. Also, like skin tissue, the collagen-ChS-C scaffold-treated wounds are more elastic, soft and pliable unlike the meshed autografting controls, which generate skin that is brittle and thick. The ChS constituent imparts the scaffold with improved mechanical and bioactive properties. The improved mechanical properties include increased elastic moduli and energy of fracture [110], more open pores [110] and decreased enzymatic degradation [111]. This leads to better cell infiltration and tissue regeneration. Coprecipitation of collagen type I with ChS-C results in improved blood compatibility of the collagen scaffolds by interfering with fibrillogenesis of collagen fibres [103, 112]. The reduced fibre formation leads to increased whole-blood clotting times and a decrease in platelet activation [103, 112]. Functionally, when the scaffolds are applied on skin wounds, the prevention of clot formation leads to better integration of the material into the native tissue. Even though these scaffolds show improved outcomes in burn patients, the neodermis that forms still lacks important parts of the skin including sweat glands and oil glands, nerves and hair follicles. Therefore, room for improvement still exists and further scientific work is needed to meet these challenges. As mentioned previously, ChS-A and ChS-C may inhibit nerve growth, while ChS-E has the opposite effect. Maybe using a different type of ChS like ChS-E could improve nerve growth and eventually lead to a tissue more closely resembling native skin. Nevertheless, this material has been very successful and this has led to the development of INTEGRA® Dermal Regeneration Template, an FDA-approved product for use in burn and reconstructive surgery.

8.8.3 Cartilage Tissue Engineering

Incorporation of ChS into scaffolds can be used to match the fixed charge density (FCD) of cartilage tissue, which will in turn mimic the streaming potentials that chondrocytes experience *in vivo* during dynamic loading. Cartilage has an FCD of 0.098–0.132 mEq/ml [113]. Thus, under dynamic loading, chondrocytes encapsulated in ChS-A-PEG hydrogels having an FCD of 0.095 mEq/ml exhibit enhanced production of GAG and collagen when compared to PEG-alone hydrogels [114, 115]. No such effects are observed under static conditions. Alternatively, in a separate study, both superficial and deep-zone chondrocytes showed enhanced collagen and proteoglycan synthesis when cultured in ChS-A-PEG hydrogels as opposed to PEG-alone hydrogels even in the absence of dynamic loading [116]. One possible explanation for the difference in outcomes is that the dynamically loaded chondrocytes were evaluated immediately after a 6-hour loading period, so the study was very short as opposed to the study performed by Hwang and co-workers, which evaluated matrix production following 3 weeks of culture. This is supported by the fact that following a short 3-day static culture of chondrocytes encapsulated in ChS-A-PEG hydrogels, a decrease in collagen synthesis was observed with an increase in ChS-A in the scaffolds, while an opposite trend was observed following a longer culture time of 7 days [86]. Therefore, in addition to mimicking the *in vivo* FCD, alternative pathways must exist that lead to an increase in proteoglycan and collagen synthesis following the incorporation of ChS in scaffolds for cartilage tissue engineering.

In addition to promoting ECM production, the presence of ChS in scaffolds leads to synthesis of tissue that resembles hyaline cartilage instead of fibrocartilage. When chondrocytes are cultured on tissue culture polystyrene surfaces, they usually undergo dedifferentiation such that they exhibit a fibroblastic phenotype. Their morphology becomes flat and spread out, and the ECM produced contains mostly collagen type I instead of the predominant collagen type II observed in articular cartilage. When bovine chondrocytes are cultured on ChS-A-containing membranes instead, the cells remain rounded and the secreted variant of collagen is mostly type II [87]. An increase in collagen type II and aggrecan expression, as well as very low expression of collagen type X and collagen type I, was also observed when rat articular chondrocytes were cultured on membranes containing ChS-C [117]. Additionally, when cultured in the presence of collagen type I or collagen type II, canine chondrocytes produce better hyaline cartilage on collagen type II matrices [118]. Therefore, incorporating ECM molecules present in cartilage tissue into tissue engineering scaffolds guides the seeded cells or infiltrating cells to replace the matrix with tissue that more closely resembles native cartilage. In effect, the regenerated tissue should last longer than temporary fibrocartilage produced by conventional methods such as microfracture.

8.9 Conclusion

The different variants of ChS can exhibit differences in biological activity towards the same type of cell. Alternatively, the same variant of ChS can have a different and sometimes opposite effect on different types of cells. A lot of work has been done and is under way to elucidate the mechanisms behind these effects of ChS. Thus far, the results show that ChS is able to change intracellular signalling pathways by binding to and presenting growth factors to cells. ChS has also been shown to interact directly with the CD-44 receptor expressed on the surface of some cells, consequently altering intracellular signalling. The GAG has shown beneficial effects for treating arthritic conditions and is also able to improve the biological activity of biomaterials for applications such as heart valve engineering, wound healing and cartilage regeneration.

References

1. P.A. Mourao, *Arthritis and Rheumatism*, 1988, **31**, 8, 1028.
2. D.C. Seldin, N. Seno, K.F. Austen and R.L. Stevens, *Analytical Biochemistry*, 1984, **141**, 1, 291.
3. N. Volpi and P. Tarugi, *Journal of Biochemistry*, 1999, **125**, 2, 297.
4. N. Volpi, M. Petrini, A. Conte, P. Valentini, T. Venturelli, L. Bolognani and G. Ronca, *Experimental Cell Research*, 1994, **215**, 1, 119.
5. N. Volpi, *Biomaterials*, 1999, **20**, 15, 1359.
6. A. Hjerpe, C.A. Antonopoulos and B. Engfeldt, *Journal of Chromatography*, 1979, **171**, 339.
7. A.D. Theocharis, D.A. Theocharis, G. De Luca, A. Hjerpe and N.K. Karamanos, *Biochimie*, 2002, **84**, 7, 667.
8. Y. Zhang, I. Schmack, D.G. Dawson, H.E. Grossniklaus, A.H. Conrad, Y. Kariya, K. Suzuki, H.F. Edelhauser and G.W. Conrad, *Investigative Ophthalmology & Visual Science*, 2006, **47**, 6, 2390.
9. V.M. Mania, A.G. Kallivokas, C. Malavaki, A.P. Asimakopoulou, J. Kanakis, A.D. Theocharis, G. Klironomos, G. Gatzounis, A. Mouzaki, E. Panagiotopoulos and N.K. Karamanos, *IUBMB Life*, 2009, **61**, 4, 447.
10. C. Ueoka, N. Kaneda, I. Okazaki, S. Nadanaka, T. Muramatsu and K. Sugahara, *Journal of Biological Chemistry*, 2000, **275**, 48, 37407.
11. M. Ishii and N. Maeda, *Glycobiology*, 2008, **18**, 8, 602.
12. N. Volpi, I. Sandri and T. Venturelli, *Carbohydrate Research*, 1995, **279**, 193.
13. G.M. Campo, A. Avenoso, S. Campo, G. Nastasi, P. Traina, A. D'Ascola and A. Calatroni, *Experimental Biology and Medicine*, 2008, **233**, 6, 741.
14. G.M. Campo, A. Avenoso, S. Campo, A. D'Ascola, P. Traina and A. Calatroni, *Osteoarthritis and Cartilage*, 2008, **16**, 12, 1474.
15. X.L. Yue, S. Lehri, P. Li, V. Barbier-Chassefiere, E. Petit, Q.F. Huang, P. Albanese, D. Barritault, J.P. Caruelle, D. Papy-Garcia and C. Morin, *Cell Death and Differentiation*, 2009, **16**, 5, 770.

16. M.A. Raymond, A. Desormeaux, P. Laplante, N. Vigneault, J.G. Filep, K. Landry, A.V. Pshezhetsky and M.J. Hebert, *The FASEB Journal*, 2004, **18**, 6, 705.

17. P. Laplante, M.A. Raymond, G. Gagnon, N. Vigneault, A.M. Sasseville, Y. Langelier, M. Bernard, Y. Raymond and M.J. Hebert, *Journal of Immunology*, 2005, **174**, 9, 5740.

18. S. Lerouge, A. Major, P.L. Girault-Lauriault, M.A. Raymond, P. Laplante, G. Soulez, F. Mwale, M.R. Wertheimer and M.J. Hebert, *Biomaterials*, 2007, **28**, 6, 1209.

19. N.J. Cartel and M. Post, *American Journal of Physiology: Lung Cellular and Molecular Physiology*, 2005, **288**, 2, L285.

20. Y. Sato, K. Nakanishi, Y. Tokita, H. Kakizawa, M. Ida, H. Maeda, F. Matsui, S. Aono, A. Saito, Y. Kuroda, M. Hayakawa, S. Kojima and A. Oohira, *Journal of Neurochemistry*, 2008, **104**, 6, 1565.

21. G.M. Campo, A. D'Ascola, A. Avenoso, S. Campo, A.M. Ferlazzo, C. Micali, L. Zanghi and A. Calatroni, *Glycoconjugate Journal*, 2004, **20**, 2, 133.

22. N. Canas, T. Valero, M. Villarroya, E. Montell, J. Verges, A.G. Garcia and M.G. Lopez, *Journal of Pharmacology and Experimental Therapeutics*, 2007, **323**, 3, 946.

23. G.M. Campo, A. Avenoso, S. Campo, A. D'Ascola, P. Traina, D. Sama and A. Calatroni, *Journal of Applied Toxicology*, 2008, **28**, 4, 509.

24. R. Albertini, G. De Luca, A. Passi, R. Moratti and P.M. Abuja, *Archives of Biochemistry and Biophysics*, 1999, **365**, 1, 143.

25. O.M. Toledo and P.A. Mourao, *Biochemical and Biophysical Research Communications*, 1979, **89**, 1, 50.

26. A.M. Tovar, D.C. Cesar, G.C. Leta and P.A. Mourao, *Arteriosclerosis, Thrombosis and Vascular Biology*, 1998, **18**, 4, 604.

27. J.E. Christner and J.R. Baker, *Analytical Biochemistry*, 1990, **184**, 2, 388.

28. L.E. Cardoso and P.A. Mourao, *Arteriosclerosis and Thrombosis: A Journal of Vascular Biology*, 1994, **14**, 1, 115.

29. M. Ida, T. Shuo, K. Hirano, Y. Tokita, K. Nakanishi, F. Matsui, S. Aono, H. Fujita, Y. Fujiwara, T. Kaji and A. Oohira, *Journal of Biological Chemistry*, 2006, **281**, 9, 5982.

30. E.L. Shipp and L.C. Hsieh-Wilson, *Chemistry & Biology*, 2007, **14**, 2, 195.

31. P. Zou, K. Zou, H. Muramatsu, K. Ichihara-Tanaka, O. Habuchi, S. Ohtake, S. Ikematsu, S. Sakuma and T. Muramatsu, *Glycobiology*, 2003, **13**, 1, 35.

32. C.H. Wu, C.S. Ko, J.W. Huang, H.J. Huang and I.M. Chu, *Journal of Materials Science: Materials in Medicine*, 2010, **21**, 2, 725.

33. E. Fthenou, A. Zafiropoulos, P. Katonis, A. Tsatsakis, N.K. Karamanos and G.N. Tzanakakis, *Journal of Cellular Biochemistry*, 2008, **103**, 6, 1866.

34. L. Li, C.H. Heldin and P. Heldin, *Journal of Biological Chemistry*, 2006, **281**, 36, 26512.

35. G.P. Dowthwaite, J.C. Bishop, S.N. Redman, I.M. Khan, P. Rooney, D.J. Evans, L. Haughton, Z. Bayram, S. Boyer, B. Thomson, M.S. Wolfe and C.W. Archer, *Journal of Cell Science*, 2004, **117**, Part 6, 889.

36. A.J. Hayes, D. Tudor, M.A. Nowell, B. Caterson and C.E. Hughes, *Journal of Histochemistry and Cytochemistry*, 2008, **56**, 2, 125.

37. S. Varghese, N.S. Hwang, A.C. Canver, P. Theprungsirikul, D.W. Lin and J. Elisseeff, *Journal of the International Society for Matrix Biology*, 2008, **27**, 1, 12.

38. S.A. Oberlender and R.S. Tuan, *Development*, 1994, **120**, 1, 177.

39. A.M. Delise and R.S. Tuan, *Developmental Dynamics*, 2002, **225**, 2, 195.

40. Y. Luo, I. Kostetskii and G.L. Radice, *Developmental Dynamics*, 2005, **232**, 2, 336.

41. M.T. Bayliss, D. Osborne, S. Woodhouse and C. Davidson, *Journal of Biological Chemistry*, 1999, **274**, 22, 15892.

42. W.E. Johnson, B. Caterson, S.M. Eisenstein, D.L. Hynds, D.M. Snow and S. Roberts, *Arthritis and Rheumatism*, 2002, **46**, 10, 2658.

43. H. Wang, Y. Katagiri, T.E. McCann, E. Unsworth, P. Goldsmith, Z.X. Yu, F. Tan, L. Santiago, E.M. Mills, Y. Wang, A J. Symes and H.M. Geller, *Journal of Cell Science*, 2008, **121**, 18, 3083.

44. J. Melrose, S. Roberts, S. Smith, J. Menage and P. Ghosh, *Spine*, 2002, **27**, 12, 1278.

45. C.I. Gama, S.E. Tully, N. Sotogaku, P.M. Clark, M. Rawat, N. Vaidehi, W.A. Goddard III, A. Nishi and L.C. Hsieh-Wilson, *Nature Chemical Biology*, 2006, **2**, 9, 467.

46. A.M. Clement, K. Sugahara and A. Faissner, *Neuroscience Letters*, 1999, **269**, 3, 125.

47. S. Nadanaka, A. Clement, K. Masayama, A. Faissner and K. Sugahara, *Journal of Biological Chemistry*, 1998, **273**, 6, 3296.

48. S.S. Deepa, Y. Umehara, S. Higashiyama, N. Itoh and K. Sugahara, *Journal of Biological Chemistry*, 2002, **277**, 46, 43707.

49. M. Asada, M. Shinomiya, M. Suzuki, E. Honda, R. Sugimoto, M. Ikekita and T. Imamura, *Biochimica et Biophysica Acta*, 2009, **1790**, 1, 40.

50. A.M. Clement, S. Nadanaka, K. Masayama, C. Mandl, K. Sugahara and A. Faissner, *Journal of Biological Chemistry*, 1998, **273**, 43, 28444.

51. Z. Li, W.S. Hou and D. Bromme, *Biochemistry*, 2000, **39**, 3, 529.

52. Z. Li, W.S. Hou, C.R. Escalante-Torres, B.D. Gelb and D. Bromme, *Journal of Biological Chemistry*, 2002, **277**, 32, 28669.

53. Z. Li, Y. Yasuda, W. Li, M. Bogyo, N. Katz, R.E. Gordon, G.B. Fields and D. Bromme, *Journal of Biological Chemistry*, 2004, **279**, 7, 5470.

54. Z. Li, M. Kienetz, M.M. Cherney, M.N. James and D. Bromme, *Journal of Molecular Biology*, 2008, **383**, 1, 78.

55. J. Selent, J. Kaleta, Z. Li, G. Lalmanach and D. Bromme, *Journal of Biological Chemistry*, 2007, **282**, 22, 16492.

56. Y. Yasuda, Z. Li, D. Greenbaum, M. Bogyo, E. Weber and D. Bromme, *Journal of Biological Chemistry*, 2004, **279**, 35, 36761.

57. D. Uebelhart, M. Malaise, R. Marcolongo, F. de Vathaire, M. Piperno, E. Mailleux, A. Fioravanti, L. Matoso and E. Vignon, *Osteoarthritis and Cartilage*, 2004, **12**, 4, 269.

58. A. Kahan, D. Uebelhart, F. De Vathaire, P.D. Delmas and J.Y. Reginster, *Arthritis and Rheumatism*, 2009, **60**, 2, 524.

59. B. Mazieres, B. Combe, A. Phan Van, J. Tondut and M. Grynfeltt, *Journal of Rheumatology*, 2001, **28**, 1, 173.

60. D. Uebelhart, E.J. Thonar, P.D. Delmas, A. Chantraine and E. Vignon, *Osteoarthritis and Cartilage*, 1998, **6**, Supplement A, 39.

61. L.B. Lazebnik and V.N. Drozdov, *Terapevticheskii Arkhiv*, 2005, **77**, 8, 64.

62. L.I. Alekseeva, L.I. Benevolenskaia, E.L. Nasonov, N.V. Chichasova and A.N. Kariakin, *Terapevticheskii Arkhiv*, 1999, **71**, 5, 51.

63. B.A. Michel, G. Stucki, D. Frey, F. De Vathaire, E. Vignon, P. Bruehlmann and D. Uebelhart, *Arthritis and Rheumatism*, 2005, **52**, 3, 779.

64. N. Volpi, *Journal of Pharmacy and Pharmacology*, 2009, **61**, 10, 1271.

65. K.D. Rainsford, *Journal of Pharmacy and Pharmacology*, 2009, **61**, 10, 1263.

66. V.C. Mow, A. Ratcliffe and A.R. Poole, *Biomaterials*, 1992, **13**, 2, 67.

67. D. Guo, L. Ding and G.A. Homandberg, *Inflammation Research*, 2009, **58**, 3, 161.

68. G.A. Homandberg, G. Davis, C. Maniglia and A. Shrikhande, *Osteoarthritis and Cartilage* 1997, **5**, 6, 450.

69. R. Pichika and G.A. Homandberg, *Inflammation Research*, 2004, **53**, 8, 405.

70. G.A. Homandberg, F. Hui, C. Wen, C. Purple, K. Bewsey, H. Koepp, K. Huch and A. Harris, *The Biochemical Journal*, 1997, **321**, 3, 751.

71. Y. Dang, A.A. Cole and G.A. Homandberg, *Osteoarthritis and Cartilage*, 2003, **11**, 7, 538.

72. K. Imada, H. Oka, D. Kawasaki, N. Miura, T. Sato and A. Ito, *Biological & Pharmaceutical Bulletin*, 2010, **33**, 3, 410.

73. G.M. Campo, A. Avenoso, S. Campo, A. D'Ascola, P. Traina, D. Sama and A. Calatroni, *Journal of Cellular Biochemistry*, 2009, **106**, 1, 83.

74. F. Ronca, L. Palmieri, P. Panicucci and G. Ronca, *Osteoarthritis Cartilage*, 1998, **6**, Supplement A, 14.

75. C.T. Bassleer, J.P. Combal, S. Bougaret and M. Malaise, *Osteoarthritis and Cartilage*, 1998, **6**, 3, 196.

76. G.M. Campo, A. Avenoso, S. Campo, A. D'Ascola, P. Traina, D. Sama and A. Calatroni, *Innate Immunity*, 2008, **14**, 4, 233.

77. G.M. Campo, A. Avenoso, S. Campo, P. Traina, A. D'Ascola and A. Calatroni, *Archives of Biochemistry and Biophysics*, 2009, **491**, 1/2, 7.

78. K. Yudoh, T. Nguyen, H. Nakamura, K. Hongo-Masuko, T. Kato and K. Nishioka, *Arthritis Research & Therapy*, 2005, **7**, 2, R380.

79. I. Strehin, Z. Nahas, K. Arora, T. Nguyen and J. Elisseeff, *Biomaterials*, 2010, **31**, 10, 2788

80. A. Baici and P. Bradamante, *Chemico-Biological Interactions*, 1984, **51**, 1, 1.

81. N. Volpi, *Chemico-Biological Interactions*, 1997, **105**, 3, 157.

82. D. Huang, *Journal of Cell Biology*, 1974, **62**, 3, 881.

83. N.B. Schwartz and A. Dorfman, *Connective Tissue Research*, 1975, **3**, 2, 115.

84. M. David-Raoudi, B. Deschrevel, S. Leclercq, P. Galera, K. Boumediene and J.P. Pujol, *Arthritis and Rheumatism*, 2009, **60**, 3, 760.

85. D.A. Wang, S. Varghese, B. Sharma, I. Strehin, S. Fermanian, J. Gorham, D.H. Fairbrother, B. Cascio and J.H. Elisseeff, *Nature Materials*, 2007, **6**, 5, 385.

86. S.J. Bryant, J.A. Arthur and K.S. Anseth, *Acta Biomaterialia*, 2005, **1**, 2, 243.

87. V.F. Sechriest, Y.J. Miao, C. Niyibizi, A. Westerhausen-Larson, H.W. Matthew, C.H. Evans, F.H. Fu and J.K. Suh, *Journal of Biomedical Materials Research*, 2000, **49**, 4, 534.

88. T.C. Flanagan, B. Wilkins, A. Black, S. Jockenhoevel, T.J. Smith and A.S. Pandit, *Biomaterials*, 2006, **27**, 10, 2233.

89. F. Wang, Z. Li, M. Khan, K. Tamama, P. Kuppusamy, W.R. Wagner, C.K. Sen and J. Guan, *Acta Biomaterialia*, 2010, **6**, 6, 1978.

90. M. Jensen, P. Birch Hansen, S. Murdan, S. Frokjaer and A.T. Florence, *European Journal of Pharmaceutical Sciences*, 2002, **15**, 2, 139.

91. S.C. Wang, B.H. Chen, L.F. Wang and J.S. Chen, *International Journal of Pharmaceutics*, 2007, **329**, 1/2, 103.

92. M.K. Yoo, K.Y. Cho, H.H. Song, Y.J. Choi, J.W. Kwon, M.K. Kim, J.H. Lee, W.R. Wee and C.S. Cho, *Drug Development and Industrial Pharmacy*, 2005, **31**, 4/5, 455.

93. S. Cai, Y. Liu, X. Zheng Shu and G.D. Prestwich, *Biomaterials*, 2005, **26**, 30, 6054.

94. Y. Liu, S. Cai, X.Z. Shu, J. Shelby and G.D. Prestwich, *Wound Repair and Regeneration*, 2007, **15**, 2, 245.

95. A. Pirouzmanesh, S. Herretes, J.M. Reyes, O. Suwan-Apichon, R.S. Chuck, D.A. Wang, J.H. Elisseeff, W.J. Stark and A. Behrens, *Archives of Ophthalmology*, 2006, **124**, 2, 210.

96. I. Strehin, W.M. Ambrose, O. Schein, A. Salahuddin and J. Elisseeff, *Journal of Cataract and Refractive Surgery*, 2009, **35**, 3, 567.

97. J.M. Reyes, S. Herretes, A. Pirouzmanesh, D.A. Wang, J.H. Elisseeff, A. Jun, P.J. McDonnell, R.S. Chuck and A. Behrens, *Investigative Ophthalmology and Visual Science*, 2005, **46**, 4, 1247.

98. S.H. Yang, P.Q. Chen, Y.F. Chen and F.H. Lin, *Artificial Organs*, 2005, **29**, 10, 806.

99. W.C. Hsu, M.H. Spilker, I.V. Yannas and P.A. Rubin, *Investigative Ophthalmology and Visual Science*, 2000, **41**, 9, 2404.

100. I.V. Yannas, J.F. Burke, D.P. Orgill and E.M. Skrabut, *Science*, 1982, **215**, 4529, 174.

101. R. Muggli and H.R. Baumgartner, *Thrombosis Research*, 1973, **3**, 6, 715.

102. L.F. Brass and H.B. Bensusan, *Journal of Clinical Investigation*, 1974, **54**, 6, 1480.

103. F.H. Silver, I.V. Yannas and E.W. Salzman, *Thrombosis Research*, 1978, **13**, 2, 267.

104. F.J. O'Brien, B.A. Harley, I.V. Yannas and L.J. Gibson, *Biomaterials*, 2005, **26**, 4, 433.

105. K.R. Kirker, Y. Luo, S.E. Morris, J. Shelby and G.D. Prestwich, *Journal of Burn Care & Rehabilitation*, 2004, **25**, 3, 276.

106. A.H. Park, C.W. Hughes, A. Jackson, L. Hunter, L. McGill, S.E. Simonsen, S.C. Alder, X.Z. Shu and G.D. Prestwich, *Otolaryngology: Head and Neck Surgery*, 2006, **135**, 6, 877.

107. M.E. Gilbert, K.R. Kirker, S.D. Gray, P.D. Ward, J.G. Szakacs, G.D. Prestwich and R.R. Orlandi, *The Laryngoscope*, 2004, **114**, 8, 1406.

108. K.R. Kirker, Y. Luo, J.H. Nielson, J. Shelby and G.D. Prestwich, *Biomaterials*, 2002, **23**, 17, 3661.

109. J.F. Burke, I.V. Yannas, W.C. Quinby, Jr., C.C. Bondoc and W.K. Jung, *Annals of Surgery*, 1981, **194**, 4, 413.

110. I.V. Yannas and J.F. Burke, *Journal of Biomedical Materials Research*, 1980, **14**, 1, 65.

111. I.V. Yannas, J.F. Burke, C. Huang and P.L. Gordon, *Polymer Reprints*, 1975, **16**, 2, 209.

112. F.H. Silver, I.V. Yannas and E.W. Salzman, *Journal of Biomedical Materials Research*, 1979, **13**, 5, 701.

113. C.C. Wang, X.E. Guo, D. Sun, V.C. Mow, G.A. Ateshian and C.T. Hung, *Biorheology*, 2002, **39**, 1/2, 11.

114. I. Villanueva, S.K. Gladem, J. Kessler and S.J. Bryant, *Journal of the International Society for Matrix Biology*, 2010, **29**, 1, 51.

115. I. Villanueva, B.J. Klement, D. von Deutsch and S.J. Bryant, *Biotechnology and Bioengineering*, 2009, **102**, 4, 1242.

116. N.S. Hwang, S. Varghese, H.J. Lee, P. Theprungsirikul, A. Canver, B. Sharma and J. Elisseeff, *FEBS Letters*, 2007, **581**, 22, 4172.

117. Y.L. Chen, H.P. Lee, H.Y. Chan, L.Y. Sung, H.C. Chen and Y.C. Hu, *Biomaterials*, 2007, **28**, 14, 2294.

118. S. Nehrer, H.A. Breinan, A. Ramappa, S. Shortkroff, G. Young, T. Minas, C.B. Sledge, I.V. Yannas and M. Spector, *Journal of Biomedical Materials Research*, 1997, **38**, 2, 95.

9 Micro- and Nano-particles Based on Polysaccharides for Drug Release Applications

Cătălina Anişoara Peptu and Marcel Popa

9.1 Introduction

The enhancement of the drug treatment efficacy and the elimination of the undesired side effects caused by drug overdoses or by their aggressiveness to both normal and affected cells are nowadays an important issue of chemotherapy. Two general directions have been developed to surpass these major drawbacks:

- Synthesis of new drugs with higher specificity and minimum side effects
- Development of controlled/sustained drug delivery systems based on polymers

In the case of polymer-based drug delivery systems, there are several mechanisms that ensure the desired effect:

- Diffusion through hydrogels (films or micro/nanoparticles) or membranes (micro/nanocapsules)
- Erosion of polymer matrix containing the drug (micro/nanoparticles)
- Hydrolysis of chemical bonds established between the drug and polymeric supports (drug-polymer conjugates)

The above mechanisms are mainly controlled by zero- or first-order release kinetics. Evidently, the association of the active principle with polymers is not randomly done, but certain restrictions are being imposed: the polymer is required to be biocompatible, biodegradable, to present high reactivity towards the drug under mild conditions (temperature, and so on) and not to form toxic degradation products. The administration paths and the nature of the disease that needs to be treated are also very important.

Natural polymers are excellent candidates for drug delivery systems due to their abilities to meet the previously mentioned requirements. Among the natural polymers, polysaccharides occupy a special position in the top of research interests due to their undoubtable qualities related to biocompatibility, biodegradability, hydrophilicity, nontoxicity, adhesivity and so on.

Different formulations of drug delivery systems are already there on the market, such as films, powders, tablets, insert, implants, conjugates, micro/nanoparticles (including micro/nanocapsules), and for different administration paths, such as parenteral, intravenous, oral, transdermal, rectal, vaginal and respiratory. In some cases, the particulated systems are guided from the exterior towards the targets (magnetic particles), with an evident higher efficiency of the treatment.

The products of micro/nanotechnology have proven to be capable of revolutionising modern medicine, as shown by the numerous results reported in the last decade in scientific literature. The drug delivery field is the most evident beneficiary of this research technology as the micro/nanoparticles are versatile regarding the targeting of the affected tissues, access at cellular level and the controlled release of the biological active principle.

So far, the classic way to present the research results in this area was to address the formulation/preparation techniques or the material involved (polymers, drugs, and so on). The present work proposes a different approach, relating the polysaccharide-based micro/nanoparticulated systems to the treatment of various classes of diseases and to medical diagnostics.

Although polysaccharides have been intensively studied with respect to their use in micro/nanotechnology for obtaining controlled drug delivery systems, it is clear that the properties/qualities of these natural polymers are not entirely exploited; researchers are engaged in ongoing research to solve various issues arising in human medicine by using micro/nanoparticles based on polysaccharides.

It is hoped that specialists in pharmaceutical industry, academic community, graduate/undergraduate students in the field of medical bioengineering, chemical engineering and pharmaceutical science and all other people interested in this fascinating research area will find this review useful.

Further on, the authors will present the most important types of nano/microparticles based on polysaccharides suitable for drug delivery, for different classes of medical applications:

- Ophthalmic applications
- Cancer applications

- Infectious diseases
- Diabetes therapy
- Respiratory diseases
- Diagnostics

9.2 Applications of Polysaccharide Nano/Microparticles in Ophthalmologic Therapy

Nanomedicine offers potentially safe and successful treatment regimens for ocular disorders especially when polysaccharides are used, because micro/nanoparticles based on this kind of natural polymers are able to persist at the surface of the eye to give prolonged drug delivery. Many efforts have been focused on increasing the corneal penetration of drugs with the final goal of improving the therapeutic outcome of treatments of different ocular diseases. These attempts involve the use of colloidal drug delivery systems, such as liposomes, biodegradable nanoparticles and nanocapsules. One of the main disadvantages of this class of materials is represented by the short residence time of the colloidal carrier systems in the ocular mucosa, an important aspect for the therapy of extraocular diseases, such as keratoconjunctivitis sicca or dry eye disease. Consequently, the researchers are looking at designing a mucoadhesive carrier system with improved drug delivery properties to the ocular surface, which would be a promising step towards the management of external ocular diseases.

Chitosan (CS) is a polysaccharide with characteristic properties such as adhesivity, biodegradability and low toxicity. It has been shown to enhance drug penetration across not only cell monolayers, but also corneal epithelia [1]. CS by itself is not suitable for permeabilising the cornea because this polymer is insoluble in the tear fluid, and hence inactive at the physiological pH of 7.4. The research group of Zambito and Di Colo have synthesised *N*-trimethylchitosan (TMC), a polycationic derivative of CS, soluble in the tear fluid at physiological pH, by partially quaternising the primary amino group of CS; TMC significantly increased *in vivo* transcorneal absorption rate of ofloxacin in rabbits [2] and the permeability of the highly lipophilic dexamethasone across the excised rabbit cornea [3].

In order to increase the hydrophobicity of CS, Yuan and co-workers [4] have prepared cholesterol-modified CS, and loaded cyclosporine A (CyA) into the CS and cholesterol 3-hemisuccinate conjugate (CS-CH) nanoparticles by a self-aggregated method; the results demonstrated good retention ability. CS-CH conjugates were prepared by using Lee's modified method [5] and were further used for the preparation of self-aggregates using the diafiltration method. The CS-CH micelles obtained were spherical in shape and between 50 and 200 nm. From the *in vivo* test point of view, the CS-CH nanoparticles showed a good spreading over the entire precorneal area immediately after the topical administration, then part of the suspension drained into the lachrymal duct and from there into the lachrymal sac. To evaluate the ability of nanoparticles as drug carriers, CyA was incorporated as a hydrophobic model drug. The drug-loaded CS-CH nanoparticles were prepared by dialysing the dimethyl sulfoxide (DMSO)/water solution of CyA and CS-CH against physiological saline. The CyA was released in a biphasic way, characterised by an initial rapid release period followed by a slower release phase. The burst effect was observed in the first 4 hours, in which 60% of the initial drug was released from nanoparticles. After this initial effect, CyA was released in a continuous way for up to 48 hours, reaching percentage of cumulative release close to 95%. CS-CH nanoparticulated aggregates have been proved to be a potential carrier of hydrophobic drug for the treatment of external ocular diseases.

Again, the same group of authors have studied the possibility of using rapamycin (RAPA)-loaded CS/poly(lactic acid) (PLA) nanoparticles for immunosuppression in corneal transplantation [6]. In corneal transplantation, which is one of the most common allografts performed, one of the possible events is graft rejection. Topical drugs have not been found efficacious in suppressing graft rejection, which is usually due to their insolubility and the difficulty in achieving clinically effective drug concentrations in the cornea and anterior chamber; however, some novel immunosuppressive agents such as RAPA have been found to be effective. The CS/cholesterol-PLA nanoparticles were prepared by nanoprecipitation, and the conjugate between CS and cholesterol was used in order to create an amphiphilic compound; thus the nanoparticles can self-aggregate, with a hydrophobic microenvironment inside and with a certain affinity for RAPA. The addition of PLA significantly improved the drug-loading efficiency of the nanoparticles (compared to the nanoparticles loaded with CyA presented above) due to the stronger hydrophobic interaction between RAPA and PLA. The incorporation of PLA into CS-CH particles leads to a much slower sustained release of RAPA, and the hydrophobicity of PLA restricts the permeation of water into the particles, preventing the diffusion of RAPA out of particles.

The ocular distribution of CS-CH/PLA nanoparticles (the nanoparticles were radiolabelled with ^{99m}technetium (^{99m}Tc)), evaluated by single photon emission computed tomography image analysis and scintillation counter, showed that nanoparticles congregated at the conjunctival sac; only a small part of the applied dose was drained into the lachrymal sac, while the major fraction of the instilled dose remained on the ocular surface where it separated into two parts, one on the inner canthus and the other on the outer canthus. The animals treated with CS-CH/PLA nanoparticles showed higher remaining radioactivities on corneal and conjunctival epithelia than those treated with a suspension of RAPA. The nanoparticles proved to have an excellent immunosuppressive effect compared to RAPA eye drops.

Motwani and co-workers [7] have investigated mucoadhesive CS-sodium alginate nanoparticles as a new vehicle for the prolonged topical ophthalmic delivery of the antibiotic gatifloxacin. The main argument for using the combination of two different polysaccharides (CS as polycation and alginate as polyanion) was that the interaction between biodegradable cationic and anionic biopolymers leads to the formation of polyionic hydrogels with good characteristics for drug entrapment and delivery. The authors obtained nanoparticles with dimensions between 62 and 193 nm and loaded with gatifloxacin (a fourth-generation fluoroquinolone with a broad-spectrum antibacterial activity) used in the treatment of ocular infections, resulting in a successful formulation in the form of CS-alginate nanoreservoir system. This new formulation is intended to be a viable alternative to conventional eye drops due to its ability to sustain the drug release and for its ease of administration because of reduced dosing frequency.

Tuovinen and co-workers [8] prepared calcein-loaded starch acetate microparticles by using modified water-in-oil-in-water technique for targeting the retinal pigment epithelium (RPE) of the eye. The cellular uptake of microparticles and degradation of starch acetate by cultured human RPE cell line (D407) were analysed. The prepared microparticles showed an encapsulation efficiency for calcein of 18%; they present a spherical shape with a smooth surface and their size is between 5 and 22 μm. The cellular uptake was analysed by using confocal microscopy and flow cytometry and it was shown that certain percentages of RPE cells took up the particles; the viability of RPE cells was not affected by calcein loading.

Polysaccharides are often used in order to improve the properties of other controlled drug release entities. For example, Li and co-workers [9] have investigated the *in vitro* and *in vivo* properties of liposomes coated with low-molecular-weight chitosan (LCH), and their potential use in ocular drug delivery.

The coating of liposomes by CS modified the liposome surface charge and increased their particle size, but the drug (diclofenac sodium salt) encapsulation was not affected. After coating, the liposome displayed a prolonged *in vitro* drug release profile. The ocular bioadhesion characteristics were evaluated *in vivo* in rabbits by precorneal retention tests, and LCH-coated liposome displayed a significantly prolonged retention compared with noncoated liposome or drug solution. The CS-coated liposomes showed a potential penetration-enhancing effect for transcorneal delivery of the drug. The authors also performed an ocular tolerance study, and found no irritation or toxicity being caused by continued administration of LCH-coated liposome during the

7 days of analysis. In conclusion, the coating of liposomes by CS significantly modified the properties of liposomes bringing numerous advantages for ocular drug delivery.

The authors of this chapter have contributed to the development of micro/nanoparticles for ophthalmic drug release by studying the possibility of using the combination of polysaccharides (CS) with proteins (gelatin) for preparing micro/nanoparticles for the controlled release of adrenalin (a hydrosoluble drug that is usually applied for maintaining mydriasis during eye surgery) [10]. Particles were prepared by an original method consisting of a two-step crosslinking (ionic and covalent) technique applied in a reverse emulsion system. Sodium sulfate was used as a nontoxic ionic crosslinker and small amounts of glutaraldehyde were used as a covalent crosslinker. The prepared particles were loaded with adrenalin and tested from the point of view of biocompatibility, toxicity, swelling, loading and release characteristics. The adrenalin-loaded microparticles were tested on human eye to determine the adhesiveness, and it was shown that there was no elimination by tear flow.

9.3 Applications of Polysaccharide Nano/Microparticles in Cancer Therapy

Micro/nanotechnology generally is applied to cancer in two main areas: the development of vectors/carriers, such as micro/nanoparticles, which can be loaded with drugs or imaging agents and then targeted to tumours, and nanosensor devices for detecting the biological signatures of cancer [11].

Polysaccharides have been found useful in both areas of applications in cancer therapy. They have been used as native or modified polysaccharides, as combination of two or more polysaccharides and as combinations of polysaccharides with other natural or synthetic polymers.

Novel hybrid nanoparticles consisting of hyaluronic acid (HA) and iron oxide and with an average diameter between 80 and 160 nm were synthesised and characterised for the first time by Kumar and co-workers [12], and it was demonstrated that nanoparticles are capable of delivering peptides to cells with almost full efficiency. The HA nanoparticles (HA-NP) were prepared by self-assembling microemulsion techniques; HA and a dihydrazide (Sigma-Aldrich) were mixed, homogenised and incubated with the crosslinking reagent in an aqueous solution. For the preparation of the hybrid Fe_2O_3-HA particles, the HA-NP were interacted with colloidal Fe_2O_3 fluid and then a certain amount of acid solution was added to the functionalised solution under stirring. The final mixture was left for 24 hours to allow the formation of nanoparticles. To test the potential of these particles and to deliver proteins for cancer therapy, atrial natriuretic peptide was encapsulated within the HA-NP. The human alveolar type II epithelial cell line A549 (adenocarcinomic human alveolar basal epithelial cells) and HEK 293 cells (human embryonic kidney 293 cells; ATCC) were used for nanoparticle-mediated peptide delivery, the reason for using these cells being that such cells might have an important role in preventing metastasis. The results demonstrate that both HA and HA-Fe_2O_3 hybrid nanoparticles are capable of delivering the peptides to almost all (100%) cells and that the particles delivered the peptides to the nucleus of cells.

HA-NP, formed by the self-assembly of hydrophobically modified HA derivatives, were prepared to investigate their physicochemical characteristics and fates in tumour-bearing mice after systemic administration [13]. For this, amphiphilic HA-5b-cholanic acid conjugates (HA-CA conjugates) were synthesised by conjugation of hydrophobic bile acid (5b-cholanic acid) to the hydrophilic HA backbone through amide formation. For the preparation of HA-NP, amphiphilic HA-CA conjugates were dissolved in a phosphate-buffered saline (PBS) and the solution was sonicated three times using a probe-type sonicator and then the solution was filtered through a membrane filter. In order to monitor receptor-mediated cellular uptake of HA-NP, Cy5.5 (dye)-labelled nanoparticles were incubated with SCC7 cancer cells (squamous cell carcinoma) overexpressing CD44 (receptor) or normal fibroblast cells (CV-1) for 90 minutes. In the confocal microscopy studies, strong fluorescent signals were detected in the cytoplasm for all tested nanoparticles, indicating that HA-NP are readily taken up by SCC7 cancer cells. In this study, the authors have shown that HA-NP selectively target the tumour tissue, passively accumulate via the enhanced permeation and retention effect and actively target CD44 by the strong receptor binding affinity of HA to CD44. Self-assembled HA-NP could be used as drug carriers and/or imaging agents for cancer therapy.

Duceppe and Tabrizian [14] have prepared nanoparticles based on two natural polysaccharides (LCH and HA) which are at the same time polyelectrolytes and studied their potential as carriers for gene delivery, especially as nonviral nanocarriers with high transfection efficiency. The microparticles were obtained by a simple method consisting of simultaneous filtering of the two polymer solutions, resulting in particles between 140 and 160 nm. The transfection efficiency of the system was determined by fluorescent microscopy and fluorescence-activated cell sorting analysis using HEK 293T cells; all studied polyplexes presented high transfection efficiency. Cell transfection conditions (incubation time, pH and concentration of DNA) play a major role in the efficiency of transfection and the system has shown great potential as specific cell targeting system with cells that have a high occurrence of CD44 receptors, such as cancer cells, due to the presence of HA in its formulation.

The HA-coated poly(butyl cyanoacrylate) (PBCA) nanoparticles were prepared by a radical polymerisation technique of butyl cyanoacrylate initiated by cerium ions in the presence of HA [15]. The nanoparticles were designed as an anticancer drug delivery carrier using paclitaxel (PTX) as an anticancer model drug. The cellular uptake of the noncoated PBCA nanoparticles was compared with the uptake of the HA-coated PBCA nanoparticles in Sarcoma-180 (S-180) and HEK 293 cells. For the two types of nanoparticles, the anticancer efficiency of PTX-loaded nanoparticles was investigated in S-180-bearing mice. The prepared particles present a spherical shape, size between 200 and 400 nm and a negative zeta potential which is not influenced by the addition of HA. The PTX release studies showed that by adding more HA, the release will be slower and the initial burst effect will be reduced. Uptake of HA-PBCA nanoparticles by S-180 cells was significantly higher than the uptake of unmodified nanoparticles and the modification of HA at the nanoparticles' surface increased its affinity towards S-180 cells, which facilitates the targeting effect of PTX-loaded HA-PBCA nanoparticles to the tumour.

CS nanoparticles containing the mitotic inhibitor drug PTX were prepared by a solvent evaporation and emulsification crosslinking method with trisodium citrate as the crosslinking agent [16]. The nanoparticles presented a fine spherical shape with smooth surfaces, without aggregation and with an average size of 116 ± 15 nm. The cell uptake efficiency of CS nanoparticles by cancer cells was analysed by using confocal fluorescent microscopy. Plain CS nanoparticles were shown to induce A2780 cancer cell (human ovarian carcinoma) apoptosis, which was further enhanced by the cytotoxicity of PTX; CS nanoparticles have been shown to function as not only potential anticancer drug carriers but also anticancer agents.

The antitumoral effect of ionically crosslinked CS nanoparticles on BEL7402 cells in nude mice was investigated by Qi and co-workers [17]. The nanoparticles with size between 40 and 100 nm were prepared by co-acervation between positively charged CS and negatively charged sodium tripolyphosphate (TPP). Human hepatoma cells (BEL7402) were implanted subcutaneously into the right flank of male athymic BALB/c nude mice. Each treatment group received once daily oral administration of CS or CS nanoparticles with different particle size (40, 70 and 100 nm). Nanoparticles with small particle size and high surface charge exhibited a strong dose- and time-dependent antitumour activity on human hepatoma BEL7402 cells *in vitro*. CS nanoparticles showed higher tumour-inhibitory effects than CS *in vivo*; the particle size of CS nanoparticles had a great influence on antitumour activity. Consequently, CS nanoparticles ionically stabilised could be a potential candidate in the treatment of hepatocellular carcinoma.

Recently, Bilensoy and co-workers [18] compared three types of nanoparticles based on the following polymers and combinations from the point of view of chemotherapeutical activity of the encapsulated mitomycin C (MMC): CS, CS and polycaprolactone (PCL) and PCL with polylysine (PLL). The nanoparticles were prepared by different techniques as follows: CS particles were prepared by using ionotropic gelation technique with TPP as crosslinker; CS/PLL-coated poly-E-caprolactone nanoparticles were obtained by coating the PCL nanoparticles already formed by nanoprecipitation; nanoparticle diameters varied between 180 and 340 nm depending on the polymer used for preparation (the smallest particles were those based on PLL-PCL) and zeta potential values demonstrated positive charge. The highest encapsulation efficiency (about 21%) was obtained for CS and PLL-PCL nanoparticles. CS-PCL was the most efficient formulation for the uptake of both hydrophilic and hydrophobic fluorescent markers for all cell lines used in this study (mouse bladder cancer (MB49), normal epithelial (G/G) cell lines). A very important conclusion is that MMC-loaded polysaccharide-containing nanoparticles showed higher toxicity against cancer cells compared to MMC solution.

In their efforts to develop a versatile carrier system for gene transfer into mammalian cells that meets the demands of safety, biocompatibility and enhanced stability, Gupta and Gupta [19] proposed some hydrogel nanoparticles of pullulan with a hydrophilic core that can encapsulate water-soluble materials like DNA for intracellular delivery [19]. Void pullulan nanoparticles were prepared inside the inner aqueous core of the reverse micelles formed by dissolving the surfactant sodium bis(2-ethylhexyl) sulfosuccinate in *n*-hexane followed by covalent crosslinking with glutaraldehyde. The average size was determined by measuring the size of around 150 particles; the particles as seen by transmission electron photographs were found to be monodispersed with diameters of 45 ± 80 nm. The viability of COS-7 (cell line derived from kidney cells of the African green monkey) and HEK cells after incubation with pullulan nanoparticles was determined with the 3-(4,5-dimethylthiazol-2-yl)-2,5-diphenyltetrazolium bromide (MTT) assay [colorimetric assays for measuring the activity of enzymes that reduce MTT to formazan dyes, giving a purple colour; a main application allows assessment of the viability (cell counting) and the proliferation of cells (cell culture assays)]. It can also be used to determine the cytotoxicity of potential medicinal agents and toxic materials, since those agents would stimulate or inhibit cell viability and growth. The cytotoxicity of nanoparticles was found to be increased in relation to increased concentration of pullulan. After 24 hours, the COS-7 cells incubated with pullulan nanoparticles showed a viability of around 80% at a concentration of 20.0 mg/ml relative to control cells; the maximum transfection occurred at 250 μg/ml pullulan. Then, the transfection efficiency of the pullulan nanoparticles was evaluated in COS-7 and HEK 293 cells and compared with the commercial reagent Lipofectamine 2000. The efficacy of transfection *in vitro* in HEK and COS cells demonstrated cell type dependence, with COS cells found to have higher gene expression compared to HEK 293 cells. The β-gal expression (β-galactosidase, also called beta-gal or β-gal, is a hydrolase enzyme that catalyses the hydrolysis of β-galactosides into monosaccharides) in COS-7 cells by pullulan nanoparticle-mediated gene transfer was found to be at a level comparable to that with commercially available Lipofectamine 2000. The cytotoxicity of the pullulan nanoparticles as assessed by MTT assay shows that these nanoparticles could be used as gene carriers without any apparent cytotoxicity.

Photodynamic therapy (PDT) is a treatment that uses special drugs, called photosensitising agents, along with light to kill cancer cells. The drugs work only after they have been activated or 'turned on' by certain kinds of light. PDT is also called photoradiation therapy, phototherapy, or photochemotherapy. It was first used to treat cancer over 100 years ago. Depending on the part of the body being treated, the photosensitising agent is either injected into the bloodstream through a vein or put on the skin. Over a certain amount of time the drug is absorbed by the cancer cells. Then light is applied to the area to be treated. The light causes the drug to react with oxygen, which forms a chemical that kills the cancer cells. PDT may also work by destroying the blood vessels that feed the cancer cells and by alerting the immune system to attack the cancer [20].

The period of time between when the drug is given and when the light is applied is called the drug-to-light interval. It can be anywhere from a couple of hours to a couple of days and depends on the drug used. PDT is a rapidly growing area of medical treatment. The diseases that can be successfully treated by PDT include skin cancer, brain tumours, tumours under the surface of the skin and tumours located on the lining of internal organs. PDT is less invasive than other therapies and often produces a better cosmetic outcome, with little or no scarring. PDT involves the use of light-activated dyes (photosensitisers) that localise in target cells (e.g., in tumours) but not in normal, healthy cells [21]. Photosensitisers utilise energy from treatment light to produce a cytotoxic oxygen species which kills cancerous or diseased cells. This toxic oxygen species is not a radical but is actually an excited state of oxygen. The excited state is more reactive than usual oxygen, and the atoms are in a different quantum spin state than is normally the case. However, there are some difficulties with the use of a photosensitiser, such as water insolubility and low selectivity for the target site, which often limit clinical applications in PDT. Self-quenching polysaccharide-based nanogels synthesised from pullulan/folate-pheophorbide-a (Pheo-A) conjugates were investigated by Bae and Na [22] for their potential to reduce the phototoxicity of photosensitisers in normal tissue and to enhance the efficacy of tumour treatment. The pullulan/folate conjugate was easily synthesised by one-step chemistry and spontaneously formed a self-organised nanogel in an aqueous environment. Self-organised nanogels from pullulan/folate conjugate (PFP samples) were prepared by a dialysis method, which prevented the uncontrolled and rapid precipitation of polymer during the self-assembly process. The size and size distribution of pullulan/folate conjugate and PFP nanogels in PBS were measured by dynamic light scattering. The size of pullulan/folate nanogels was higher than 500 nm, but the mean diameter of nanogels prepared from PFP was approximately 170 nm, which is smaller than the critical size required for recognition by the reticuloendothelial system. The intracellular localisation and distribution of Pheo-A in the PFP nanogel as a function of time were studied and were found to be different from that of free Pheo-A; the results indicate that the nanogel was retained in cellular compartments and Pheo-A remained attached to PFP. These studies, the fluorescence studies and the photocytotoxicity of nanogels lead to the conclusion that self-quenching PFP nanogels may contribute to the development of a new generation of carriers for enhanced PDT treatment of tumours.

Pullulan was hydrophobised to pullulan acetate (PA; by replacing the hydroxyl groups of the glucose unit with acetate groups to produce a hydrophobically modified pullulan) and used in order to prepare nanoparticles by dialysis method for epirubicin (EPI) (antitumoral model drug) encapsulation, reduce its toxic effects against normal cells and increase its therapeutic activity [23]. The particles are between 185 and 423 nm and present a round shape and a smooth surface. Exposure of KB cells (a cell line derived from a human carcinoma of the nasopharynx, used as an assay for antineoplastic agents) to free EPI and EPI-loaded PA nanoparticles revealed that in both cases the drug gains access to the cell nucleus, but the route and the kinetics of uptake are different. A higher cytotoxicity was found for EPI-loaded PA nanoparticles in comparison with free EPI, which could be ascribed to the internalisation of the nanoparticles by the cells as shown in the confocal microscopy images.

9.4 Applications of Polysaccharide Nano/Microparticles in Infectious Diseases Therapy

Infectious disease is one of the major threats to global health care. Current antibiotic therapy is too long and burdensome for most infected patients, resulting in many patients not completing the required drug regimen, which will finally lead to multiple drug-resistant infections. Advanced drug delivery techniques can be utilised to decrease the length of treatment time and to simplify therapy, resulting in increased compliance and effective treatment of disease.

Polysaccharides are intensely used for creating carriers that will prevent the side effects of classical formulations of drugs. Liu and co-workers [24] prepared amoxicillin (α-amino-hydroxybenzylpenicillin - a semisynthetic, orally absorbed, broad-spectrum antibiotic) mucoadhesive microspheres using ethylcellulose as the matrix and carbopol 934P as a mucoadhesive polymer for the potential use in treating gastric and duodenal ulcers, associated with *Helicobacter pylori*. There are two main reasons for the incomplete eradication of *H. pylori* during the administration of amoxicillin by the classical way: the first reason is the short residence time of antimicrobial agents in the stomach, due to which effective antimicrobial concentration cannot be achieved in the gastric mucous layer or epithelial cell surfaces where *H. pylori* exists [25]; and the second reason could be the degradation of amoxicillin in gastric acid [26].

The microparticles with a round shape and a smooth surface were prepared by an emulsification/evaporation method and their size varied between 400 and 1000 μm. The *in vitro* release test revealed that about 90% of amoxicillin was released within 4 hours at pH 1.0. The *in vivo* evaluation of mucoadhesiveness showed that almost 64% of microspheres still remained in the rat stomach 4 hours after administration, which meant that amoxicillin could be released almost completely when most microspheres still resided in the stomach. One of the most important conclusions of the authors was that mucoadhesive microspheres can protect amoxicillin from being degraded.

In order to localise antibiotics at the *H. pylori* infection site on the gastric epithelium for improving the efficacy of anti-*H. pylori* agents, Lin and co-workers [27] have developed a novel cationic nanoparticle delivery system based on the natural polysaccharides CS and heparin. The nanoparticles with an average size between 100 and 300 nm were prepared by using a simple ionic gelation technique. The reason for using CS and heparin is that in the pH range 1.2–6.5, CS and heparin are ionised and can form polyelectrolyte complexes, which result in a matrix structure with a spherical shape. Outside this pH range (e.g., pH 7.0), the nanoparticles become unstable and break apart, because at pH 7.0, CS is deprotonated, causing the collapse of the nanoparticles. The pH-sensitive nanoparticles that are stable at pH 1.2–2.5 (simulating gastric acid) can protect drugs from destruction by gastric acids. The *in vivo* results clearly indicate that the nanoparticles localised to the spaces in the gastric villi of the mouse model; the nanoparticles also infiltrated the cell-cell junctions and interacted locally with *H. pylori* infection sites in the intercellular spaces.

Malaria is a mosquito-borne infectious disease caused by a eukaryotic protist of the genus *Plasmodium*. Each year, there are approximately 350–500 million cases of malaria, killing between one and three million people. Carrier systems, such as nanoparticles, present a possible approach to improve the delivery properties of antisense oligodeoxynucleotides (ODN) (an interesting alternative to traditional antimalarial drugs) and the complexation of ODN into nanoparticles enhances stability against enzymatic degradation [28]. Foger and co-workers [29] have incorporated 30mer antisense ODN against malarial topoisomerase II gene in CS nanoparticles, in order to study the possibility of applying nanotechnology against *Plasmodium falciparum* (the most virulent of the four human malarial parasites).

Phosphorothioate ODN were complexed with depolymerised CS by mixing aliquots of CS, dissolved in acetic acid (pH 5.5 adjusted with 1 M NaOH), with solutions of ODN, dissolved in double-distilled sterile filtered water. The hydrodynamic diameter of the nanoparticles measured by photon correlation spectroscopy was around 50 nm; the amount of CS and also the sequence of the ODN did not practically affect the mean diameter of the particles.

The nanoparticles obtained from the complex between CS and phosphorothioate ODN were analysed from many points of view and the most interesting conclusions for the subject described in this chapter are the following:

- Nanoparticles presented a more pronounced sequence-specific antisense effect compared to free ODN due to the sustained release of ODN from CS nanoparticles.
- Nanoparticles protected the oligonucleotides against nuclease degradation - a major requirement for *in vivo* use of antisense technology.
- CS nanoparticles may not harm erythrocytes under *in vivo* conditions. This represents one of the conditions for *in vivo* use of nanoparticles against malaria.

Polysaccharides were found to be useful in the treatment of hepatitis B, which is an infectious illness caused by hepatitis B virus (HBV). Failure of immunisation against HBV infection is due to the following factors: improper storage or administration of the vaccine, advanced age of the individual, chronic liver disease, immunosuppression and resistance of the virus. Alginate-coated CS nanoparticles were investigated for the first time by Borges and co-workers [30] as adjuvant for subcutaneous vaccination with the recombinant hepatitis B surface antigen (HBsAg). The nanoparticles were obtained by an ionic crosslinking using sodium sulfate as crosslinker, and loaded with HBsAg by incubating a solution of HBsAg with a suspension of particles in phosphate buffer of pH 7.4. Alginate-coated nanoparticles were prepared by mixing equal volumes of the HBsAg-loaded nanoparticle suspension and a solution of sodium alginate with magnetic stirring. The major advantage of this particular system is that the antigen is encapsulated under nonstressful conditions and, consequently, the biological properties of the antigen are expected to remain intact. The adjuvant effect of the alginate-coated CS nanoparticles was demonstrated in this study by the production of high anti-HBsAg IgG titres with a predominance of Th2-type antibodies (IgG1NIgG2a). All studied to the CS formulations studied that contain CpGODN have been shown to have a great potential for the improvement of the existing HBV vaccines, being useful for the treatment of chronic hepatitis B, where Th-1 cellular immune response induction is required.

9.5 Applications of Polysaccharide Nano/Microparticles in Diabetes Therapy

Diabetes is a group of metabolic diseases in which a person has high blood sugar, either because the body does not produce enough insulin or because cells do not respond to the insulin that is produced. One of the most important difficulties that are met in diabetes treatment is represented by the poor gastrointestinal uptake of poorly absorbable drugs such as insulin. Polysaccharides (like those that are contained in black tea) may benefit people with diabetes because they help retard absorption of glucose, but also polysaccharides like starch, CS, alginates and glucomannan are intensively studied for being used as carriers for diabetic drugs. However, the oral delivery of many diabetic drugs, like insulin which is a peptide, still remains an unresolved challenge because of the large size, hydrophilicity and instability. The potential utility of CS, related to its mucoadhesive properties and to the fact that it can facilitate drug absorption by localising drug concentration around absorptive cells [31], led Ma and co-workers [32] to use CS nanoparticles, prepared by an ionotropic gelation method, as a carrier for the oral delivery of insulin. The sizes of the particles are in the nanometric range (129–3030 nm) and they have a positive zeta potential with values between 13 and 48. Test samples were administered intragastrically to male Wistar rats with blood glucose levels above 250 mg/dl and a marked and sustained lowering of the serum glucose levels was observed in the diabetic rats 10 hours after the oral administration of the insulin-loaded nanoparticles; the samples were demonstrated to be effective in lowering the serum glucose levels of streptozotocin-induced diabetic rats; besides, one of the samples tested was able to maintain the rat serum glucose level at prediabetic levels for more than 11 hours.

Alginate-dextran nanoparticles, prepared by emulsion dispersion/*in situ* triggered gelation, were reported few years ago by Reis and co-workers [33]; their method resulted in a unimodal size distribution. The authors have shown that by controlling the conditions under which the water-in-oil emulsion is produced, dispersed droplet size and thus the resulting particle size can be controlled. Insulin-loaded nanoparticles, presenting a high encapsulation efficiency (82.5 ± 3.3% of the initial amount of insulin formulated), were tested using an '*in vitro*' assay in order to measure the biological activity of insulin after a subcutaneous administration of released insulin to diabetic rats. The tests demonstrated that bioactivity was largely retained in the peptide-loaded nanospheres and that nanospheres were able to protect and preserve protein stability during particle formulation, recovery and insulin release in phosphate buffer.

Very recently, insulin-loaded nanoparticles based on positively charged CS and negatively charged arabic gum (a complex mixture of polysaccharides and glycoproteins) were synthesised using ionic gelation between the two polysaccharidic compounds and the effect of different variables on nanoparticle preparation was investigated [34].

The insulin-loaded nanoparticles had a small size, positive charge and median association efficiency (AE) and the authors suggested that the CS and arabic gum concentrations are the most important parameters that affect the AE of nanoparticles.

The release of insulin from nanoparticles followed more than one mechanism: possibly diffusion, dissolution and relaxation of the polymer chains.

Another example of using two polysaccharides for obtaining insulin-loaded nanoparticles is represented by the work of Sonaje and co-workers [35]. They have prepared an enteric-coated capsule loaded with freeze-dried nanoparticles for oral delivery of insulin, the general concept being that the enteric-coated capsule remains intact in the highly acidic environment of the stomach and dissolves rapidly in the neutral environment of the small intestine. Consequently, this capsule could prevent the disintegration of nanoparticles in the stomach, increasing the amount of intact nanoparticles released to the proximal segment of the small intestine. This could be one of the successful ways for increasing the bioavailability of insulin. Well-tolerated nanoparticles (around 100 nm in size) from the toxicological point of view were prepared by ionic crosslinking of CS and poly(γ-glutamic acid) using TPP and magnesium sulfate simultaneously.

The '*in vivo*' tests for determining the glucose levels were performed on Wistar rats with glycaemia higher than 3 g/l the blood samples were collected at different time intervals after dosing from the tail veins of rats prior to drug administration. The hypoglycaemic effect was not reduced after oral administration of the capsule filled with the free-form insulin, indicating the poor oral absorption of insulin in the absence of an appropriate delivery system. Oral administration of the Eudragit L100-55-coated capsule filled with the freeze-dried nanoparticles produced a significant hypoglycaemic effect. The hypoglycaemia after oral administration of the capsule containing insulin-loaded nanoparticles produced a slower but prolonged reduction in blood glucose levels compared with the effect produced by the administration of insulin in free form. The results of this study could be exploited in the future by developing this special formulation for the oral delivery of insulin.

Another interesting idea developed by Jain and co-workers [36] was to prepare insulin-loaded starch nanoparticles with good mucoadhesive properties for transnasal insulin delivery taking into account that they have higher surface area to cover the well-vascularised nasal mucosa providing a great concentration gradient. Starch was chosen for nanoparticle preparation because, like other polysaccharides such as CS and cellulose, it is well known that it could significantly increase the systemic absorption of polypeptides like insulin across the nasal mucosa [37]. Different insulin-loaded nanoparticles were prepared and compared for their size and *in vitro/in vivo* release and performance; the method used for nanoparticle synthesis was the chemical crosslinking of starch using two different crosslinkers – epichlorohydrin and phosphorus oxychloride ($POCl_3$) – and with two different techniques – gel method and emulsion method.

The preparation method and the type of crosslinker are very important for the final size of the nanoparticles, which was between 194 and 997 nm:

- Particles prepared by the emulsion method were more uniform in size distribution with a narrow polydispersity index and had a smaller mean size compared to nanoparticles obtained by the gel method.
- The nanoparticles crosslinked with epichlorohydrin in place of $POCl_3$ had a smaller size.

The '*in vitro*' insulin release study revealed another important difference: the nanoparticles prepared by the gel method released around 80% of insulin in 12 hours, whereas particles prepared by emulsion crosslinking released faster (85–90% in 12 hours). The high release rate is considered useful for drug molecules with short '*in vivo*' half-life such as insulin.

9.6 Applications of Polysaccharide Nano/Microparticles in Respiratory Diseases Therapy

Because nanocarrier systems may be administered to the airways easily, a number of respiratory diseases may be approached using nanoparticles: obstructive lung diseases, genetic disorders affecting the airways, infectious diseases including tuberculosis and cancer [38].

Definitely, the most frequently studied polysaccharide as carrier for controlled release of drugs used in respiratory therapy is CS, especially due to its properties such as high bioadhesiveness and the ability to enhance the permeability of the nasal mucosa [39–41].

Thus, besides offering a substrate for the adsorptive deposition or conjugation of bioactive ligands, CS microparticles feature intrinsic bioadhesive and permeation-enhancing properties, which make them particularly suitable for nasal and pulmonary administration.

Nanoparticles prepared by the ionotropic gelation of CS with the counteranion TPP, in which the positively charged amino groups of CS interact with the negatively charged TPP, were investigated by Florea and co-workers [42] in order to determine the compatibility of the nanoparticle-containing inhalation delivery system with the Calu-3 (a human lung epithelial cell line) and A549 respiratory cell lines in terms of cytotoxicity and cell layer permeability. Cell viability assays are generally used in order to evaluate the safety of nanoparticulated material. Even though in other studies

CS was found to present a toxic effect against respiratory cells [42], it was shown by the authors of this study using the MTT assay, which allows quantification of the metabolic activity of a population of cells, that the CS/TPP formulations exhibited a low cytotoxicity in cell lines of human origin from airway and alveolar regions of the pulmonary tract.

Asthma, from the Greek Άσθμα (ásthma), meaning gasp, is a common chronic inflammatory disease of the airways characterised by variable and recurring symptoms, reversible airflow obstruction and bronchospasm. One of the most widely prescribed antiasthmatic agents is theophylline, but the use of theophylline is complicated by the fact that it presents important side effects, such as nausea, headache and cardiac arrhythmias [43]. Based on the idea that adsorption of theophylline to mucoadhesive nanoparticles would improve theophylline absorption by the bronchial epithelium and enhance its anti-inflammatory effects, the research group of Lee have proposed thiolated CS as support polysaccharide for nanoparticle preparation [44]. The addition of thiol groups to CS chains was performed in order to increase the mucoadhesiveness and permeation of CS nanoparticles without affecting their biodegradability. After the chemical modification with thioglycolic acid for obtaining the thiolated CS, the nanoparticles were prepared by ionic crosslinking with TPP, the new system being then characterised from mucoadhesivity point of view by using the mucin test (as the amount of mucin adsorbed by a certain amount of CS nanoparticles in a certain time period; mucin is the major component of the mucus that coats the cells lining the surfaces of the respiratory tract). Compared with unmodified CS nanoparticles, the ones prepared with thiolated CS presented higher mucoadhesiveness due to the formation of disulfide bonds between thiol groups on CS and cysteine-rich subdomains of mucus glycoproteins.

The authors of this very interesting study observed that the administration of CS or thiolated CS nanoparticles to the sensitised and challenged mice (mice allergic to ovalbumin) did not produce anti-inflammatory effects. The mice treated with theophylline-loaded thiolated CS nanoparticles, however, showed a considerable reduction in pulmonary inflammation, decreased epithelial damage, reduced goblet cell hyperplasia and fewer infiltrating inflammatory cells in the interstitial and peribronchovascular regions (compared to treatment with unmodified ('pure') CS nanoparticles, theophylline plus unmodified CS and theophylline itself).

Recently, Ampuero and co-workers [45] prepared heparin-loaded nanoparticles based on the combination of two polysaccharides (CS and HA) for the treatment of asthma. Heparin is a macromolecular drug (highly sulfated glycosaminoglycan) generally used as a strong anticoagulant but its inhalation is effective in preventing acute bronchoconstrictor responses and airway hyperresponsiveness. Heparin-associated nanoparticles with sizes between 150 and 200 nm and with an AE around 70% were prepared by ionotropic gelation technique using TPP due to the ability of cationic CS to form a gel after contact with negatively charged molecules such as HA, TPP and heparin. One of the interesting conclusions of this study is that the nanoparticulates may prolong the anti-asthmatic effect by improving the effect of a conventional heparin formulation because of slow drug release. The *ex vivo* experiments have been performed in order to evaluate the capacity of heparin to prevent histamine release in rat mast cells and the tests indicated that both free and encapsulated drug exhibited the same dose-response behaviour.

9.7 Applications of Polysaccharide Nano/Microparticles in Diagnostics

In the last decade, nanotechnology has been found to be a successful alternative for solving major problems in diagnostics and imaging. Multifunctional nanoparticles, which incorporate diagnostic properties (such as quantum dots (QD), magnetic, metallic, polymeric and silica nanoparticles) and/or therapeutic properties (such as magnetic and metallic nanoparticles), have already been developed. Nanoparticles due to their small size can be tailored to have specific or multiple functions and can be used for investigating and pursuing an in-depth understanding of the mechanisms involved in biochemical processes [46]. Their unique characteristics such as high surface/volume ratio or size-dependent optical/magnetic properties are extremely different from those of their bulk materials and show promise in the clinical field for disease diagnosis [47, 48].

Concerning the contribution of polysaccharides to the development of this important area of medicine, the scientific literature reports a great interest for this kind of products probably due to their natural sources and their low toxicity, biocompatibility and biodegradability.

Hollow nanospheres based on CS were prepared very recently to be used as ultrasound-induced imaging agents by Liu and co-workers [49]; perfluoropentane was used as an imaging gas to be filled into the CS hollow nanospheres. Reaction of CS with the 4-isocyanato-4′-(3,3-dimethyl-2,4-dioxo-azetidino)diphenylmethane-modified silica nanoparticles led to silica-CS core-shell nanoparticle formation, and the removal of the silica cores with hydrofluoride generates CS hollow nanospheres.

In order to test the ability of the CS hollow nanospheres to be applied in ultrasound imaging techniques and ultrasound-induced cavitation gene delivery, an ultrasound-induced imaging test with perfluoropentane as an imaging gas was performed. The ultrasound-induced images that resulted for the CS hollow spheres demonstrated that they are able to trap perfluoropentane gas and the gas-infused CS hollow nanospheres demonstrated ultrasound imaging abilities.

Spherical microparticles made of crosslinked alginate gel and microcapsules composed of an oil-in-water emulsion where the continuous aqueous phase is crosslinked to an alginate gel matrix were prepared by Reilly and Shapley [50] as contrast agents for the study of the flow properties of fluids using nuclear magnetic resonance imaging (MRI). The authors have demonstrated that combined spin-spin (T2) relaxation and diffusion contrast can be used to identify, among rigid polymer particles, plain alginate beads and alginate emulsion beads.

It is interesting that the properties of several types of particles, such as the alginate emulsion beads, the plain alginate beads and the rigid polystyrene particles, were compared by packing them into vials and imaging them together, while surrounded by deionised water. The alginate beads are produced by simple ionic gelation. In order to obtain alginate emulsion beads, the sodium alginate solution and the oil-in-water emulsion were added together and equilibrated overnight; the emulsion was then dropped into a bath with calcium chloride.

The diffusion weighted imaging produces a sharp contrast between the two types of alginate beads, due to the restricted diffusion inside the embedded oil droplets of the alginate emulsion beads. When a suspension of alginate emulsion beads is imaged in an abrupt annular expansion flow, the emulsion beads are able to be clearly distinguished from the surrounding fluid and rigid polystyrene particles, through either T2 relaxation or diffusion contrast. This property could be exploited in the future and it will be possible to use the alginate emulsion beads as tracer particles.

The work of Wang and co-workers [51] focuses on the differentiation of embryonic stem (ES) cells into insulin-producing cells by application of one of the alginate encapsulation techniques. It seems that the ES cells can be encapsulated within the alginate beads, maintaining a high level of cell viability. Alginate beads were prepared by using an extrusion/crosslinking method; the cell-alginate mixture was passed through a plastic syringe in a bath containing calcium chloride. The alginate beads with an average dimension of 1.5 mm presented a porous structure of the bead matrix and the pore size is influenced by the concentration of the initial alginate solution. In order to promote differentiation of encapsulated ES cells into insulin-producing cells, the alginate beads were incubated with specific growth factors in order to induce differentiation towards pancreatic lineage. The differentiation efficiency was evaluated by determining the ability of cells to adhere to the tissue culture surface; the released cells maintain their ability to adhere to the tissue culture (gelatin-coated tissue culture) surfaces and form cellular aggregates.

In order to measure the glucose-responsive insulin secretion, after the completion of the differentiation process, the enzyme-linked immunosorbent assay was applied and it was observed that insulin secretion, after the high-glucose challenge, was significantly higher from alginate beads (three-dimensional culture), relative to classical two-dimensional culturing conditions.

Another interesting and important research direction in the field of diagnostics is represented by the use of magnetic micro/nanoparticles. Saboktakin and his research group [52] have prepared and characterised superparamagnetic nanoparticles coated with carboxymethyl starch (CMS) for MRI technique. First they prepared a conjugate between CMS and superparamagnetic iron oxide (SPIO) by a chemical reaction between starch and monochloroacetic acid. Afterwards, the CMS-iron oxide particles were synthesised and analysed as drug delivery system for parenteral administration of a model drug 5-aminosalicyclic acid (mesalamine). The CMS-coated SPIO nanoparticles (SPION) of about 10 nm diameter having a core-shell structure with magnetic core and polymeric shell have been successfully prepared, with the nanoparticles presenting a good solubility and stability in ferrofluid. CMS was used as a coating polysaccharide in order to avoid the aggregation between SPION in physiological medium and also to enable SPION to be delivered to tumoral tissue, suggesting the possibility of using CMS-coated SPION as a contrast agent for cancer diagnosis.

Park and co-workers [53] have designed a nanoparticulate intratumoral radioisotope carrier based on PA, which was synthesised in advance via dialysis. The PA dissolved in DMSO was dialysed against aqueous solutions of different ionic strength (IS), and later the dialysate was filtered through a nanometric filter (0.45 μm) to obtain particles with a diameter between 50 and 130 nm.

In order to verify the idea of intratumoral delivery of radioisotopes, ^{99m}Tc was used as a model radioisotope because it is generally used in thyroid and blood-brain barrier imaging. The tests were performed on 6-week-old male BALB/c mice weighing 20–30 g and which had been inoculated with tumour cells. After a certain period of time, mice were randomly selected and ^{99m}Tc-labelled PA nanoparticles were injected into the tumour sites, the nanoparticles presenting a high degree of ^{99m}Tc labelling efficiency (approximately 98%). The authors have also shown that the retention rate (%RR) of the ^{99m}Tc-labelled PA nanoparticles was much higher than that of the free ^{99m}Tc ($p < 0.05$), due to the IS-sensitivity of PA nanoparticles.

The final conclusion is that self-assembled PA nanoparticles with IS-sensitivity under physiological conditions may constitute an alternative way for achieving maximal radioisotope efficiency regarding the intratumoral administration.

Hong and co-workers [54] have prepared biocompatible ferrofluid based on dextran-coated Fe_3O_4 magnetic nanoparticles (MNP) using a novel one-step method of coprecipitation consisting of the addition of hydrazine hydrate as reducing agent and precipitator. The authors investigated the effects of hydrazine hydrate, the weight ratio of dextran to MNP and the molecular weight of dextran on the dispersibility of MNP in water. Dextran was found to be responsible for the decrease in Fe_3O_4 MNP aggregation and for the stabilisation of the colloidal suspension in water. It has been noticed that the weight ratio of dextran to MNP and the molecular weight of dextran have great effect on the dispersibility of nanoparticles, the size of modified Fe_3O_4 MNP, coating efficiency and magnetic properties of the suspension. Another interesting conclusion is that on increasing the weight ratio of dextran to MNP and the molecular weight of dextran, the stability, particle size and coating efficiency also increased, but the saturation magnetisation decreased. In order to analyse the biodistribution and the biotransportation of ferrofluid in different organs, the biocompatible ferrofluid was intravenously injected into rabbits and the iron content in blood and organs was measured at different times. MRI analysis of liver, marrow and lymph revealed that the magnetic resonance signal intensity of these organs notably decreased after being intensified by ferrofluid. In the case of pre-existing tumours, the signal intensity of tumours remained constant after injection. The use of MNP proved to be very useful and reliable in functional molecular imaging for biomedical research and clinical diagnosis.

Polysaccharides have been found useful in the field of luminescent semiconductor QD, which in turn have been used to label biomolecules (biological detection), in other bioapplications such as gene expression studies and in medical diagnostics based on optical coding technology. For example, Tan and his research group [55] developed an easy technique to produce ultrafine nanometer-sized CS nanoparticles encapsulating multicolour QD for the previously mentioned applications. CdSe/ZnS QDS are encapsulated in the CS for improving their biocompatibility, forming monodisperse CS nanoparticles in a single step with a spherical shape and an average size around 60 nm. The authors have effectuated fluorescence spectra of the CS/QD nanoparticles made with different green:red QD ratios in the synthetic solutions and the results have suggested that the fluorescence emission spectra of the various multicolour CS/QD nanoparticles changed according to the ratios between the green and red QD.

After CS encapsulation, *in vitro* cytotoxicity test with MTT on primary myoblast cells indicated that the cytotoxicity of the QD was greatly reduced and fluorescence confocal microscopy studies proved that nanoparticles were small enough to be internalised into the myoblast cells.

References

1. S. Majumdar, K. Hippalgaonkar and M.A. Repka, *International Journal of Pharmaceutics*, 2008, **348**, 175.
2. G. Di Colo, S. Burgalassi, Y. Zambito, D. Monti and P. Chetoni, *Journal of Pharmaceutical Science*, 2004, **93**, 2851.
3. Y. Zambito, C. Zaino, S. Burchielli, V. Carelli, M.F. Serafini and G. Di Colo, *Journal of Drug Delivery Science and Technology*, 2007, **17**, 19.
4. X.B. Yuan, H. Li and Y.B. Yuan, *Carbohydrate Polymers*, 2006, **65**, 337.
5. K.Y. Lee, I.C. Kwon and Y.H. Kim, *Journal of Controlled Release*, 1998, **51**, 213.
6. X.B. Yuan, Y.B. Yuan, W. Jiang, J. Liu, E.J. Tian, H.M. Shun, D.H. Huang, X.Y. Yuan, H. Li and J. Sheng, *International Journal of Pharmaceutics*, 2008, **349**, 1/2, 241.
7. S.K. Motwani, S. Chopra, S. Talegaonkar, K. Kohli, F.J. Ahmad and R.K. Khar, *European Journal of Pharmaceutics and Biopharmaceutics*, 2008, **68**, 513.
8. L. Tuovinen, E. Ruhanen, T. Kinnarinen, S. Rönkkö, J. Pelkonen, A. Urtti, S. Peltonen and K. Järvinen, *Journal of Controlled Release*, 2004, **98**, 3, 407.
9. N. Li, C. Zhuang, M. Wang, X. Sun, S. Nie and W. Pan, *International Journal of Pharmaceutics*, 2009, **379**, 1, 131.
10. C.A. Peptu, G. Buhus, M. Popa, A. Perichaud and D. Costin, *Journal of Bioactive and Compatible Polymers*, 2010, **25**, 1, 98.
11. *Nature Reviews Drug Discovery*, 2007, **6**, 174.
12. A. Kumar, B. Sahoo, A. Montpetit, S. Behera, R.F. Lockey and S.S. Mohapatra, *Nanomedicine: Nanotechnology, Biology and Medicine*, 2007, **3**, 2, 132.
13. K.Y. Choi, H. Chung, K.H. Min, H.Y. Yoon, K. Kim, J.H. Park, I.C. Kwon and S.Y. Jeong, *Biomaterials*, 2010, **31**, 1, 106.
14. N. Duceppe and M. Tabrizian, *Biomaterials*, 2009, **30**, 13, 2625.

15. M. He, Z. Zhao, L. Yin, C. Tang and C. Yin, *International Journal of Pharmaceutics*, 2009, **373**, 1/2, 21, 165.

16. F. Li, J. Li, X. Wen, S. Zhou, X. Tong, P. Su, H. Li and D. Shi, *Materials Science and Engineering: C*, 2009, **29**, 8, 2392.

17. L. Qi, Z. Xu and M. Chen, *European Journal of Cancer*, 2007, **43**, 1, 184.

18. E. Bilensoy, C. Sarisozen, G. Esendagli, A.L. Dogan, Y. Aktas¸ S. Murat and N.A. Mungane, *International Journal of Pharmaceutics*, 2009, **371**, 170.

19. M. Gupta and A.K. Gupta, *Journal of Controlled Release*, 2004, **99**, 157.

20. S. Brown, E.A. Brown and I. Walker, *Lancet Oncology*, 2004, **5**, 497.

21. C. Hopper, *Lancet Oncology*, 2000, **1**, 212.

22. B. Bae and K. Na, *Biomaterials*, 2010, **31**, 6325.

23. H. Zhang, F. Gao, L. Liu, X. Li, Z. Zhou, X. Yang and Q. Zhang, *Colloids and Surfaces B: Biointerfaces*, 2009, **71**, 1, 19.

24. Z. Liu, W. Lu, L. Qian, X. Zhang, P. Zeng and J. Pan, *Journal of Controlled Release*, 2005, **102**, 135.

25. J.C. Atherton, A. Cockayne, M. Balsitis, G.E. Kirk, C.J. Hawley and R.C. Spiller, *Gut*, 1995, **36**, 670.

26. A.T. Axon, *Scandinavian Journal of Gastroenterology*, 1994, **29**, 16.

27. Y.H. Lin, C.H. Chang, Y.S. Wu, Y.M. Hsu, S.F. Chiou and Y.J. Chen, *Biomaterials*, 2009, **30**, 3332.

28. M. Junghans, J. Kreuter and A. Zimmer, *Nucleic Acids Research*, 2000, **28**, 45.

29. F. Foger, W. Noonpakdee, B. Loretz, S. Joojuntr, W. Salvenmoser, M. Thaler and A.B. Schnurch, *International Journal of Pharmaceutics*, 2006, **319**, 139.

30. O. Borges, M. Silva, A. de Sousa, G. Borchard, H.E. Junginger and A. Cordeiro-da-Silva, *International Immunopharmacology*, 2008, **8**, 1773.

31. C.M. Lehr, J.A. Bouwstra, E.H. Schacht and H.E. Junginger, *International Journal of Pharmaceutics*, 1992, **78**, 43.

32. Z. Ma, T.M. Lim and L.Y. Lim, *International Journal of Pharmaceutics*, 2005, **293**, 271.

33. C.P. Reis, A.J. Ribeiro, S. Houng, F. Veiga and R.J. Neufeld, *European Journal of Pharmaceutical Sciences*, 2007, **30**, 392.

34. M.R. Avadi, A.M.M. Sadeghi, N. Mohammadpour, S. Abedin, F. Atyabi, R. Dinarvand and M.R. Tehrani, *Nanomedicine: Nanotechnology, Biology and Medicine*, 2010, **6**, 58.

35. K. Sonaje, Y.J. Chen, H-L. Chen, S.P. Wey, J.H. Juang, H.N. Nguyen, C.W. Hsu and K.J. Lin, *Biomaterials*, 2010, **31**, 3384.

36. A.K. Jain, R.K. Khar, F.J. Ahmed and P.V. Diwan, *European Journal of Pharmaceutics and Biopharmaceutics*, 2008, **69**, 426.

37. L. Ryden and P. Edman, *International Journal of Pharmaceutics*, 1992, **83**, 1,

38. U. Pisona, T. Welte, M. Giersigc and D.A. Groneberg, *European Journal of Pharmacology*, 2006, **533**, 1–3, 341.

39. A. Vila, A. Sanchez, M. Tobio, P. Calvo and M.J. Alonso, *Journal of Controlled Release*, 2002, **78**, 1–3, 15.

40. L. Illum, N.F. Farraj and S.S. Davism, *Pharmaceutical Research*, 1994, **11**, 8, 1186.

41. I.M. van der Lubben, J.C. Verhoef, G. Borchard and H.E. Junginger, *European Journal of Pharmaceutical Science*, 2001, **14**, 3, 201.

42. B.I. Florea, M. Thanou, H.E. Junginger and G. Borchard, *Journal of Controlled Release*, 2006, **110**, 2, 353.

43. C.L. Emerman, R.M. Nowak, M.C. Tomlanovich, S. Yanari, D. Sarkar and J.A. Anderson, *The American Journal of Emergency Medicine*, 1983, **1**, 1, 12.

44. D.W. Lee, S.A. Shirley, R.F. Lockey and S.S. Mohapatra, *Respiratory Research* 2006, 7, 112.

45. F.A. Oyarzun-Ampuero, J. Brea, M.I. Loza, D. Torres and M.J. Alonso, *International Journal of Pharmaceutics*, 2009, **381**, 2, 122.

46. S. Bamrungsap, Y.F. Huang, M.C. Estevez and W. Tan, Clinical Laboratory International.

http://www.cli-online.com/featured-articles/nanoparticles-for-molecular-diagnostics-and-therapy/index.html

47. K.Y. Kim, *Nanomedicine: Nanotechnology, Biology and Medicine*, 2007, **3**, 2, 103.

48. F. Goya, V. Grazu and M.R. Ibarra, *Current Nanoscience*, 2008, **4**, 1, 16.

49. Y.L. Liu, Y.H. Wu, W.B. Tsai, C.C Tsai, W.S. Chen and C.S. Wu, *Carbohydrate Polymers*, 2010.

50. H.J.H. Reilly and N.C. Shapley, *Journal of Magnetic Resonance*, 2007, **188**, 168.

51. N. Wang, G. Adams, L. Buttery, F.H. Falcone and S. Stolnik, *Journal of Biotechnology*, 2009, **144**, 304.

52. M.R. Saboktakin, A. Maharramov and M.A. Ramazanov, *Carbohydrate Polymers*, 2009, **78**, 292.

53. K.H. Park, H.C. Song, K. Na, H.S. Bom, K.H. Lee, S. Kim, D. Kang and D.H. Lee, *Colloids and Surfaces B: Biointerfaces*, 2007, **59**, 16.

54. R.Y. Hong, B. Feng, L.L. Chen, G.H. Liu, H.Z. Li, Y. Zheng and D.G. Wei, *Biochemical Engineering Journal*, 2008, **42**, 290.

55. W.B. Tan, N. Huang and Y. Zhang, *Journal of Colloid and Interface Science*, 2007, **310**, 464.

Acknowledgements

This work was supported by the project PERFORM-ERA 'Postdoctoral Performance for Integration in the European Research Area' (ID-57649), financed by the European Social Fund and the Romanian Government.

10 Carbohydrate-Containing Dendrimers in Biomedical Applications

Jaroslav Sebestik, Milan Reinis and Jan Jezek

10.1 Introduction

The more than 14,500 published references (*Web of Science*, 22 October 2010) about dendrimers and their applications clearly show the merits and perspectives of this class of compounds. There are more than 1100 papers about glycodendrimers. Presently, there are about 300 papers about dendrimers published in the last 3 months (October-December 2010). Many reviews have appeared recently [1–37]. Dendrimers are being studied in all areas of chemistry, nanotechnology, biology and medicine, and in many other fields. We could say that the scope of dendrimers is fast dendrimering [22]. We have already explained the basic terms and nomenclature of peptide and glycopeptide dendrimers [21–23]. This review is limited to glyco- and glycopeptide dendrimers with emphasis on their biological activity. Since some synthetic principles, physicochemical properties, and biological and biomedical applications can be seen in different types of dendrimers, we have included these as examples and inspiring hints that are also usable in the glycopeptide dendrimer field.

The 'glue' of the dendrimer field is the cluster effect. From our point of view, dendrimers are not strictly limited to regular geometrical branching, but generally to branched compounds, where the cluster effect plays an important role. Roy and co-workers published the first paper about glycopeptide dendrimers in 1993. They obtained a compound with strong inhibitory properties against the flu virus haemagglutinin (HA). A history of dendrimers is covered by many reviews [9, 38, 39].

Applications of dendrimers were found in many areas, such as drug delivery [2, 4, 8, 9, 11, 14, 15, 17, 18, 25, 26, 28, 29, 31, 32, 39–57]; gene carriers and vectors [4, 9, 11, 14, 15, 17, 18, 25, 40, 51, 57]; anticancer polymeric nanomedicines and nanocarriers [2, 8, 9, 15, 18, 25, 26, 31, 39, 41–43, 50, 51, 53, 58]; contrast agents for molecular imaging [2, 9, 15, 18, 30, 39, 44, 59–61]; cancer diagnosis and therapy [2, 8, 14, 15, 23, 26–31, 39, 42–45, 50, 52, 53, 56, 61]; immune response modulators [3, 12, 20, 27, 55, 62]; vaccines against infectious diseases and cancer [3, 12, 20, 23, 27–29, 33, 48, 54, 55, 62–64]; and anti-infective and anti-inflammatory drugs [8, 20, 23, 26, 28, 29, 42, 43, 48, 53, 55, 63].

Their highly branched, multivalent nature and molecular architecture make dendrimers ideal tools for a variety of tissue engineering applications [33, 65], components in scaffolds that mimic natural extracellular matrices.

10.2 Sugar Code – A Key to Biodiversity

Permutations of the linear structure of peptide and nucleotide sequences are the only source of their coding capacity. For carbohydrates [13, 66, 67], four additional parameters dramatically increase their coding capacity: (i) positions of linkage points (e.g., 1→2, 1→3, 1→4 or 1→6); (ii) anomeric position (α or β; glycogen/starch in comparison with cellulose differs only in this parameter); (iii) ring size (pyranose or furanose); and (iv) introduction of branches [66, 67]. For complete characterisation of the saccharide structure, the sequence and all listed parameters have to be defined. α-Lactose is not simply galactosyl-glucose (Gal-Glc), but (1→4). Structural determination of saccharides is more complicated than that of nucleotides or peptides [66, 67]. This is the reason carbohydrates and their dendrimers possess much higher diversity [13, 66, 67] than the dendrimers of proteins or nucleic acids.

The size of information that can be coded by nucleic acids, proteins and carbohydrates was evaluated [13, 21, 66–69] for trimers and hexamers. Nucleotides coded by the four pyrimidine and purine bases form 64 (4^3) trimeric and 4096 (4^6) hexameric isomers. Peptides coded by 20 amino acids provide 8000 trimeric (20^3) and 64,000,000 (20^6) hexameric structures. Due to more flexible branching of hexamers, sugars possess 9,000,000 trimeric and 1.44×10^{15} hexameric isomers. This makes carbohydrates the best high-density coding system. This language of carbohydrates was named glycocode [21, 70], sugar code [13, 66, 67, 71], and it uses sugar building blocks – the third alphabet of life, thus, the variability of structures and size of information carried by glycopeptide dendrimers or glycodendrimers are much higher than those of corresponding peptide dendrimers.

10.3 The Dendrimeric State

10.3.1 Physicochemical Properties and General Consequences

From many definitions of dendrimers, we have chosen a lyrical approach [72], which defines dendrimers as 'a jungle of entangled branches traversed by winding trails which lead to sweet fruits and bright blossoms'. Walking down these trails, the thicket's interior can be approached and one can also track a way out. The thicket represents regularly branched, thickly packed structures, whereas the trails stand for empty space and channels filled by solvent. The blossoms and fruits represent electrochemically, photochemically or synthetically addressable species. They can also be catalytically active sites. The motions to and from the trails can be viewed as transport processes.

In the past, three classes of macromolecular architectures (i.e., linear, crosslinked and branched) were widely recognised for formation of slightly polydisperse products of different molecular weights [52, 73]. Since 1984, the 'dendritic state' has been recognised as a new, fourth class of macromolecular architecture, which can be divided into five subclasses: random hyperbranched polymers, dendrigrafts, dendrons, dendrimers, and tecto(dendrimers) or megamers [21, 26, 38, 39, 56, 73, 74]. The quantised precision of dendrons/dendrimers allows these entities to be viewed as nanoscale monomer-type building blocks [5, 75]. They are suitable for the construction of regio-crosslinked dendrimers referred to as 'megamers'. The synthesis of dendrimers proceeds smoothly affording monodisperse, structure-controlled macromolecular systems similar to those observed in nature [21, 26, 39, 74, 76]. Dendrimers (dendrimeric polymers) with polydispersities of $M_w/M_n \sim 1.0005–1.05$ were routinely obtained in multigram to kilogram scale starting from commercially available chemicals [21, 39, 74]. As a result of multistep synthesis, the obtained dendrimeric material is always a mixture of both ideal and nonideal structures [49, 76].

Due to the limitations or a complete lack of interpenetration and the presence of a large number of terminal groups, the physicochemical properties of dendrimers are different from those of classical polymers [26, 49]. The basic properties of dendrimers can be summarised as follows [5, 8, 39, 61, 74]:

1. The nanoscale dimensions are similar to the size of biomacromolecules such as proteins and DNA.

2. An immense variety of surface groups is available for the conjugation of molecules.

3. Suppressed immunogenicity is typical for PEGylated dendrimers.

4. Encapsulation of small molecule drugs, metals or probes occurs in the interior void space.

5. Biocompatibility is a property of lower generations of dendrimers with anionic or neutral polar surface groups.

With increasing generations, the number of terminal groups increases exponentially. The overall density of the dendrimer molecules increases as well. As a consequence, the flexibility of the dendrimer and accessibility of all functional groups decrease. Rigidity and overall density increased by high level of symmetry lead to limited interpenetration between molecules. Due to the high density of terminal groups, penetration of dendrimers is forbidden and interactions appear solely on the surface of the dendrimers [5, 49].

The solvent plays a decisive role in the folding of the dendrimers. Compatibility of a solvent with terminal groups leads to their expansion, whereas incompatibility causes their burial [8, 49]. Important molecular machines can be based on pH-sensitive dendrimers [19], which can be controlled by simple perturbation by acids and bases.

Branching of peptides multiple antigenic peptide provides resistance towards proteolysis by trypsin and α-chymotrypsin [77] and paves the way for biological applications of the peptides. The spheroidal shape of dendrimers collapses during drying, and thus the structure of dendrimers differs in the solid and solution phases [5, 8, 39, 49, 74]. The symmetry and spherical shape of dendrimers prevent their crystallisation, and they remain in the amorphous phase. Conjugation to crosslinked polymers leads to improvement of mechanical properties of dendrimers. In the molten state, dendrimers behave as ideal Newtonian liquids [5, 8, 39, 49, 74].

Biocompatibility (or toxicity) of dendrimers has been reviewed [2, 3, 8, 14, 18, 22, 26, 39, 50, 51, 53, 55–57, 64]. Useful techniques for screening biological properties of dendrimers have also been critically evaluated [7, 40].

10.3.2 Dendrimeric Effects

10.3.2.1 Glycocluster

A sterical arrangement of two and more glycotopes, for example, in the form of dendron or dendrimer, leads to the amplification of a desired property. In comparison with the individual contributions, the achieved amplification is a few orders of magnitude higher. For more details, see **Section 10.3.2.3** [1, 20, 21, 24, 63, 78–93].

10.3.2.2 Multivalency

In order to attain biologically useful affinities between carbohydrate ligands and their protein receptors, the concept of multivalency has to be applied [1, 55, 79, 94–98]. Multiple copies of the carbohydrate epitopes on an appropriate scaffold (polymeric, molecular, dendritic) lead to multivalent displays mimicking the nature of affinity enhancement. This affords higher affinities than that can be expected from the sum of the individual interactions. This was termed 'cluster effect'. Many reviews cover the cooperativity and multivalency in dendrimer and supramolecular chemistry [20, 66, 99].

Precise design and strong individual interactions are not necessary for recognition of a protein surface, because the same effect can be achieved by the sum of weak multivalent interactions [33].

10.3.2.3 Cluster Effect

Since isolated interactions between saccharides and peptides or proteins are weak, nature compensates by clustering multiple copies of carbohydrate ligands and their receptors. This stronger cooperative binding is known as the 'cluster effect' or 'multivalent effect' [1, 21, 63, 64, 66, 78, 89, 92, 94, 97, 100]. Roy and co-workers [100] proved this concept by studies of dendrimer-lectin interactions. The hexadecavalent multiple antigenic glycopeptide (MAG) bound 10^6 times stronger than a monosialoside. The cluster effect was confirmed by various groups [24, 55, 89, 93, 94, 98, 101–107]. Three explanations of the cluster effects were proposed: intramolecular, intermolecular and steric stabilisation. With some exaggeration, we can say that in the case of cluster effect 1 × 8 is a few orders of magnitude higher than 8 × 1 [21]. The optimal enhancement depends on many parameters. In some cases, for the given activity, a valency of 4 is better than 16, in other cases it can be the opposite (steric reasons) [80]. Generally, tetravalent dendrimers provide the best results. For each combination of dendrimer and its application, the valency must be optimised separately. Library-based optimisation led to a compound with 440-fold higher potency than that of L-fucose [108] (see also **Section 10.5**).

Besides 'cluster effect' or 'multivalent effect', other terms, such as a multivalent glycotope [109] have also been used.

10.3.2.4 Macromolecular Effect

The multivalency or cluster effect could not provide an explanation for all types of interactions. According to Lindhorst, understanding fimbriae-mediated bacterial adhesion might require at least two different points of view [110]. Neither the knowledge of crystal structure of FimH nor its interpretation in the sense of a classical 'multivalency effect' could provide a plausible explanation for the observed inhibition of bacterial adhesion. The inhibition of bacterial adhesion to the glycocalyx or a glycocalyx mimetic should be rather explained by a 'macromolecular effect' [110, 111].

10.4 Synthesis of Dendrimers: Convergent and Divergent Approaches

There are two major strategies of dendrimer synthesis (**Figure 10.1**) [6, 9, 11, 39, 63, 74]. In a 'divergent method', the growth of a dendron originates from a core site (**Figure 10.1: 1**). Monomeric modules are built in a radial, branch-upon-branch motif in accord with certain dendritic rules and principles. In contrast, a 'convergent method' starts from blocks that will create the dendrimer shell (**Figure 10.1:** 5) inwards to a reactive focal point, forming a single reactive dendron (7). A reaction of several dendrons with a multifunctional core will lead to a final dendrimer structure (**4**). Both convergent and divergent syntheses were simplified by 'Lego', 'click' and ligation chemistries [37].

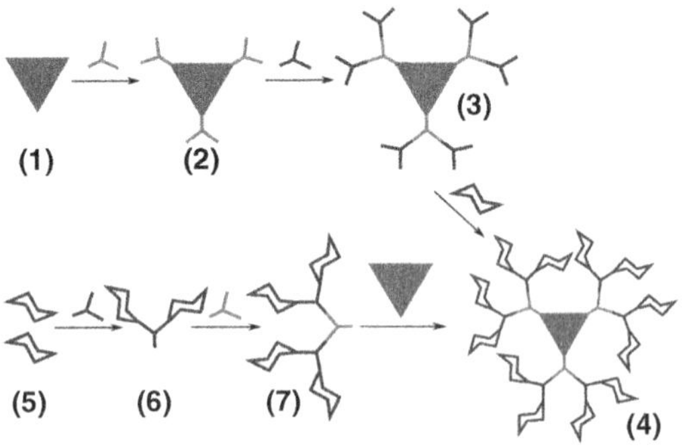

Figure 10.1 Glycodendrimer (**4**) synthesis by divergent and convergent approaches [9, 11, 20, 39, 63, 74, 78]. The core (**1**) and glycans (5) are used as starting points in divergent and convergent syntheses, respectively. Divergent synthesis is analogous with stepwise peptide synthesis. The differences between the product and impurities are small, therefore their behaviour is similar and the separation is more difficult. Convergent synthesis resembles ligation strategy; the building blocks are bigger and more sophisticated. The differences between the product and impurities are bigger and therefore the purification process is easier in comparison with the divergent strategy

Divergent strategy played a guiding role in dendrimer synthesis owing to its major advantages, such as fast synthesis, cheap reagents, exponential growth and possibility to prepare large dendrimers [9, 11, 20, 26, 29, 49, 57, 63, 74, 76, 78, 112]. The key disadvantage is purification of compounds prepared by these strategies, because they are contaminated by thousands of deletion compounds with charge, polarity, hydrophilicity, molecular weight and so on, which are very similar to the desired product. Another limitation is the steric hindrance of higher generation dendrimers, which can prevent couplings of desired building blocks and can cause major defects on the surface of dendrimer. The growing glycodendrimer must be purified in every generation in order to avoid cumulative defects from failed couplings [9, 20, 26, 29, 49, 57, 63, 74, 76, 78, 112].

In conclusion, the divergent approach is a simple way for synthesis of dendrimers using either chemical ligation or coupling of sugars and dendrimers with linking bridges.

The main advantages of the convergent approach are monodispersity of the dendrimer, the possibility to attach different types of dendrons to one dendrimer and much simpler purification of the product [9, 11, 20, 26, 29, 49, 57, 63, 74, 76, 78, 112–114]. Since the by-products differ from the desired structure more significantly, the purification of the dendrimer is simpler. Steric restraints can suppress the attachment of large dendrons to the central core. Both strategies have theirs pros and cons, and thus the choice of desired synthesis of dendrimers is influenced by their structure and generation [20, 26, 49, 57, 63, 74, 76, 78, 112, 114].

The synthesis of glycodendrimers by the convergent approach led to higher sugar-loading efficacy than that in the divergent one [112].

Glycocluster synthesis by native chemical ligation (NCL) was described by Wehner and Lindhorst [115]. NCL of mannoside-derived thioesters and a mannosidic cysteine derivative was carried out. This enables protecting-group-free synthesis of small glycopeptide clusters. These glycoclusters can be dimerised via disulfide bridge formation. The thiol group provides a site for chemoselective modification of prepared glycoclusters.

Globular carbosilane dendrimers with mannose (Man) groups at the shell were synthesised and their toxicity in dendritic cells (DC) was evaluated [116]. A synthetic strategy was based on hydrosilylation reaction of allyl tetraacetylmannose with carbosilane dendrimers containing monohydrosilane end groups and the subsequent deacetylation reaction. Dendrimer toxicities were evaluated in DC by 3-4,5-dimethylthiazol-2-yl-2,5-diphenyltetrazolium bromide (MTT) assay. Their biocompatibility was very good.

10.5 Dendrimeric Libraries

Interactions between carbohydrate receptors and carbohydrates or their mimetics suffer from low binding affinity. Glycopeptides are superior ligands with greater affinity for a receptor than the natural carbohydrate ligand [22]. Glycopeptide and oligosaccharide libraries are the means of discovery of new leads. This topic has been reviewed [9, 22, 28, 52, 63, 78, 117–121] with different glycotopes (TN, TF, mannose, GlcNAc, L-fucose, etc.), resins (TentaGel, PEGA, POEPOP, SPOCC, etc.) and applications. The libraries were prepared by syntheses in solution, on the solid phase, in the active site of enzymes and in the living cells [22, 119–121].

Selection of inhibitors preventing biofilm formation of *Pseudomonas aeruginosa* [108, 118, 122–124] was studied by glycopeptide and oligosaccharide libraries. *P. aeruginosa* is a human pathogenic bacterium, which produces an L-fucose-specific lectin, LecB. LecB plays an important role in tissue attachment and the formation of biofilms. The glycopeptide dendrimer library [122], which consisted of 15,625 members with α-C-L-fucosyl residues at the N-termini, was synthesised on TentaGel and tested for binding to L-fucose-specific lectins. A potent ligand against *Ulex europaeus* lectin and against *P. aeruginosa* lectin PA-IIL (latter = 0.14 μM) was identified as (L-Fuc-α-CH_2-CO-Lys-Pro-Leu)$_4$-(Lys-Phe-Lys-Ile)$_2$-Lys-His-Ile-NH_2 (IC_{50} = 11 μM). The same group described a larger (390,625-member) self-encoded glycopeptide dendrimer library [108, 123, 124], which contained leads for inhibition of biofilm formation in the *P. aeruginosa* with formulae (L-Fuc-α-p-O-C_6H_4-CO-Lys-Ala-Asp)$_4$-(Lys-Ser-Gly-Ala)$_2$-Lys-His-Ile-NH_2 (IC_{50} = 0.11 μM) and (L-Fuc-α-CH_2-CO-Lys-Pro)$_8$-(Lys-Leu-Phe)$_4$-(Lys-Lys-Ile)$_2$-Lys-His-Ile-NH_2 (IC_{50} = 0.025 μM). The latter lead was 440-fold stronger than L-fucose [108]. It may be suitable for development of adhesion and biofilm formation inhibitors of certain pathological bacteria.

Other examples of dendrimer libraries can be found in Reference [1].

10.6 Dendrimers in Drug Solubilisation and Delivery

Many compounds cannot be developed into drugs because of their poor solubility, hydrophobicity and/or limited cell membrane permeability [2, 8, 53, 125]. A crucial task for drug delivery is to understand and enhance the drug solubility and bioavailability. Dendrimeric structure and properties are useful for drug solubilisation. The solubilising properties are attributed to hydrogen bonding and ionic and hydrophobic interactions. Dendrimers with hydrophobic interior and hydrophilic exterior can serve as water-soluble containers of hydrophobic drugs [8, 125]. Enhanced solubilisation of drugs can be achieved by simple encapsulation (**Figure 10.2: 8, 9**), electrostatic interaction (**Figure 10.2: 10**) and/or covalent conjugation (**Figure 10.2: 11, 12**) [2, 9, 15, 26, 41–43, 50, 58]. Dendrimeric vehicles for drug delivery not only improve the drug bioavailability but they also protect drug molecules against enzymatic degradation in the body [8, 15]. Various effects on drug delivery and solubilisation by dendrimers have been studied [8, 9, 15, 25, 26, 31, 49–51, 53, 57].

There is a list of dendrimer applications in drug delivery and solubilisation: anticancer drugs (camptothecin, carboplatin, chlorambucil, cisplatin, dimethoxycurcumin, doxorubicin, etoposide, 5-fluorouracil, methotrexate, oxaliplatin and paclitaxel), antidepressants (venlafaxine), antifungal drugs (amphotericin B), antihaemorrhagic drugs (nimodipine), antihistamines (famotidine), anti-inflammatory drugs (diclofenac, diflunisal, ibuprofen, indomethacin, ketoprofen, mefenamic acid, methylprednisolone, naproxen, nifedipine, phenylbutazone and piroxicam) and antimicrobial drugs (artemether, niclosamide, nadifloxacin, penicillin V, prulifloxacin and sulfamethoxazole). Excellent reviews about drug solubilisation by dendrimers have been published recently [8, 15, 29, 53, 126–129].

An antimalarial drug chloroquine phosphate was more safely delivered in the form of galactose-coated peptide dendrimers than in uncoated form [130].

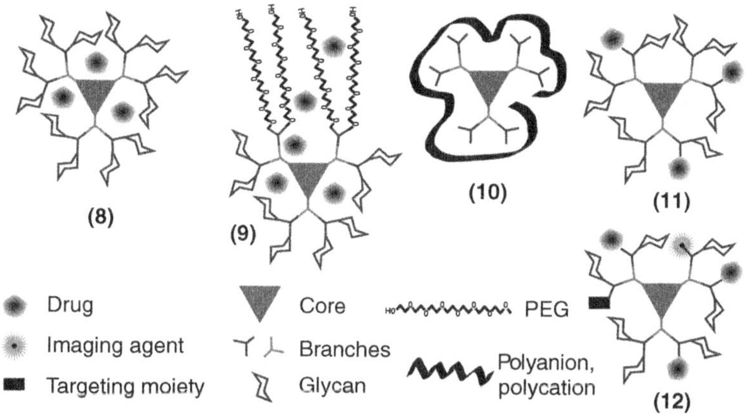

Figure 10.2 Drug solubilisation by dendrimers [15]. The solubilisation proceeds through a number of mechanisms or their combination: the drug can be encapsulated in the interior cavities of the dendrimer molecule (**8**); partly in the interior and between the polyethylene glycol (PEG) chains as in (**9**); or bound by electrostatic forces between polyanion or polycation, which surround the dendrimer, with the corresponding cationic or anionic dendrimer, respectively (**10**). An insoluble drug is bound to a hydrophilic glycodendrimer by a chemical bond (**11**) and attachment of the imaging agent and targeting moiety to **11** leads to the most sophisticated example (**12**)

Carboxymethylchitosan/polyamidoamine (PAMAM) dendrimer nanoparticles were utilised for dexamethasone delivery to bone marrow stromal cells in macroporous hydroxyapatite scaffolds. Intracellular dexamethasone release promoted ectopic bone formation in a rat model [131].

Controlled drug release of an anticancer drug chlorambucil was achieved by hyperbranched polyamidoamines (HPMA) containing beta-cyclodextrin (beta symbol-CD) [127].

Dendrimers with β-CD core with 17-β-estradiol serve as a selective extranuclear oestrogen source [132]. These compounds are potential tools for the studies on the different biological effects of oestrogen during activation of their respective nuclear and extranuclear receptors.

Dendrimeric *N*-acetyl-galactosyl blood-clearing reagent was used for clearance of endogenous avidin as a pretreatment to increase the specificity and efficacy of the B-cell lymphoma therapy with 4 anti-CD20 single-chain Fv fragments and streptavidin fragments followed by radiolabelled DOTA-biotin derivative (bis-biotin) reacting with mutated streptavidin [133].

Dendrimer-drug conjugates can be applied by almost every administration method. The prolonged lifetime and more stable level of the active substance are two key advantages of drug-dendrimer conjugates over drug-dendrimer complexes and traditional drug dosing (**Figure 10.3**) [26, 43].

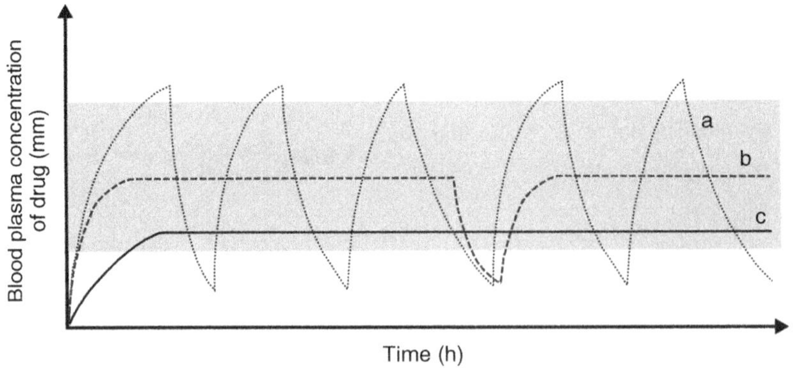

Figure 10.3 Pharmacokinetics during treatment with [26, 43] traditional drugs (a), drug-dendrimer complexes (b) and drug-dendrimer conjugates (c).The light-grey area represents effective and harmless treatment. During traditional dosing (a), spikes of drugs above maximal tolerable levels can occur for a short interval. Safe and continuous release of drug from covalent conjugate is possible (c)

The star PAMAM-*b*-poly(ε-caprolactone)-*b*-poly(gluconamidoethyl methacrylate) block copolymers [128] formed nanoparticles, which served as carriers of nimodipine and binders of concanavalin A (ConA). The aim of this study was the development of a drug against haemorrhage.

Carbohydrate-functionalised cyclodextrins and liposomes for hepatocyte-specific targeting were designed, synthesised and evaluated [134]. Glycan-binding receptors are attractive targets for cell-specific drug and gene delivery. A liver-specific delivery can be achieved using the C-type lectin asialoglycoprotein receptor (ASGPR). ASGPR is exclusively expressed by parenchymal hepatocytes. The authors designed and developed synthesis of carbohydrate-functionalised β-CD and liposomes for hepatocyte-specific delivery. Fluorescent glycodendrimers functionalised with β-CD were used for targeting of ASGPR. Liposomes containing a terminal GalNAc residue bind to ASGPR. Gal/GalNAc-derivatised liposomes were incorporated by endocytosis of the HepG2 human hepatocellular carcinoma cells. β-CD and liposomes with terminal Man or GlcNAc residues were incorporated to a lesser extent. Gal-functionalised β-CDs were excellent molecular carriers for doxorubicin delivery *in vitro* into hepatocytes. This delivery led to induced apoptosis of cancer cells.

10.7 Dendrimers in Gene Delivery

Viral and nonviral systems are used for delivery of nucleic acids to particular target sites [4, 51, 57]. In contrast to viral vectors, the nonviral ones are safer and more flexible. However, they have lower efficiency. Defined architecture and a high ratio of multivalent surface moieties to molecular volume are the principal properties that make these materials very important for the development of synthetic (nonviral) vectors for nucleic acid delivery [57].

Dendrimers form complexes with many types of nucleic acids such as plasmid DNA, antisense oligonucleotides and RNA. These complexes protect the nucleic acids against degradation. The cationic dendrimers interact with the anionic backbone of nucleic acids mainly electrostatically. The net positive charge of the dendrimer nucleic acid complex determines the transfection efficiency. However, highly cationic complexes are extremely cytotoxic. The properties of these complexes can be tuned by many factors such as concentration of dendrimer amines and nucleic acid phosphates, salt concentration, stoichiometry and bulk solvent properties such as pH and buffer strength [4, 57]. Delivery of small-interference RNA (siRNA) has been covered by an excellent review [135].

PAMAM-G2 dendrimers conjugated with α-CD and Lac-α-CD were used as a novel artificial vector for gene delivery, which is selective for hepatocytes [136]. Lactose coating of dendrimers suppressed their cytotoxicities.

Conjugates of α-CD with PAMAM-G3 served as a carrier of short hairpin RNA (shRNA) expressing plasmid DNA (shpDNA) [137]. The conjugates shielded shpDNA against degradation by DNase I.

Dendrimers with β-CD cores were also used for gene delivery [138, 139]. Bis(guanidinium)-tetrakis-(β-CD) tetrapod-containing dendrimers [140] were successfully applied as efficient nonviral carriers for transfection of DNA and siRNA to human embryonic lung fibroblasts.

Since dendriplex complexes with nucleic acids serve as transfection reagents, knowledge of their transfection efficiencies and cytotoxicities is crucial. For transfection efficacy of reporter genes, standard techniques, such as luciferase reporter assay, β-galactosidase assay or green fluorescent protein microscopy, have been summarised [7]. The cytotoxicity is usually determined by standard MTT test.

10.8 Carbohydrate Interactions of Glycopeptide Dendrimers

The cluster of highly branched glycoconjugate macromolecules, which forms an envelope around every cell, is called a cell's glycocalyx [10, 86, 141–144]. This layer is the principal player in molecular recognition and interactions of protein receptors, such as lectins, selectins and their carbohydrate ligands. Many biological processes, such as immune response, fertilisation, cell-cell recognition, metastasis, microbial adhesion, inflammation and other disease states of a cell or tissue, are strongly influenced by interactions with the glycocalyx. Glycodendrimers are excellent tools for studies of these biological processes occurring on cell surfaces. A glycocalyx model based on glyco-self-assembled monolayers (glyco-SAM) is suitable for binding assays using surface plasmon resonance (SPR) [143].

An excellent high-throughput assay based on microarrays for optimisation of glycodendrimer-lectin binding affinity has been described in Reference [145]. It covers the influence of multivalency (generation of dendrimer), type of sugar and type of lectin.

10.8.1 Bacteria

Due to enhanced resistance of bacteria towards common antimicrobial agents, the search for new effective drugs is a challenge for scientists today. The resistant microorganisms cause several diseases that have a dramatic impact on mortality, disability and economics. A set of infectious diseases is initiated by the adhesion of bacterial lectins to host cells glycoconjugates. *Escherichia coli* and several other enterobacteria use Type 1 fimbriae – 30 kDa lectin-like subunit FimH – for mannose-mediated specific binding during anchoring to the host [10, 63, 86, 96, 146, 147]. Plant lectins, particularly ConA, *Dioclea grandiflora lectins* and pea lectins, bind mannopyranosides such as high-mannose oligosaccharides [1, 20, 24, 63, 78, 91]. On pathogen surface, multipresentation and oligomerisation of lectin domains are responsible for multivalency.

Precise understanding of the molecular basis of the adhesion phenomena is necessary in order to find ways to develop new antibiotics [20, 24, 63, 91, 96]. Glycomimetics play the role of inhibitors of cellular recognition and thus suppress adhesion and colonisation of host tissues by pathogens (**Figure 10.4**) [20, 24, 63, 86, 91, 146]. In order to enhance very weak interactions of saccharides, glycodendrimers are employed (see cluster effect in **Section 10.3.2**).

The design and applications of glycodendrimeric mannosylated inhibitors of fimbriae-mediated adhesion have been reviewed [1, 20, 24, 63, 86, 91]. Antibacterial vaccines based on dendrimeric glycopeptide have also been reviewed [12].

Mannosylation of spacered glucose and oligosaccharide (α,α-trehalose, β-melibiose, raffinose) cores provided carbohydrate-centred (octopus) mannosides [110]. An interpretation of their relative inhibitory potency (RIP) of type 1 fimbriae-mediated

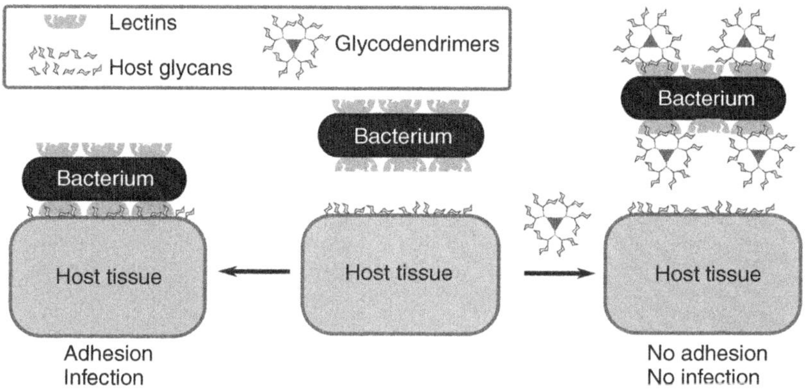

Figure 10.4 How glycodendrimers can prevent host colonisation by bacteria [91, 96]: the interaction of bacterial lectins with host glycans starts the adhesion process leading to infection (left side). When a suitable glycodendrimer with affinity to bacterial lectins is added, it saturates the surface of bacteria and no adhesion (and no infection) of the host tissue takes place (right side)

adhesion used a new macromolecular rather than multivalency effect. To study multivalent interactions with glycocalyx constituents, the same authors [146] prepared clustered glycomimetics with 12 α-mannose units as model compounds. In antiadhesive properties of type 1 fimbriated *E. coli*, RIP of this dendrimer – 190 – was ranked above methyl α-mannoside (RIP = 1).

Nonavalent cluster mannosides have shown very poor if any inhibitory activities of *E. coli* type 1 fimbriae-mediated adhesion [86]. For better understanding of glycocalyx interactions, trivalent MAG with L-fucose and mannose were anchored on gold surface via thiofunctionalised alkane and alkane-oligoethylene glycol spacers [142]. Syntheses of other glycocalyx mimetics have also been described [141].

Touaibia and Roy [148] developed new mannosylated dendrimers as potential drug candidates for gastrointestinal and urinary tract infections caused by *E. coli*. They elaborated a short and efficient strategy for the first synthesis of 'Majoral-type' [149] multivalent glycodendrimers with α-mannopyranosides covalently bound onto a cyclotriphosphazene scaffold assembled using single-step Sonogashira and 1,3-cycloaddition click chemistry (**Figure 10.5**). The complete series (valencies 6–18 units) of dendrimers was tested by a kinetic turbidimetric assay with well-established ConA. The fastest and more complete formation of the insoluble crosslinked lattice was achieved with the decavalent dendrimer (**Figure 10.5**: **15**) containing alkyne spacer. Higher flexibility of dodecamer and octadecamer provided lower crosslinking potencies than **15**. Significant crosslinking enhancement was explained by more favourable extended intersugar distances, which facilitate the penetration into the active site of the carbohydrate and permit a higher protein crosslinking ability [148].

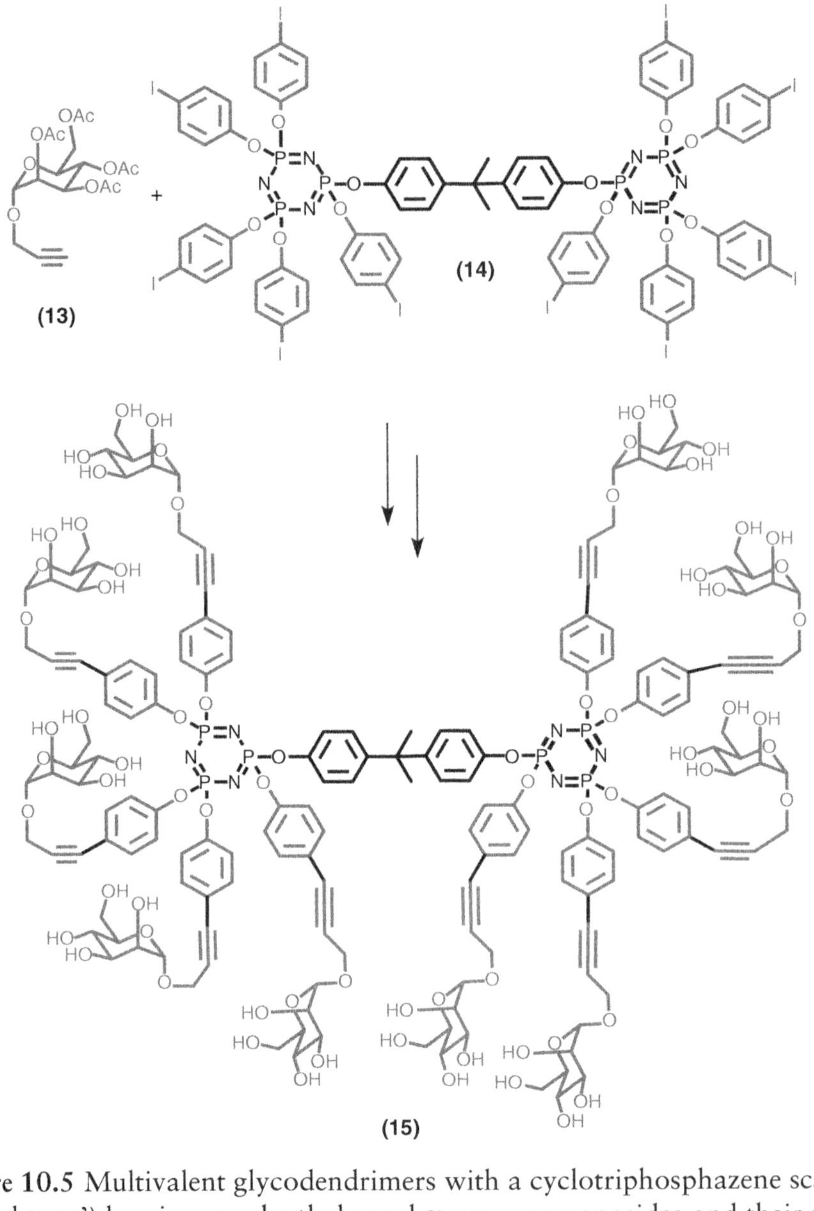

Figure 10.5 Multivalent glycodendrimers with a cyclotriphosphazene scaffold ('Majoral-type') bearing covalently bound α-mannopyranosides and their synthesis using single-step Sonogashira coupling [148]

These α-mannopyranoside dendrimers can serve as probes or effectors of biological processes involving multivalent carbohydrate-binding proteins.

A new approach in glycodendrimer synthesis has been elaborated [144]. In this iterative synthesis of spaced glycodendrons as oligomannoside mimetics, a 3,6-diallylated carbohydrate is utilised as core molecule; hydroboration-oxidation is the activating step; and glycosylation with branched and unbranched sugar trichloroacetimidates is used for dendritic growth. The six glycodendrons were tested by enzyme-linked immunosorbent assay as potential inhibitors of *E. coli* type 1 fimbriae-mediated bacterial adhesion. The data obtained were discussed with regard to spacer characteristics and sugar valency.

Galactose containing di-, tetra- and octavalent dendrimers were prepared via click chemistry. Their inhibition of cholera toxin binding is as strong as in the case of the natural ganglioside GM1 oligosaccharide [150]. This result is important in the therapy and detection of cholera toxin.

Click chemistry with dendritic scaffolds and extended arms was used for synthesis of multivalent GM1os and GM2os dendrimers [151]. These dendrimers showed strong multivalent binding to cholera toxin B-subunit, with an excellent value of at least 380,000-fold higher binding for octavalent GM1os dendrimer than that for monovalent GM1os derivatives. These results can be exploited for development of very sensitive sensor applications.

Glycosylated propargylated pentaerythritol phosphodiester oligomers with 4, 6, 8 and 10 L-fucose residues were prepared [111]. Their binding to the L-fucose-specific bacterial lectin (PA-IIL) was measured by an enzyme-linked lectin amplification competition assay. The IC_{50} values found are 10–20 times better in comparison with that of monovalent L-fucose. This fact cannot be explained by a cluster effect, because the relative activity per carbohydrate is only of 2 for every glycocluster, regardless of the number of L-fucose residues. The recently described 'macromolecular' effect [110] is most probably responsible for the increased binding of these glycoclusters.

The ability of intracellular pathogens (*Mycobacterium tuberculosis* and *Leishmania*) to hide within immune cells made the development of effective therapies particularly challenging. Song and co-workers [152] presented that changes of surface oligosaccharides on synthetic probes were sufficient to block IL-12 production by macrophages and thus dampen innate immune responses.

Dendrimer technology could be also used for sensitive bacteria detection. Fluorescent HPAMAM with mannose groups were proved to bind *E. coli* cell surface [153]. Bacteria could be detected at concentrations higher than 10^2 cfu/ml.

10.8.2 Viruses

The most severe influenza pandemic was the Spanish flu in 1918–1919, which caused the deaths of at least 20 million people. The pathogenicity of influenza virus is high

[154, 155]. Almost a quarter of a million patients with flu are hospitalised every year in the United States. About 62 million people are expected to die worldwide, when a new type of pandemic flu strikes. On the surface of influenza A viruses, two unique glycoproteins are localised, namely, HA and neuraminidase (NA). These glycoproteins play a decisive role in infection and replication. To avoid the first contact of the virus and a host cell, the first approach has focused on HAs, which participate in infection of influenza virus to host cells. HAs of influenza A viruses bind to a specific receptor – sialyl lactose (Neu5Ac-α-(2$\rightarrow$3)-Gal-β-(1$\rightarrow$4)-Glc). This event starts the adhesion of influenza viruses to the surface of the host cell [154, 155]. Dendrimers with sialic acid as synthetic inhibitors of influenza virus have been reviewed [156].

In the search for novel influenza sialidase inhibitors, Japanese authors [154] prepared 12 types of sialylated carbosilane dendrimers with thioglycosidic linkage that are resistant to hydrolysis by the sialidases. Dendrimers with 3, 4, 6 and 12 sialyl residues and with different spacer patterns, that is, aliphatic linkage, ether and amide linkages, were synthesised. All of the ether- and amide-elongated compounds had inhibitory potencies for the influenza sialidases in the micromolar range.

The same authors [155] prepared a series of carbosilane dendrimers functionalised with Neu5Ac-α-(2$\rightarrow$3)-Gal-β-(1$\rightarrow$4)-Glc moieties. These glycodendrimers were systematically tested for anti-influenza virus activity. Their biological activities depended on the form of their core frame. The dumbbell(1)6-amide type glycodendrimer showed particularly high inhibitory activities against human influenza viruses [A/PR/8/34 (H1N1) and A/Aichi/2/68 (H3N2)]. The results of this study showed that dumbbell-shaped dendrimers were the most suitable core scaffolds. Improved inhibitory activities in the micromolar level were obtained. These glycodendrimers are promising therapeutic agents for influenza disease.

High-mannose clusters on the surface of HIV envelope glycoprotein gp120 serve for targeting of the C-type lectin on dendritic cells (dendritic cell-specific intracellular adhesion molecule-3-grabbing non-integrin (DC-SIGN)). In order to develop carbohydrate-based antiviral agents, mimics of the cluster presentation of oligomannosides on the virus surface were designed [157]. A multivalent gold nanoparticle (manno-GNP) library containing truncated $Man_9GlcNAc_2$ was prepared and screened for inhibitors of DC-SIGN binding to gp120. These DC-SIGN ligands can also interfere with the early steps of other infections via specific recognition of associated glycans. Different thiol-containing spacers serving for attachment of the (oligo)mannosides to the gold surface were prepared. SPR was used to study their inhibition potency towards DC-SIGN binding to gp120. Complete inhibition of the binding in the micro- to the nanomolar range was achieved by the tested manno-GNP. Millimolar concentration was required by the corresponding monovalent mannosides.

Manno-GNP containing the disaccharide α-Man*p*-(1→2)-α-Man*p* were the best inhibitors of gp120 binding to DC-SIGN. Their activity was roughly 20,000-fold higher in comparison with that of the corresponding monomeric disaccharide.

Polylysine dendritic sulfated cellobiose blocked HIV-1 replication due to electrostatic interactions between negatively charged sulfated groups and positively charged gp120 [158]. The cluster effect was important for the biological activity since nonbranched sulfated oligosaccharides have little anti-HIV-1 activity.

Water-soluble hyperbranched β-galactosylceramide containing dendritic polymers bind HIV-1 gp120 and mimic multivalent glycosphingolipid (GSL) display on cell surface (specific lipid rafts) required for assembly of the HIV-1 entry complex [159].

Poly(propyleneimine) dendrimers or mannosylated fifth-generation poly(propyleneimine) dendrimers increased 3TC antiretroviral activity against HIV-1. This approach is focused on targeting reservoirs such as quiescent CD4+ cells, mononuclear phagocytic cells such as macrophages or DC and so on, central nervous system and male genital tract compartments [160].

A series of catanionic multivalent analogues of GalCer have been described [161]. Dendrimers based on cyclotriphosphazene core with phosphonic acid in the branches formed noncovalent complexes with *N*-hexadecylamino lactitol moieties. These supramolecular systems showed anti-HIV1 activity. They have submicromolar IC_{50} in a cell-based HIV-infection model and also a high general cytotoxicity.

First-generation gallic acid triethylene glycol (GATG) dendrimers endowed with a large hydrophobic moiety (benzoate) blocked HIV capsid formation *in vitro* [162]. One of the tested dendrimers hampered *in vitro* formation of the human immunodeficiency virus capsid. This finding suggests the possibility that dendrimers can be developed as anti-HIV-1 therapeutic agents targeting capsid assembly.

The HIV-1 external envelope glycoprotein recognises specific GSL, which are present on the surface of target immune cells as alternate cell surface receptors. The globotriose and 3′-sialyllactose carbohydrate head groups present on two GSL were conjugated with a dendrimer [163]. These conjugates inhibited Env-mediated membrane fusion and HIV-1 infection of T-cell lines and primary peripheral blood mononuclear cells (PBMC) by T-cell line adapted viruses or primary isolates, with IC_{50} ranging from 0.1 to 7.4 μg/ml. These entry inhibitors thus may represent a new class of anti-HIV-1 drugs.

The topic of dendrimeric glycopeptide antiviral vaccines has been reviewed [12, 23, 164].

10.8.3 Cancer

DC have a unique ability to display tumour epitopes via the major histocompatibility complex class I pathway inducing cytotoxic CD8+ T-lymphocyte responses. They therefore play a key role in the development of therapeutic cancer vaccines. The mannose receptor and DC-SIGN are membrane lectins located on the surface of DC. They recognise oligosaccharides containing mannose and/or L-fucose and mediate sugar-specific endocytosis of synthetic oligolysine-based glycoclusters.

Srinivas and co-workers [165] synthesised glycocluster conjugates containing a CD8+ epitope of the Melan-A/Mart-1 melanoma antigen. The glycocluster-Melan-A conjugates were prepared by coupling different glycosynthons, for example, oligosaccharyl-pyroglutamyl-β-alanine derivatives containing either a dimannoside (Man-α-(1→6)-Man) or lactoside, or Lewis oligosaccharides Le^x and Le^a, to Melan-A 16-40 peptide Ac-G^{16}HGHSYTTAE*E^{26}LAGIGILTV^{35}*-ILGVL^{40}KKKK (Ac, for acetyl) containing a variant of the HLA-A2-restricted CD8+ epitope (26–35) *ELAGIGILTV* instead of EAAGIGILTV, with a tetralysine tail on the C terminus. Flow cytometry and confocal microscopy have shown that fluorescent Melan-A glycoclusters with either dimannoside or Lewis oligosaccharide were taken up by DC and concentrated in acidic vesicles, contrary to lactoside glycopeptides, which were not at all taken up. Tight binding of the dimannoside and Lewis-Melan-A conjugates to MR and DC-SIGN was shown by SPR. The lactose derivative was inactive. These conjugates with DC call forth a CD8+ T-lymphocyte response and represent promising tools for the antitumour vaccines.

Mannose containing dendrimeric porphyrins, which interact with mannose-specific receptors at the surface of cancer cells, were designed for photodynamic therapy [166].

A 65,536-member one-bead-one-compound combinatorial library of β-Gal containing glycopeptide dendrimers provided a lead compound against Jurkat cancer cells with LD_{50} = 1.5 μM [167].

Tetra- and hexa-valent mannoside inhibitors of the pro-apoptotic, antiproliferative and cell surface clustering effects of ConA with pentaerythritol scaffolds have been synthesised by click chemistry [168]. Their influence on membrane type 1-matrix metalloproteinase (MT1-MMP) functions in marrow-derived mesenchymal stem cells (MSC) was investigated. MSC morphological changes caused by ConA were reversed by the tetra- and hexavalent mannosides. The mannosides serve as antagonists of ConA-induced caspase-3 activity and proMMP-2 activation. Antiproliferative and pro-apoptotic reversal of ConA on the MT1-MMP/glucose-6-phosphate transporter signalling axis were observed. MT1-MMP pleiotropic functions in cell survival, proliferation and extracellular matrix degradation are the targets of these glycoclusters; therefore these glycoclusters are useful for anticancer therapy [168].

The application of dendrimeric glycopeptides in cancer detection and prevention including anticancer vaccines has been reviewed [15, 23, 30, 169, 170].

10.8.4 Other Examples

Cellular recognition and adhesion is mediated by carbohydrate-carbohydrate interactions (CCIs) between cell surface glycans. Multivalent glycoconjugates based on G4 PAMAM dendrimers were prepared [171]. Their interaction with Langmuir monolayers containing GM3 was investigated. The CCI is adversely affected by excessive carbohydrate valency. In the presence of calcium ions, the interaction of the GM3 monolayer with lactose-functionalised dendrimers is selective. These results are the first example of the use of glycodendrimers as model systems for studying CCI. These glycodendrimers can also find application as targeted diagnostic and antimetastatic agents and may serve as useful agents for probing CCI *in vivo*.

Dendrimers with up to eight α-mannose residues were synthesised by click chemistry [93]. The attachment of the dendrimer core to aluminium oxide chips was done via a spacer. A real time binding of the fluorescent lectins ConA and GNA to the glycodendrimer chips was observable. In a single experiment, the multivalency enhancement or cluster effect of the binding event could be observed. The difference between the small effect for ConA and the large one for GNA was explained by the 12 times higher density of GNA binding sites than those of ConA. Chips coated with these dendrimers serve for screening of multivalency effects. Kinetic and thermodynamic data of binding and inhibition events are easily accessible.

Peptoid glycopeptide dendrimers containing four dansyl groups as fluorescence label were synthesised by Jin and co-workers [172]. The trisaccharide building block was formulated using major antigenic epitope of pectic polysaccharides from the root of the plant *Bupleurum falcatum* L. The roots serve as traditional Chinese and Japanese medicine for the treatment of autoimmune diseases, chronic hepatitis and nephrotic syndrome.

Glycodendrimer-lectin interactions were studied by a novel, digital, single-operation analytical method [173]. Tris(bipyridine)ruthenieum(II) complexes bearing 2, 4, 6 or 18 mannose or galactose units were used as molecular logic circuits working as fluorescence devices. These devices responded to changes of pH and concentration of *N,N'*-4,4'-bis(benzyl-2-boronic acid)bipyridinium dibromide and different lectins (ConA, GNA and asialoglycoprotein). The output of the devices was indicated by the relative change in fluorescence and quantum yield. This led to a single-step, high-throughput method for quick screening of carbohydrate-lectin interactions.

Fluorescent, large, monodisperse G4 MAG capped with 16 mannose residues were internalised into DC in a manner consistent with mannose-receptor-mediated uptake as monitored by confocal microscopy [174].

Upconverting lanthanide Ln^{3+}-doped nanoparticles conjugated with glycodendrimers were developed [175]. These particles are able to recognise lectins and they can serve for an assay based on luminescence resonance energy transfer (LRET). PAMAM dendrimers were adsorbed on the surface of these nanoparticles by direct ligand exchange. The surface modification was further tightened by thiourea linkage formation between the amine surface and *p*-isothiocyanatophenyl α-mannopyranoside. These water-dispersible and biocompatible mannose-coated PAMAM conjugates were used to recognise ConA conjugated with tetramethylrhodamine via LRET from the nanoparticles. Nanoparticles play the role of energy donors to the labelled lectin molecules, which act as energy acceptors. The energy transfer phenomenon of the nanoparticles was induced by excitation at 980 nm, to which the ConA conjugate responded by emissions at 585 nm. These particles are novel sensors for monitoring of the binding interaction between lectins and sample in aqueous solution.

The ability of G0 to G5 lysine MAG with benzhydrylamine at the core and with 2–64 mono-, di- and tri-α-mannopyranosyl residues to induce DC maturation was studied; however, no significant trends were observed [176].

Dendrimers and their derivatives can serve as potential therapeutic tools in regenerative medicine strategies [177] due to the great potential of their applications in tissue engineering and the central nervous system.

10.9 Conclusion

Glyco and glycopeptide dendrimers are outstanding coding systems with extremely high density. This is a key to their enormous functional variability, which paves the avenue for many applications in analytical chemistry, biological (anticancer, antimicrobial, antiprion and antiviral activities; contrast agents for molecular imaging; drug and gene delivery; solubilisation effect on drugs; and so on) and forensic sciences. The introduction of modern strategies of convergent glycodendrimer synthesis can lead to a number of important discoveries. Tailor-made anticancer synthetic vaccines can emerge soon due to combination of small targeting peptides or tumour-associated antigens with dendrimeric cargos. Potentially, neurodegenerative disorders (Alzheimer's disease and prion-caused diseases) can be inhibited with cationic glycodendrimers.

References

1. Y.M. Chabre and R. Roy, *Advances in Carbohydrate Chemistry and Biochemistry*, 2010, **63**, 165.
2. D. Astruc, E. Boisselier and C. Ornelas, *Chemical Reviews*, 2010, **110**, 4, 1857.
3. P.M.H. Heegaard, U. Boas and N.S. Sorensen, *Bioconjugate Chemistry*, 2010, **21**, 3, 405.
4. T. Dutta, N.K. Jain, N.A.J. McMillan and N.H.S. Parekh, *Nanomedicine: Nanotechnology, Biology and Medicine*, 2010, **6**, 1, 25.
5. D.A. Tomalia, *Soft Matter*, 2010, **6**, 3, 456.
6. G.R. Newkome and C. Shreiner, *Chemical Reviews*, 2010, **110**, 10, 6338.
7. D. Shcharbin, E. Pedziwiatr, J. Blasiak and M. Bryszewska, *Journal of Controlled Release*, 2010, **141**, 2, 110.
8. S. Svenson, *European Journal of Pharmaceutics and Biopharmaceutics*, 2009, **71**, 3, 445.
9. B.M. Rosen, C.J. Wilson, D.A. Wilson, M. Peterca, M.R. Imam and V. Percec, *Chemical Reviews*, 2009, **109**, 11, 6275.
10. M. Lahmann, *Topics in Current Chemistry*, 2009, **288**, 17.
11. B.K. Nanjwade, H.M. Bechra, G.K. Derkara, F.V. Manvia and V.K. Nanjwade, *European Journal of Pharmaceutical Sciences*, 2009, **38**, 3, 185.
12. J. Rojo, *Anti-Infective Agents in Medicinal Chemistry*, 2009, **8**, 1, 50.
13. *The Sugar Code: Fundamentals of Glycosciences*, Ed., H-J. Gabius, Wiley-VCH Verlag GmbH, John Wiley and Sons Ltd., New York, NY, USA, 2009.
14. C.M. Paleos, L.A. Tziveleka, Z. Sideratou and D. Tsiourvas, *Expert Opinion in Drug Delivery*, 2009, **6**, 1, 27.
15. R.K. Tekade, P.V. Kumar and N.K. Jain, *Chemical Review*, 2009, **109**, 1, 49.
16. W. Wang and A.E. Kaifer, *Inclusion Polymers - Advances in Polymer Science*, 2009, **222**, 205.
17. H. Arima, and K. Motoyama, *Sensors*, 2009, **9**, 8, 6346.

18. S.F.M. van Dongen, H.P.M. de Hoog, R.J.R.W. Peters, M. Nallani, R.J.M. Nolte and J.C.M. van Hest, *Chemical Reviews*, 2009, **109**, 11, 6212.

19. K.C.F. Leung, C.P. Chak, C.M. Lo, W.Y. Wong, S. Xuan and C.H.K. Cheng, *Chemistry - An Asian Journal*, 2009, **4**, 3, 364.

20. Y.M. Chabre and R. Roy, *Current Topics in Medicinal Chemistry*, 2008, **8**, 14, 1237.

21. P. Niederhafner, J. Sebestik and J. Jezek, *Journal of Peptide Science*, 2008, **14**, 1, 2.

22. P. Niederhafner, J. Sebestik and J. Jezek, *Journal of Peptide Science*, 2008, **14**, 1, 44.

23. P. Niederhafner, M. Reinis, J. Sebestik and J. Jezek, *Journal of Peptide Science*, 2008, **14**, 5, 556.

24. A. Imberty, Y.M. Chabre and R. Roy, *Chemistry - An European Journal*, 2008, **14**, 25, 7490.

25. C.M. Paleos, D. Tsiourvas, Z. Sideratou and L. Tziveleka, *Current Topics in Medicinal Chemistry*, 2008, **8**, 14, 1204.

26. Y. Cheng, Z. Xu, M. Ma and T. Xu, *Journal of Pharmaceutical Sciences*, 2008, **97**, 1, 123.

27. I. Toth, P. Simerska and Y. Fujita, *International Journal of Peptide Research and Therapy*, 2008, **14**, 4, 333.

28. O. Renaudet, *Mini-Reviews in Organic Chemistry*, 2008, **5**, 4, 274.

29. C. Villalonga-Barber, M. Micha-Screttas, B.R. Steele, A. Georgopoulos and C. Demetzos, *Current Topics in Medicinal Chemistry*, 2008, **8**, 14, 1294.

30. J.B. Wolinsky and M.W. Grinstaff, *Advanced Drug Delivery Reviews*, 2008, **60**, 9, 1037.

31. R. Chadha, V.K. Kapoor, D. Thakur, R. Kaur, P. Arora and D.V.S. Jain, *Journal of Scientific and Industrial Research*, 2008, **67**, 3, 185.

32. V.K. Mourya, and N.N. Inamdar, *Reactive and Functional Polymers*, 2008, **68**, 6, 1013.

33. V. Martos, P. Castreno, J. Valero and J. de Mendoza, *Current Opinion in Chemical Biology*, 2008, **12**, 6, 698.

34. A. Dondoni, *Chemistry - an Asian Journal*, 2007, **2**, 6, 700.

35. R.J. Pieters, D.T.S. Rijkers and R.M.J. Liskamp, *QSAR and Combinatorial Science*, 2007, **26**, 11/12, 1181.

36. J. Iehl, I. Osinska, R. Louis, M. Holler and J.F. Nierengarten, *Tetrahedron Letters*, 2009, **50**, 19, 2245.

37. J. Sebestik, P. Niederhafner and J. Jezek, *Amino Acids*, 2011, **40**, 2, 301.

38. H. Frauenrath, *Progress in Polymer Science*, 2005, **30**, 3/4, 325.

39. S. Svenson and D.A. Tomalia, *Advanced Drug Delivery Reviews*, 2005, **57**, 15, 2106.

40. D. Shcharbin, E. Pedziwiatr and M. Bryszewska, *Journal of Controlled Release*, 2009, **135**, 3, 186.

41. M.E. Fox, F.C. Szoka and J.M.J. Frechet, *Accounts of Chemical Research*, 2009, **42**, 8, 1141.

42. Y. Cheng, J. Wang, T. Rao, X. He and T. Xu, *Frontiers in Bioscience* 2008, **13**, 1447.

43. Y. Cheng and T. Xu, *European Journal of Medicinal Chemistry*, 2008, **43**, 11, 2291.

44. I.J. Majoros, C.R. Williams and J.R. Baker, Jr., *Current Topics in Medicinal Chemistry*, 2008, **8**, 14, 1165.

45. J. Li, and X.J. Loh, *Advanced Drug Delivery Reviews*, 2008, **60**, 9, 1000.

46. K.K. Jain in *Drug Delivery Systems*, Ed., K.K. Jain, Methods in Molecular Biology, Volume 437, Humana Press, Totowa, NJ, USA, 2008, p.1.

47. K.K. Jain, *Medical Principles and Practice*, 2008, **17**, 2, 89.

48. J. Rojo and R. Delgado, *Anti-Infective Agents in Medicinal Chemistry*, 2007, **6**, 3, 151.

49. L.P. Balogh, *Advances in Experimental Medicine and Biology*, 2007, **620**, 136.

50. Y. Cheng, Y. Gao, T. Rao, Y. Li and T. Xu, *Combinatorial Chemistry and High Throughput Screening*, 2007, **10**, 5, 336.

51. C.M. Paleos, D. Tsiourvas and Z. Sideratou, *Molecular Pharmaceutics*, 2007, **4**, 2, 169.

52. T. Darbre and J.L. Reymond, *Accounts of Chemical Research*, 2006, **39**, 12, 925.

53. S. Svenson and D. Tomalia in *Nanoparticulates as Drug Carriers*, Ed., V.P. Torchilin, Imperial College Press, London, UK, 2006, p.277.

54. K.J. Doores, D.P. Gamblin and B.G. Davis, *Chemistry - An European Journal*, 2006, **12**, 3, 656.

55. P.M.H. Heegaard and U. Boas, *Recent Patents on Anti-Infective Drug Discovery*, 2006, **1**, 3, 333.

56. E.R. Gillies and J.M.J. Frechet, *Drug Discovery Today*, 2005, **10**, 1, 35.

57. C. Dufes, I.F. Uchegbu and A.G. Schatzlein, *Advanced Drug Delivery Reviews*, 2005, **57**, 15, 2177.

58. R. Tong and J. Cheng, *Polymer Reviews*, 2007, **47**, 3, 345.

59. V.V. Mody, M.I. Nounou and M. Bikram, *Advanced Drug Delivery Reviews*, 2009, **61**, 10, 795.

60. M. Longmire, P.L. Choyke and H. Kobayashi, *Current Topics in Medicinal Chemistry*, 2008, **8**, 14, 1180.

61. D.A. Tomalia, L.A. Reyna and S. Svenson, *Biochemical Society Transactions* 2007, **35**, 1, 61.

62. W. Zhong, M. Skwarczynski, Y. Fujita, P. Simerska, M.F. Good and I. Toth, *Australian Journal of Chemistry*, 2009, **62**, 9, 993.

63. R. Roy and M. Touaibia in *Carbohydrate-Protein and Carbohydrate-Carbohydrate Interactions; Comprehensive Glycoscience*, Eds., J.P. Kamerling, G.J. Boons, Y.C. Lee, A. Suzuki, N. Taniguchi and A.G.J. Voragen, Elsevier, Utrecht, The Netherlands, 2007, p.821.

64. D.E. Tsvetkov and N.E. Nifantiev, *Russian Chemical Bulletin*, International Edition, 2005, **54**, 5, 1065.

65. N. Joshi and M. Grinstaff, *Current Topics in Medicinal Chemistry*, 2008, **8**, 14, 1225.

66. H.J. Gabius, *Biochemical Society Transactions*, 2008, **36**, 1491.

67. H.J. Gabius, H.C. Siebert, S. Andre, J. Jimenez-Barbero and H. Rudiger, *ChemBioChem*, 2004, **5**, 6, 741.

68. H.J. Gabius, S. Andre, H. Kaltner and H.C. Siebert, *Biochimica et Biophysica Acta*, 2002, **1572**, 2/3, 165.

69. R.A. Laine in *Glycosciences: Status and Perspectives*, Eds., H.J. Gabius and S. Gabius, Chapman & Hall, London, UK, 1997, p.1.

70. M. Ambrosi, N.R. Cameron and B.G. Davis, *Organic and Biomolecular Chemistry*, 2005, **3**, 9, 1593.

71. D. Solis, J. Jimenez-Barbero, H. Kaltner, A. Romero, H.C. Siebert, C.W. von der Lieth, and H.J. Gabius, *Cells, Tissues, Organs*, 2001, **168**, 1/2, 5.

72. A.D. Schluter and J.P. Rabe, *Angewandte Chemie International Edition*, 2000, **39**, 5, 864.

73. L. Hartmann and H.G. Borner, *Advanced Materials*, 2009, **21**, 32/33, 3425.

74. D.A. Tomalia, *Progress in Polymer Science*, 2005, **30**, 3/4, 294.

75. D.A. Tomalia, *Chimica Oggia/Chemistry Today*, 2005, **23**, 6, 41.

76. U. Boas, J.B. Christensen and P.M.H. Heegaard, *Journal of Materials Chemistry*, 2006, **16**, 38, 3786.

77. P. Sommer, V.S. Fluxa, T. Darbre and J.L. Reymond, *ChemBioChem*, 2009, **10**, 9, 1527.

78. R. Roy, *Trends in Glycosciences and Glycotechnology*, 2003, **15**, 85, 291.

79. T.K. Lindhorst, *Topics in Current Chemistry*, 2002, **218**, 201.

80. R. Roy, and M.G. Baek, *Reviews in Molecular Biotechnology*, 2002, **90**, 3/4, 291.

81. R. Roy and J.M. Kim, *Tetrahedron*, 2003, **59**, 22, 3881.

82. M. Kohn, J.M. Benito, C.O. Mellet, T.K. Lindhorst, and J.M.G. Fernández, *ChemBioChem*, 2004, **5**, 6, 771.

83. Y. Singh, O. Renaudet, E. Defrancq and P. Dumy, *Organic Letters*, 2005, **7**, 7, 1359.

84. S.A. Kalovidouris, O. Blixt, A. Nelson, S. Vidal, W.B. Turnbull, J.C. Paulson and J.F. Stoddart, *Journal of Organic Chemistry*, 2003, **68**, 22, 8485.

85. S. Andre, H. Kaltner, T. Furuike, S. Nishimura and H.J. Gabius, *Bioconjugate Chemistry*, 2004, **15**, 1, 87.

86. A. Patel and T.K. Lindhorst, *Carbohydrate Research*, 2006, **341**, 10, 1657.

87. Y. Aoyama, *Chemistry - An European Journal*, 2004, **10**, 3, 588.

88. T. Nakai, T. Kanamori, S. Sando and Y. Aoyama, *Journal of the American Chemical Society*, 2003, **125**, 28, 8465.

89. J.J. Lundquist and E.J. Toone, *Chemical Reviews*, 2002, **102**, 2, 555.

90. K. Sato, N. Hada and T. Takeda, *Carbohydrate Research*, 2006, **341**, 7, 836.

91. M. Touaibia and R. Roy, *Mini-Reviews in Medicinal Chemistry*, 2007, **7**, 12, 1270.

92. J. Zhang, G. Pourceau, A. Meyer, S. Vidal, J.P. Praly, E. Souteyrand, J.J. Vasseur, F. Morvan and Y. Chevolot, *Biosensors and Bioelectronics*, 2009, **24**, 8, 2515.

93. H.M. Branderhorst, R. Ruijtenbeek, R.M.J. Liskamp and R.J. Pieters, *ChemBioChem*, 2008, **9**, 11, 1836.

94. W.B. Turnbull and J.F. Stoddard, *Reviews in Molecular Biotechnology*, 2002, **90**, 3/4, 231.

95. O. Ouerfelli, J.D. Warren, R.M. Wilson and S.J. Danishefsky, *Expert Review of Vaccines*, 2005, **4**, 5, 677.

96. R.J. Pieters, *Medicinal Research Reviews*, 2007, **27**, 6, 796.

97. M.L. Wolfenden and M.J. Cloninger, *Journal of the American Chemical Society*, 2005, **127**, 35, 12168.

98. P.I. Kitov and D.R. Bundle in *Carbohydrate-Based Drug Discovery,* Volume 2, Ed., C-H. Wong, Wiley-VCH GmbH & Co. KGaA, Weinheim, Germany 2003, p.541.

99. J.D. Badjic, A. Nelson, S.J. Cantrill, W.B. Turnbull and J.F. Stoddart, *Accounts of Chemical Research*, 2005, **38**, 9, 723.

100. R. Roy, D. Zanini, S.J. Meunier and A. Romanowska, *Journal of the Chemical Society, Chemical Communications*, 1993, **1993**, 24, 1869.

101. R. Roy, M.G. Baek and K. Rittenhouse-Olson, *Journal of the American Chemical Society*, 2001, **123**, 9, 1809.

102. D. Arosio, M. Fontanella, L. Baldini, L. Mauri, A. Bernardi, A. Casnati, F. Sansone and R. Ungaro, *Journal of the American Chemical Society*, 2005, **127**, 11, 3660.

103. D. Page, D. Zanini and R. Roy, *Bioorganic and Medicinal Chemistry*, 1996, **4**, 11, 1949.

104. J.B. Corbell, J.J. Lundquist and E.J. Toone, *Tetrahedron: Asymmetry*, 2000, **11**, 1, 95.

105. T.K. Dam and C.F. Brewer, *Methods in Enzymology*, 2004, **379**, 107.

106. M.G. Baek and R. Roy, *Bioorganic and Medicinal Chemistry*, 2002, **10**, 10, 11.

107. R. Roy and M.G. Baek, *Methods in Enzymology*, 2003, **362**, 240.

108. E. Kolomiets, M.A. Swiderska, R.U. Kadam, E.M.V. Johansson, K.E. Jaeger, T. Darbre and J.L. Reymond, *ChemMedChem* 2009, **4**, 4, 562.

109. T. Singh, J.H. Wu, W.J. Peumans, P. Rouge, E.J.M. van Damme, R.A. Alvarez, O. Blixt and A.M. Wu, *Biochemical Journal*, 2006, **393**, 1, 331.

110. M. Dubber, O. Sperling and T.K. Lindhorst, *Organic and Biomolecular Chemistry*, 2006, **4**, 21, 3901.

111. F. Morvan, A. Meyer, A. Jochum, C. Sabin, Y. Chevolot, A. Imberty, J.P. Praly, J.J. Vasseur, E. Souteyrand and S. Vidal, *Bioconjugate Chemistry*, 2007, **18**, 5, 1637.

112. Y. Li, Y. Cheng and T. Xu, *Current Drug Discovery Technologies*, 2007, **4**, 4, 246.

113. C. Ozawa, H. Katayama, H. Hojo, Y. Nakahara and Y. Nakahara, *Organic Letters*, 2008, **10**, 16, 3531.

114. C.P.R. Hackenberger and D. Schwarzer, *Angewandte Chemie*, International Edition, 2008, **47**, 52, 10030.

115. J.W. Wehner and T.K. Lindhorst, *Synthesis*, 2010, **2010**, 18, 3070.

116. P. Ortega, M.J. Serramia, M.A. Munoz-Fernandez, F.J. de la Mata and R. Gomez, *Tetrahedron*, 2010, **66**, 18, 3326.

117. H. Hojo and Y. Nakahara, *Biopolymers (Peptide Science)*, 2007, **88**, 2, 308.

118. T. Darbre and J.L. Reymond, *Current Topics in Medicinal Chemistry*, 2008, **8**, 14, 1286.

119. J.J. Banik, and S.F. Brady *Proceedings of the National Academy of Sciences of the United States of America*, 2008, **105**, 45, 17273.

120. T. Murase, T. Tsuji and Y. Kajihara, *Carbohydrate Research*, 2009, **344**, 6, 762.

121. C.E.P. Maljaars, S. Andre, K.M. Halkes, H.J. Gabius and J.P. Kamerling, *Analytical Biochemistry*, 2008, **378**, 2, 190.

122. E. Kolomiets, E.M.V. Johansson, O. Renaudet, T. Darbre and J.L. Reymond, *Organic Letters*, 2007, **9**, 8, 1465.

123. E.M.V. Johansson, E. Kolomiets, F. Rosenau, K.E. Jaeger, T. Darbre and J.L. Reymond, *New Journal of Chemistry*, 2007, **31**, 7, 1291.

124. E.M.V. Johansson, S.A. Crusz, E. Kolomiets, L. Buts, R.U. Kadam, M. Cacciarini, K.M. Bartels, S.P. Diggle, M. Camara, P. Williams, R. Loris, C. Nativi, F. Rosenau, K.E. Jaeger, T. Darbre and J.L. Reymond, *Chemistry and Biology*, 2008, **15**, 12, 1249.

125. Y. Shi, W. Porter, T. Merdan and L.C. Li, *Expert Opinion on Drug Delivery*, 2009, **6**, 12, 1261.

126. R.N. Prajapati, R.K. Tekade, U. Gupta, V. Gajbhiye and N.K. Jain, *Molecular Pharmaceutics*, 2009, **6**, 3, 940.

127. Y. Zhou, Z. Guo, Y. Zhang, W. Huang, Y. Zhou and D. Yan, *Macromolecular Bioscience*, 2009, **9**, 11, 1090.

128. X.H. Dai, H.D. Zhang and C.M. Dong, *Polymer*, 2009, **50**, 19, 4626.

129. L. Zhao, Y. Cheng, J. Hu, Q. Wu and T. Xu, *Journal of Physical Chemistry B*, 2009, **113**, 43, 14172.

130. P. Agrawal, U. Gupta and N.K. Jain, *Biomaterials*, 2007, **28**, 22, 3349.

131. J.M. Oliveira, N. Kotobuki, M. Tadokoro, M. Hirose, J.F. Mano, R.L. Reis and H. Ohgushi, *Bone*, 2010, **46**, 5, 1424.

132. H.Y. Kim, J. Sohn, G.T. Wijewickrama, P. Edirisinghe, T. Gherezghiher, M. Hemachandra, P.Y. Lu, R.E. Chandrasena, M.E. Molloy, D.A. Tonetti and G.R.J. Thatcher, *Bioorganic and Medicinal Chemistry*, 2010, **18**, 2, 809.

133. D.S. Wilbur, S.L. Park, M.-K. Chyan, F. Wan, D.K. Hamlin, J. Shenoi, Y. Lin, S.M. Wilbur, F. Buchegger, A. Pantelias, J.M. Pagel and O.W. Press, *Bioconjugate Chemistry* 2010, **21**, 7, 1225.

134. G.J.L. Bernardes, R. Kikkeri, M. Maglinao, P. Laurino, M. Collot, S.Y. Hong, B. Lepenies and P.H. Seeberger, *Organic and Biomolecular Chemistry*, 2010, **8**, 21, 4987.

135. J. Wang, Z. Lu, M.G. Wientjes and J.L.S. Au, *AAPS Journal*, 2010, **12**, 4, 492.

136. H. Arima, S. Yamashita, Y. Mori, Y. Hayashi, K. Motoyama, K. Hattori, T. Takeuchi, H. Jono, Y. Ando, F. Hirayama and K. Uekama, *Journal of Controlled Release*, 2010, **146**, 1, 106.

137. T. Tsutsumi, F. Hirayama, K. Uekama and H. Arima, *Journal of Pharmaceutical Sciences*, 2008, **97**, 8, 3022.

138. A. Diaz-Moscoso, L. Le Gourrierec, M. Gomez-Garcia, J.M. Benito, P. Balbuena, F. Ortega-Caballero, N. Guilloteau, C. Di Giorgio, P. Vierling, J. Defaye, C. Ortiz Mellet and J.M. Garcia Fernandez, *Chemistry - A European Journal*, 2009, **15**, 46, 12871.

139. F.J. Xu, Z.X. Zhang, Y. Ping, J. Li, E.T. Kang and K.G. Neoh, *Biomacromolecules*, 2009, **10**, 2, 285.

140. S. Menuel, S. Fontanay, I. Clarot, R.E. Duval, L. Diez and A. Marsura, *Bioconjugate Chemistry*, 2008, **19**, 12, 2357.

141. H.A. Shaikh, F.D. Sonnichsen and T.K. Lindhorst, *Carbohydrate Research*, 2008, **343**, 10/11, 1665.

142. M. Kleinert, N. Rockendorf and T.K. Lindhorst, *European Journal of Organic Chemistry*, 2004, **2004**, 18, 3931.

143. M. Kleinert, T. Winkler, A. Terfort and T.K. Lindhorst, *Organic and Biomolecular Chemistry*, 2008, **6**, 12, 2118.

144. C.D. Heidecke and T.K. Lindhorst, *Chemistry - A European Journal*, 2007, **13**, 32, 9056.

145. N.P. Pera, H.M. Branderhorst, R. Kooij, C. Maierhofer, M. van der Kaaden, R.M. Liskamp, V. Wittmann, R. Ruijtenbeek and R.J. Pieters, *ChemBioChem*, 2010, **11**, 13, 1896.

146. O. Sperling, M. Dubber and T.K. Lindhorst, *Carbohydrate Research*, 2007, **342**, 5, 696.

147. Y.M. Chabre, C. Contino-Pepin, V. Placide, T. Chieh Shiao and R. Roy, *Journal of Organic Chemistry*, 2008, **73**, 14, 5602.

148. M. Touaibia and R. Roy, *Journal of Organic Chemistry*, 2008, **73**, 23, 9292.

149. P. Servin, C. Rebout, R. Laurent, M. Peruzzini, A.M. Caminade and J.P. Majoral, *Tetrahedron Letters*, 2007, **48**, 4, 579.

150. H.M. Branderhorst, R.M.J. Liskamp, G.M. Visserb and R.J. Pieters, *Chemical Communication*, 2007, **47**, 5043.

151. A.V. Pukin, H.M. Branderhorst, C. Sisu, C.A.G.M. Weijers, M. Gilbert, R.M.J. Liskamp, G.M. Visser, H. Zuilhof and R.J. Pieters, *ChemBioChem*, 2007, **8**, 13, 1500.

152. E-H. Song, A.O. Osanya, C.A. Petersen and N.L.B. Pohl, *Journal of the American Chemical Society*, 2010, **132**, 33, 11428.

153. W. Yang, C.Y. Pan, M.D. Luo and H.B. Zhang, *Biomacromolecules*, 2010, **11**, 7, 1840.

154. J.I. Sakamoto, T. Koyama, D. Miyamoto, S. Yingsakmongkon, K.I.P.J. Hidari, W. Jampangern, T. Suzuki, Y. Suzuki, Y. Esumi, T. Nakamura, K. Hatano, D. Terunuma and K. Matsuoka, *Bioorganic & Medicinal Chemistry*, 2009, **17**, 15, 5451.

155. H. Oka, T. Onaga, T. Koyama, C.T. Guo, Y. Suzuki, Y. Esumi, K. Hatano, D. Terunuma and K. Matsuoka, *Bioorganic & Medicinal Chemistry*, 2009, **17**, 15, 5465.

156. I. Carlescu, D. Scutaru, M. Popa and C.V. Uglea, *Medicinal Chemical Research*, 2009, **18**, 6, 477.

157. O. Martinez-Avila, K. Hijazi, M. Marradi, C. Clavel, C. Campion, C. Kelly and S. Penades, *Chemistry - A European Journal*, 2009, **15**, 38, 9874.

158. S. Han, D. Yoshida, T. Kanamoto, H. Nakashima, T. Uryu and T. Yoshida, *Carbohydrate Polymers*, 2010, **80**, 4, 1111.

159. J.A. Morales-Serna, O. Boutureira, A. Serra, M.I. Matheu, Y. Diaz and S. Castillon, *European Journal of Organic Chemistry*, 2010, **2010**, 14, 2657.

160. U. Gupta and N.K. Jain, *Advanced Drug Delivery Reviews*, 2010, **62**, 4/5, 478.

161. A. Perez-Anes, C. Stefaniu, C. Moog, J.P. Majoral, M. Blanzat, C.O. Turrin, A.M. Caminade and I. Rico-Lattes, *Bioorganic and Medicinal Chemistry*, 2010, **18**, 1, 242.

162. R. Domenech, O. Abian, R. Bocanegra, J. Correa, A. Sousa-Herves, R. Riguera, M.G. Mateu, E. Fernandez-Megia, A. Velazquez-Campoy and J.L. Neira, *Biomacromolecules*, 2010, **11**, 8, 2069.

163. A.R. Borges, L. Wieczorek, B. Johnson, A.J. Benesi, B.K. Brown, R.D. Kensinger, F.C. Krebs, B. Wigdahl, R. Blumenthal, A. Puri, F.E. McCutchan, D.L. Birx, V.R. Polonis and C.L. Schengrund, *Virology*, 2010, **408**, 1, 80.

164. V. Gajbhiye, V.K. Palanirajan, R.K. Tekade and N.K. Jain, *Journal of Pharmacy and Pharmacology*, 2009, **61**, 8, 989.

165. O. Srinivas, P. Larrieu, E. Duverger, M.T. Bousser, M. Monsigny, J.F. Fonteneau, F. Jotereau and A.C. Roche, *Bioconjugate Chemistry*, 2007, **18**, 5, 1547.

166. A. Makky, J.P. Michel, S. Ballut, A. Kasselouri, P. Maillard and V. Rosilio, *Langmuir*, 2010, **26**, 13, 11145.

167. E.M.V. Johansson, J. Dubois, T. Darbre and J.L. Reymond, *Bioorganic & Medicinal Chemistry*, 2010, **18**, 17, 6589.

168. S. Fortier, M. Touaibia, S. Lord-Dufour, J. Galipeau, R. Roy and B. Annabi, *Glycobiology*, 2008, **18**, 2, 195.

169. J. Zhu, J.D. Warren and S.J. Danishefsky, *Expert Review of Vaccines*, 2009, 8, 10, 1399.

170. J. Li and J.P. Zhang, *Chinese Journal of Cancer Biotherapy*, 2010, **17**, 4, 462.

171. N. Seah, P.V. Santacroce and A. Basu, *Organic Letters*, 2009, **11**, 3, 559.

172. Y. Jin, N. Hada, J. Oka, O. Kanie, S. Daikoku, Y. Kanie, H. Yamada and T. Takeda, *Chemical and Pharmaceutical Bulletin*, 2006, **54**, 4, 485.

173. R. Kikkeri, D. Grunstein and P.H. Seeberger, *Journal of the American Chemical Society*, 2010, **132**, 30, 10230.

174. E.A.B. Kantchev, C.C. Chang, S.F. Cheng, A.C. Roche and D.K. Chang, *Organic and Biomolecular Chemistry*, 2008, **6**, 8, 1377.

175. N. Bogdan, F. Vetrone, R. Roy and J.A. Capobianco, *Journal of Materials Chemistry*, 2010, **20**, 35, 7543.

176. B.W. Greatrex, S.J. Brodie, R.H. Furneaux, S.M. Hook, W.T. McBurney, G.F. Painter, T. Rades and P.M. Rendle, *Tetrahedron*, 2009, **65**, 15, 2939.

177. J.M. Oliveira, A.J. Salgado, N. Sousa, J.F. Mano and R.L. Reis, *Progress in Polymer Science*, 2010, **35**, 9, 1163.

Abbreviations

^{99}Tc	Technetium-99
^{99m}Tc	Metastable technetium-99
AA	Abdominal aorta
ABTS	2,2′-Azino-bis(3-ethylbenzthiazoline-6-sulfonic acid)
ADAMTS	A disintegrin-like and metalloproteinase with thrombospondin
AE	Association efficiency
AEPS	Polysaccharides from *Actinidia eriantha*
AGU	Anhydroglucose unit
AI	Active ingredient
AOSMC	Aortic smooth muscle cells
APS-1	Polysaccharides from *Aloe vera*
APT	Attached proton test
APTT	Activated partial thromboplastin time
ASGPR	Asialoglycoprotein receptor
ATCC	American Type Culture Collection
β-gal	β-Galactosidase
BAD	Bcl-associated death protein
BASYC®	Bacterial synthesised cellulose
BC	Bacterial cellulose
Bcl-x_L	B-cell lymphoma leukaemia-x
BDNF	Brain-derived neurotrophic factor
bFGF	Basic fibroblast growth factor
BHT	Butylated hydroxytoluene
BSA	Bovine serum albumin

CA	Cellulose acetate
CAPS	Cassis polysaccharide
CCI	Carbohydrate-carbohydrate interaction(s)
CD	Cyclodextrin(s)
CF	Cellulose formate
cfu	Colony forming unit
CHP	Cholesteryl-bearing pullulan
$CHPNH_2$	Cholesteryl-bearing pullulan modified with amino groups
CI	Confidence interval
CL	Cotton linters
CMC	Carboxymethyl cellulose
CMD	Carboxymethyldextran
CMP	Carboxymethylpullulan
CMP-Dox	Carboxymethylpullulan-doxorubicin conjugate
CMP-SLe(x)	Carboxymethylpullulan conjugate with sialyl LewisX
CMS	Carboxymethyl starch
ConA	Concanavalin A
COSY	Correlation spectroscopy
CR3	Complement receptor 3
ChS	Chondroitin sulphate
CpGODN	CpG Oligodeoxynucleotides
CS	Chitosan
CS-CH	Chitosan and cholesterol 3-hemisuccinate conjugates
CSTR	Continuously stirred tank reactor
CTB5	Cholera toxin B-subunit
CTFA	Cellulose trifluoroacetate
CyA	Cyclosporine A
D_2O	Deuterated water
DAEB	Dried algae extract PS
DC	Dendritic cells
DEAE	Diethyle amino ethyl
DEPT	Distortionless enhancement by polarisation transfer
DFPP	Double-filtration plasmapheresis
DMAc	*N,N*-dimethylacetamide

DMF	*N,N*-diemthylformamide
DMSO	Dimethyl sulfoxide
DNA	Deoxyribonucleic acid
DO	Dissolved oxygen
DP	Degree of polymerisation
DPPH	1,1-Diphenyl-2-picrylhydrazyl
DRG	Dorsal root ganglion
DS	Degree of substitution
DTH	Delayed-type hypersensitivity
ECM	Extracellular matrix
EPI	Epirubicin
EPS	Exopolysaccharide
ERK	Extracellular signal-regulated kinase
ES	Embryonic stem
FCD	Fixed charged density
FDA	Food and Drug Administration
FDDS	Floating drug delivery system
FGF-2	Fibroblast growth factor-2
FPA	Folic acid-conjugated pullulan acetate nanoparticle
FPA/EPI	Epirubicin-loaded folic acid-conjugated pullulan acetate nanoparticle
FTIR	Fourier transform infrared
FurCl	2-Furoyl chloride
FVP	*Flammulina velutipes* PS
GAG	Glycosaminoglycan
GalNAc	*N*-acetyl-D-galactosamine
GALT	Gut-associated lymphoid tissue
GATG	Gallic acid triethylene glycol
GC-MS	Gas chromatography-mass spectrometry
G-CSF	Granulocyte colony-stimulating factor
GGP	*Glycyrrhiza glabra* polysaccharide
GlcA	Glucuronic acid
GLP	Polysaccharide from *Ganoderma lucidum*
Glycol-SAM	Glyco-self-assembled monolayers

GM-CSF	Granulocyte-macrophage colony-stimulating factor
GMPOLY	Galactomannan polymer
GPS	Glycyrrhiza polysaccharide
GSH	Glutathione
GSHPx	Glutathione peroxidase
GSL	Glycosphingolipid
H_2O_2	Hydrogen peroxide
H2BC	Heteronuclear two-bond correlation
HA	Haemagglutinin
HA-CA	Hyaluronic acid-5b-cholanic acid
HA-NP	Hyaluronic acid nanoparticle(s)
HBsAg	Hepatitis B surface antigen
HBV	Hepatitis B virus
HCT	Haematocrit
HEK 293 cells	Human embryonic kidney 293 cells
hFB	Human lung fibroblasts
HIV	Human immunodeficiency virus
HMBC	Heteronuclear multiple-bond correlation
HP	Hydroxypropyl
HPLC	High-performance liquid chromatography
HPMA	Hyperbranched polyamidoamines
HPMC	Hydroxypropylmethylcellulose
HPS1	Hizikia PS I
HSQC	Heteronuclear single-quantum correlation
HSV-1	*Herpes simplex* virus type 1
IA	Immunoadsorption
ICAM-1	Intercellular adhesion molecule 1
IFN	Interferon
IFNγ	Interferon gamma
IgA	Immunoglobulin A
IgG	Immunoglobulin G
IKK	IκB kinase
IL	Interleukin
IL-1β	Interleukin-1 beta

INEPT	Insensitive nuclei enhanced by polarisation transfer
iNOS	Inducible nitric oxide synthase
ip	Intraperitoneal
IRAK	IL-1R-associated kinase
IS	Ionic strength
ISO	International Standards Organization
ITAM	Immunoreceptor tyrosine-based activation motif
IκB-α	Inhibitory κB-α
IVD	Intervertebral disc
JNK	Jun *N*-terminal kinase
JPC	Japan Pharmacopoeia
KB	Cell line derived from carcinoma of the human nasopharynx
KBr	Potassium bromide
kpb	Kilo-base pair
KS	Keratan sulfate
LCH	Low-molecular-weight chitosan
LD_{50}	Median lethal dose
LiCl	Lithium chloride
LPS	Lipopolysaccharide
LRET	Luminescence resonance energy transfer
manno-GNP	Multivalent gold nanoparticle
MAG	Multiple antigenic glycopeptides(s)
MAPK	Mitogen-activated protein kinase
M-CSF	Macrophage colony-stimulating factor
MC	Microcrystalline cellulose
MCP-1	Monocyte chemoattractant protein-1
MDA	Malondialdehyde
MFC	Microfibrillated cellulose
MMC	Mitomycin C
MMP	Matrix metalloproteinase
M_n	Number average molecular weight
MNP	Magnetic nanoparticle
MO	Macrophage(s)
MPO	Myeloperoxidase

MR	Mannose receptor
M_r	Relative molecular weight
MRI	Magnetic resonance imaging
mRNA	Messenger ribonucleic acid
MSC	Mesenchymal stem cells
MTT	3-(4,5-Dimethylthiazol-2-yl)-2,5-diphenyltetrazolium bromide
M_w	Molecular weight
MyD88	Myeloid differentiation protein 88
N,*N*,-DEAE	*N*,*N*-diethylaminoethyl
N/A	Not available
NA	Neuraminidase
NCL	Native chemical ligation
NFAT	Nuclear factor of activated T cells
NF-κB	Nuclear factor κB
nkat	Nanokatal
NK cells	Natural killer cells
NMDA	*N*-methyl-D-aspartic acid
NMNO	*N*-methylmorpholine-*N*-oxide
NMR	Nuclear magnetic resonance
NO	Nitric oxide
NOESY	Nuclear Overhauser effect spectroscopy
OA	Osteoarthritis
ODN	Oligodeoxynucleotide
OR	Odds ratio
OVA	Ovalbumin
PA	Pullulan acetate
PAM	Polysaccharide microsphere from acetylated pullulan
PAMN	Acetylated pullulan-coated superparamagnetic iron oxide nanoparticle
PAP	Polysaccharide from *Potentilla anserine*
PASF	Porphyra acid-soluble fraction
PBCA	Poly(butyl cyanoacrylate)
PBMC	Peripheral blood mononuclear cells

PBS	Phosphate-buffered saline
PCL	Polycaprolactone
PDT	Photodynamic therapy
PE	Plasma exchange
PEG	Polyethylene glycol
PEI	Poly(ethylene imine)
PFE	Proximal femoral epiphysis
PFP	Pullulan/folate
PGE_2	Prostaglandin E_2
Pheo-A	Pheophorbide-a
PI3K	Phosphatidylinositol 3-kinase
PKC	Protein kinase C
PLA	Poly(lactic acid)
PLA_2	Phospholipase A_2
PLC	Phospholipase C
PLL	Polylysine
PNIPAM	Poly(*N*-isopropylacrylamide)
ppm	Parts per million
PS	Polysaccharide
PSS	Poly(styrene) sulfate
PTX	Paclitaxel
PVA	Poly(vinyl alcohol)
PWSF	Porphyra water-soluble fraction
QD	Quantum dot(s)
RAPA	Rapamycin
RIP	Relative inhibitory potency
PLGA	Polylactic-polyglycolic acid
PAMAM	Polyamidoamine
RM	Randomly methylated
RNA	Ribonucleic acid
RNS	Reactive nitrogen species
rpm	Revolutions per minute
ROS	Reactive oxygen species
RPE	Retinal pigment epithelium

RR	Rention rate
RSM	Response surface methodology
rVSMC	Rat vascular smooth muscle cells
SAM	Self-assembled monolayer
SEB	Sulfobutyl ether
SEC	Size-exclusion chromatography
shRNA	Short hairpin ribonucleic acid
shpDNA	Short hairpin ribonucleic acid expressing plasmid deoxyribonucleic acid
shRNA	Short hairpin ribonucleic acid
siRNA	Small-interference ribonucleic acid
SMC	Synthetically modified cellulose
SOD	Superoxide dismutase
SPG	Schizophyllan
SPIO	Superparamagnetic iron oxide
SPION	Superparamagnetic iron oxide nanoparticle
SPR	Surface plasmon resonance
SR	Scavenger receptor
SS	Sulfated pullulan
SSP	Spruce sulfite pulp
STAT	Signal transducers and activators of transcription
STMP	Sodium trimetaphosphate
Syk	Spleen tyrosine kinase
TBAF	Tetrabutylammonium fluoride
TEM	Transmission electron microscopy
TFA	Trifluoroacetic acid
TLR	Toll-like receptor(s)
TMC	*N*-trimethylchitosan
TMSC	Trimethylsilyl cellulose
TNF-α	Tumour necrosis factor-α
TOCSY	Total correlation spectroscopy
TPP	Sodium tripolyphosphate
TRAF-6	Tumour necrosis factor receptor-associated factor 6

TT	Thrombin time
UDP	Uridine diphosphate
USP-NF	United States Pharmacopoeia – National Formulary
VA	Vinyl acetate
VEC	Valve endothelial cells
VIC	Valve interstitial cells
VVM	Volume per volume per minute

Index

A

B

C

D

E

F

G

H

I

K

L

M

N

O

P

Q

R

S

T

V

W

X

Z

CPSIA information can be obtained at www.ICGtesting.com
Printed in the USA
LVOW02*0314280713

344950LV00006B/233/P

9 781847 354365